Environmental Chemistry for a Sustainable World

Volume 40

Other Publications by the Editors

Books
Environmental Chemistry
http://www.springer.com/978-3-540-22860-8

Organic Contaminants in Riverine and Groundwater Systems
http://www.springer.com/978-3-540-31169-0

Sustainable Agriculture
Volume 1: http://www.springer.com/978-90-481-2665-1
Volume 2: http://www.springer.com/978-94-007-0393-3

Book series
Environmental Chemistry for a Sustainable World
http://www.springer.com/series/11480

Sustainable Agriculture Reviews
http://www.springer.com/series/8380

Journals
Environmental Chemistry Letters
http://www.springer.com/10311

More information about this series at http://www.springer.com/series/11480

Inamuddin • Abdullah M. Asiri • Eric Lichtfouse
Editors

Conversion of Carbon Dioxide into Hydrocarbons Vol. 1 Catalysis

Springer

Editors
Inamuddin
Chemistry Department, Faculty of Science
King Abdulaziz University
Jeddah, Saudi Arabia

Abdullah M. Asiri
Chemistry Department, Faculty of Science
King Abdulaziz University
Jeddah, Saudi Arabia

Eric Lichtfouse
Laboratory Multiphase Flow in Power Engineering
Xi'an Jiaotong University
Xi'an, China

ISSN 2213-7114 ISSN 2213-7122 (electronic)
Environmental Chemistry for a Sustainable World
ISBN 978-3-030-28621-7 ISBN 978-3-030-28622-4 (eBook)
https://doi.org/10.1007/978-3-030-28622-4

This Springer imprint is published by the registered company Springer Nature Switzerland AG.
The registered company address is: Gewerbestrasse 11, 6330 Cham, Switzerland

Preface

Carbon dioxide is the most abundant among the gasses present in the air, which has a vital impact on indispensable plant and creature process, for example, photosynthesis and respiration. A large number of human activities increment the discharges of carbon dioxide into the atmosphere and contribute to the greenhouse effect, which cause health hazards not only to humans but also to wildlife. The sources of CO_2 emissions may be stationary, mobile, and natural. The principal anthropogenic sources of CO_2 emission are the deterioration of minerals and ignition of petroleum products. The fossil fuels are the basic need starting from everyday life to the survival of industrialization, leading to the emission of carbon dioxide gas in the atmosphere. The major sources of energy generation for stationary, portable, transportation, and industrial applications are in practice by the direct combustion of fossil fuels. Carbon dioxide emission from the energy generation may somehow be reduced by utilizing the renewable sources of energy such as wind, hydro, solar, and fuel cell. These sources do not emit CO_2 during their operations; therefore, they do not contribute toward the greenhouse effect. The emission of CO_2 into the atmosphere is also easy to cut down by sequestration using biological, chemical, or physical processes and the subsequent conversion of sequestrated CO_2 into chemicals or fuels as well as organic feedstocks. The conversion of sequestrated carbon dioxide into useful chemicals will prevent a worldwide temperature alteration caused by ozone layer depletion. The electrochemical reduction of CO_2 speaks much about some conceivable methods for creating chemicals or fuels by changing over carbon dioxide to feedstocks.

Conversion of Carbon Dioxide into Hydrocarbons Vol 1: Catalysis is focused on discussing the catalytic conversion of carbon dioxide into various hydrocarbons. It provides in-depth literature reviews on various types of catalyst used for photochemical, electrochemical, and thermochemical conversion of carbon dioxide into hydrocarbons such as formate, formic acid, alcohols, lower and higher hydrocarbons, as well as gasses such as hydrogen, carbon monoxide, and syngas. It is a unique book, extremely well-structured and essential resource for undergraduate and postgraduate students, faculty, R&D professionals, production chemists, environmental engineers, and industrial experts. Based on thematic topics, the book edition contains the following seven chapters:

Chapter 1 discusses the conversion of carbon dioxide into liquid hydrocarbons using cobalt-bearing catalysts by the Fischer-Tropsch process. It also summarizes the procedure used for the pretreatment and effect of support materials, structure, pressure, and ratio of the feed gas on cobalt catalyst necessary for carbon dioxide reduction. The use of CO_2 to reform CH_4 to produce syngas for the subsequent synthesis of Fischer-Tropsch liquids and electrochemical or photochemical reduction of CO_2 is discussed.

Chapter 2 summarizes the methods used for the conversion of carbon dioxide into formate/formic acid using lead/composite/oxide electrode. It emphasizes various electrode compositions, catalytic mechanisms, reactor and electrode forms, and the influence of the reaction conditions on the catalytic process.

Chapter 3 provides a deep insight toward the thermochemical conversion of carbon dioxide to carbon monoxide by reverse water-gas shift reaction over only the ceria-based catalysts. The reverse water-gas shift reaction mechanism, thermodynamics, catalyst, and the promoters are discussed.

Chapter 4 reviews recent developments, challenges, and novel approach applied for the photoconversion of carbon dioxide into sustainable fuels. The photocatalytic properties of ultraviolet and visible light photocatalysts applied to CO_2 reduction, as well as the recent advances in the design of photocatalytic systems, are discussed.

Chapter 5 discusses the homogeneous and heterogeneous catalysts used for the electrochemical reduction of carbon dioxide to methanol. Some benchmarks of metal-organic frameworks and nonmetal-organic framework catalysts for carbon dioxide reduction are also discussed.

Chapter 6 summarizes the fundamental aspect of heterogeneous photocatalytic carbon dioxide conversion using various semiconductor-based photocatalysts. Moreover, different surface modification routes adapted in photocatalytic materials are presented in details. Additionally, the influence of various experimental parameters and different types of photoreactors for carbon dioxide photoconversion are described with applications.

Chapter 7 highlights an approach to fulfill the challenges that occur during the electrocatalytic production of methanol from carbon dioxide. It reviews the effect of applied catalytic and electrolyte material, electrode, and electrochemical cell structure and utilized operational parameters on process performance.

Jeddah, Saudi Arabia Inamuddin
Jeddah, Saudi Arabia Abdullah M. Asiri
Xi'an, China Eric Lichtfouse

Contents

1 Conversion of Carbon Dioxide into Liquid Hydrocarbons Using Cobalt-Bearing Catalysts 1
Afsaneh Khajeh, Lijun Wang, and Abolghasem Shahbazi

2 Conversion of Carbon Dioxide Using Lead/Composite/Oxide Electrode into Formate/Formic Acid 25
Xiaowei An, Akihiro Yoshida, Abuliti Abudula, and Guoqing Guan

3 Thermochemical Conversion of Carbon Dioxide to Carbon Monoxide by Reverse Water-Gas Shift Reaction over the Ceria-Based Catalyst 43
Joshua Gorimbo and Diane Hildebrandt

4 Photocatalytic Systems for Carbon Dioxide Conversion to Hydrocarbons 63
Amel Boudjemaa and Nabila Cherifi

5 Electrochemical Reduction of Carbon Dioxide to Methanol Using Metal-Organic Frameworks and Non-metal-Organic Frameworks Catalyst 91
Fayez Nasir Al-Rowaili and Aqil Jamal

6 Photocatalytic Conversion of Carbon Dioxide into Hydrocarbons 133
Pramila Murugesan, Sheeba Narayanan, and Matheswaran Manickam

7 Electrocatalytic Production of Methanol from Carbon Dioxide 165
Esperanza Ruiz Martínez and José María Sánchez Hervás

Index 209

Contributors

Abuliti Abudula Graduate School of Science and Technology, Hirosaki University, Hirosaki, Japan

Fayez Nasir Al-Rowaili Research and Development Center, Saudi Aramco, Dhahran, Saudi Arabia
Department of Chemical Engineering, King Fahd University of Petroleum and Minerals, Dhahran, Saudi Arabia

Xiaowei An Energy Conversion Engineering Laboratory, Institute of Regional Innovation (IRI), Hirosaki University, Aomori, Japan
Graduate School of Science and Technology, Hirosaki University, Hirosaki, Japan

Amel Boudjemaa Centre de Recherche Scientifique et Technique en Analyses Physico-Chimiques (CRAPC), Tipaza, Algeria

Nabila Cherifi Centre de Recherche Scientifique et Technique en Analyses Physico-Chimiques (CRAPC), Tipaza, Algeria

Joshua Gorimbo Institute for the Development of Energy for African Sustainability (IDEAS) Research Unit, College of Science, Engineering and Technology (CSET), University of South Africa (UNISA), Johannesburg, South Africa

Guoqing Guan Energy Conversion Engineering Laboratory, Institute of Regional Innovation (IRI), Hirosaki University, Aomori, Japan

José María Sánchez Hervás Unit for Sustainable Thermochemical Valorisation, CIEMAT, Madrid, Spain

Diane Hildebrandt Institute for the Development of Energy for African Sustainability (IDEAS) Research Unit, College of Science, Engineering and Technology (CSET), University of South Africa (UNISA), Johannesburg, South Africa

Aqil Jamal Research and Development Center, Saudi Aramco, Dhahran, Saudi Arabia

Afsaneh Khajeh Nanoengineering Department, North Carolina Agricultural and Technical State University, Greensboro, NC, USA

Matheswaran Manickam Department of Chemical Engineering, National Institute of Technology, Tiruchirappalli, Tamil Nadu, India

Esperanza Ruiz Martínez Unit for Sustainable Thermochemical Valorisation, CIEMAT, Madrid, Spain

Pramila Murugesan Department of Chemical Engineering, National Institute of Technology, Tiruchirappalli, Tamil Nadu, India

Sheeba Narayanan Department of Chemical Engineering, National Institute of Technology, Tiruchirappalli, Tamil Nadu, India

Abolghasem Shahbazi Chemical, Biological and Bioengineering, North Carolina Agricultural and Technical State University, Greensboro, NC, USA

Lijun Wang Chemical, Biological and Bioengineering, North Carolina Agricultural and Technical State University, Greensboro, NC, USA

Akihiro Yoshida Energy Conversion Engineering Laboratory, Institute of Regional Innovation (IRI), Hirosaki University, Aomori, Japan

Chapter 1
Conversion of Carbon Dioxide into Liquid Hydrocarbons Using Cobalt-Bearing Catalysts

Afsaneh Khajeh, Lijun Wang, and Abolghasem Shahbazi

Contents

1.1 Introduction 2
1.2 Hydrogenation of CO_2 into Hydrocarbons over a Cobalt Catalyst in a Fischer-Tropsch Process 3
1.2.1 Fischer-Tropsch Reactions 3
1.2.2 Bimetallic Cobalt and Iron Catalysts 5
1.2.3 Promoters on Cobalt-Based Catalysts for CO_2 Hydrogenation 7
1.2.4 Effect of the Supports and Structure of Cobalt-Based Catalysts 8
1.2.5 Pretreatment of Cobalt Catalysts 11
1.2.6 Effect of Pressure and Ratio of the Feed Gas 11
1.3 Hydrogenation of CO_2 over a Cobalt Catalyst in a Solution 12
1.4 CO_2 Reforming of CH_4 over a Cobalt Catalyst 12
1.5 Electrochemical Reduction of CO_2 over a Cobalt Catalyst 15
1.5.1 Electrochemical Reduction of CO_2 15
1.5.2 Cobalt-Based Electrocatalysts for Reduction of CO_2 into Formate 16
1.5.3 Cobalt Phthalocyanines and Cobalt Porphyrins for CO_2 Reduction 17
1.6 Photocatalytic Reduction of CO_2 over a Cobalt Catalyst 19
1.7 Conclusions 19
References 20

Abstract There is an increasing interest in carbon dioxide (CO_2) capture and subsequent conversion into value-added chemicals and fuels. Cobalt-based catalysts are widely used in Fischer-Tropsch synthesis of liquid fuels from syngas ($CO + H_2$) as they have high activity, good selectivity, and superior stability for the hydrogenation of CO to long-chain hydrocarbons. However, the hydrogenation of CO_2 into

A. Khajeh
Nanoengineering Department, North Carolina Agricultural and Technical State University, Greensboro, NC, USA

L. Wang (✉) · A. Shahbazi
Chemical, Biological and Bioengineering, North Carolina Agricultural and Technical State University, Greensboro, NC, USA
e-mail: lwang@ncat.edu; ash@ncat.edu

Inamuddin et al. (eds.), *Conversion of Carbon Dioxide into Hydrocarbons Vol. 1 Catalysis*, Environmental Chemistry for a Sustainable World 40,
https://doi.org/10.1007/978-3-030-28622-4_1

long-chain hydrocarbons involves both reverse water-gas shift reaction and Fischer-Tropsch synthesis. As a cobalt catalyst lacks the ability to catalyze the reverse water-gas shift reaction, an active component for catalyzing reverse water-gas shift reaction such as iron and copper can be incorporated to a cobalt-based catalyst for CO_2 hydrogenation. Proper catalyst supports, structures, and pretreatment can further improve the performance of cobalt-based catalysts for CO_2 hydrogenation. The pressure and composition of the feed gas were found to have significant effects on the distribution of Fischer-Tropsch synthesized hydrocarbons during CO_2 hydrogenation.

There are several other chemical processes that have been studied to produce chemicals and fuels from CO_2 over a cobalt-based catalyst, which include hydrogenation of CO_2 in a basic solution, the use of CO_2 to reform CH_4 to produce syngas for the subsequent synthesis of Fischer-Tropsch liquids, and electrochemical or photochemical reduction of CO_2. Various catalyst supports and promoters have been studied to improve the performance of cobalt-based catalysts for those chemical processes.

Keywords Carbon dioxide · Hydrogenation · Fischer-Tropsch · Electrochemical reduction · Photochemical · Fuels · Hydrocarbons · Alcohols · Long-chain hydrocarbons · Organic acids · Catalyst

1.1 Introduction

Capture, storage, and utilizations of carbon dioxide (CO_2) have attracted much attention in recent decades in order to mitigate its irreparable damages to the atmosphere (Olah et al. 2009; Riduan and Zhang 2010). Several methods including hydrogenation of CO_2, reforming of CH_4 with CO_2, and electrochemical reduction of CO_2 in a solution or photocatalytic reduction of CO_2 have been studied to convert CO_2 into value-added chemicals and fuels such as methane, alcohols, long-chain hydrocarbons, and formic acid. Methanol, higher alcohols, dimethyl ether, and long-chain hydrocarbons are considered as alternative liquid fuels for internal combustion engines. Formic acid is one of the essential and main chemical feedstocks used to produce sustainable clean electricity in fuel cells, paper manufacture, pharmaceuticals, and antibacterial and deicing agent in several industries. Formate salts such as HCOONa and HCOOK are widely used as heat-transfer agents in oil and gas drilling (Olah et al. 2009).

Hydrogenation of CO_2 in a Fischer-Tropsch process or an aqueous solution into value-added chemicals and fuels has been considered as a promising and beneficial approach to reduce and control the emission of CO_2 into the atmosphere (Porosoff et al. 2017; Patel et al. 2018). A Fischer-Tropsch synthetic process has widely been used to catalytically convert syngas ($CO + H_2$) into a variety of fuels and chemicals such as methanol, ethanol, gasoline, diesel, lubricant, and olefins. Cobalt and iron catalysts are two common types of catalysts used in a Fischer-Tropsch synthesis process. A Fischer-Tropsch synthetic process can also be used to hydrogenate CO_2 to liquid fuels. CO_2 hydrogenation over a Fischer-Tropsch catalyst mainly produces light saturated hydrocarbons compared to CO hydrogenation. An iron- or cobalt-based Fischer-Tropsch catalyst requires different compositions of feed gases, particularly CO_2 content that was industrially reduced by an acid-gas removal process (Raftati et al. 2017). Most of the studies for CO_2 hydrogenation have been focused on the Fischer-Tropsch synthesis with iron-based catalysts (Rafati et al. 2015, 2017), while few studies have been done using cobalt-based catalysts in the Fischer-Tropsch synthesis (Iglesia et al. 1993; Guttel et al. 2008; Dry 1996, 2002). The activity and selectivity of catalysts are extremely dependent on the nature of the active components, supports, particle size, and number of active sites present on their surface. Stabilizing and attaining high dispersion of the active phase on a catalyst support can be controlled through the interaction of metal catalyst compounds with an oxide support (Vance and Bartholomew 1983; Goodman et al. 2017). Promoters, mostly alkali metals are another vital contributor to alter the selectivity of a catalyst (Zhang et al. 2015a, b; Wei et al. 2016; Visconti et al. 2017). Different preparation and pretreatment conditions also affect the active sites on the surface of a catalyst for various activity and selectivity toward the production of specific products (Gnanamani et al. 2015).

Several other approaches have been studied to convert CO_2 into chemicals and fuels. CO_2 can also be catalytically hydrogenated to formate in a basic solution. Cobalt nanoparticles supported on nitrogen-rich graphitic carbon was reported to be active in hydrogenation of CO_2 in a KOH solution to produce formate (Patel et al. 2018). Reforming of CH_4 with CO_2 can produce syngas for the subsequent Fischer-Tropsch synthesis of liquid fuels (Zhang et al. 2007). Electrochemical or photochemical reduction of CO_2 in an aqueous solution is another approach to convert CO_2 to value-added products such as carbon monoxide, ethylene, methanol, ethanol, and formic acid (Aga et al. 1997; Zhao et al. 2009). Bonin et al. (2017) provided a review on the recent advances on electrochemical and photochemical reduction of CO_2 with Fe and Co metal-based catalysts.

The objective of this chapter was to review the state of the art of conversion of CO_2 into hydrocarbons over cobalt-based catalysts.

1.2 Hydrogenation of CO_2 into Hydrocarbons over a Cobalt Catalyst in a Fischer-Tropsch Process

1.2.1 Fischer-Tropsch Reactions

CO_2 can be hydrogenated to liquid fuels over cobalt or iron catalysts in a Fischer-Tropsch synthesis process. Conversion of CO_2 to hydrocarbons consists of two consecutive reactions. The CO_2 is first reduced to CO with H_2 in a reverse water-gas shift reaction (Reaction 1.1). The mixture of CO and H_2 that is called syngas is then converted into hydrocarbons through a Fischer-Tropsch process (Reaction 1.2) (Porosoff et al. 2016):

$$CO_2 + H_2 \rightarrow CO + H_2O \qquad \Delta H_{300^\circ C} = 38\ kJ\ mol^{-1} \tag{1.1}$$

$$nCO + 2nH_2 \rightarrow -(CH_2)_n - +nH_2O \qquad \Delta H_{300^\circ C} = -166\ kJ \cdot mol^{-1} \tag{1.2}$$

$$CO_2 + 4H_2 \rightarrow CH_4 + 2H_2O \qquad \Delta H_{25^\circ C} = -165\ kJ \cdot mol^{-1} \tag{1.3}$$

One of the main challenges for the hydrogenation of CO_2 into hydrocarbons in a Fischer-Tropsch process is the development of catalysts with high activity and selectivity. Cobalt and iron are both of great interest in Fischer-Tropsch synthesis of hydrocarbons from syngas (CO + H_2) (Reaction 1.2). Cobalt-based catalysts are widely used in Fischer-Tropsch synthesis of liquid fuels from syngas as they have high activity, selectivity, and stability for the hydrogenation of CO to long-chain hydrocarbons. For producing long-chain hydrocarbons, a cobalt-based catalyst is preferred over an iron-based catalyst because of its higher stability and ability for the growth of long-chain hydrocarbons (Schulz 1999).

Cobalt and iron Fischer-Tropsch catalysts behave differently in the presence of CO_2 (Riedel et al. 1999). Most of the studies for CO_2 hydrogenation have been focused on the Fischer-Tropsch synthesis with iron-based catalysts (Rafati et al. 2015), while few studies have been done using cobalt-based catalysts in the CO_2 hydrogenation (Guttel et al. 2008; Iglesia et al. 1993; Dry 1996, 2002).

Iron-based catalysts have proved to be the most successful to simultaneously catalyze the reverse water-gas shift reaction (Reaction 1.1) for the conversion of CO_2 to CO and the Fischer-Tropsch synthesis (Reaction 1.2) for the further conversion of CO to hydrocarbons (Saeidi et al. 2014; Rafati et al., 2015). Therefore, for a Fischer-Tropsch iron catalyst, CO_2 does not affect the catalyst selectivity, and CO_2 can also be converted to hydrocarbons according to the equilibrium constraints of reverse water-gas shift reaction (Rafati et al. 2017).

As cobalt exhibits low activity for catalyzing RWGS reaction, the hydrogenation reaction of CO_2 over a cobalt-based catalyst is shifted toward methanation (Reaction 1.3) (Riedel et al. 1999; Gnanamani et al. 2016). It was reported that CO_2 was easily hydrogenated over an unpromoted Co/γ-Al_2O_3 catalyst and its conversion rate was faster than that of CO. However, methane is a dominant product in CO_2 hydrogenation, and long-chain hydrocarbons are dominant products during CO hydrogenation (Visconti et al. 2016). Zhang et al. (2002) conducted a comparative study of the hydrogenation of CO and CO_2 and their mixture on a cobalt catalyst. They found that the CO_2 hydrogenation was slow in the presence of CO and the hydrogenation of CO_2 only produced more than 70% of methane. They also found that the catalyst deactivated more rapidly in the hydrogenation of CO than CO_2 (Zhang et al. 2002).

The different performances of iron and cobalt catalysts in the hydrogenation of CO_2 and CO are caused by their different active phases during Fischer-Tropsch synthesis. The iron carbide phases contribute to the high activity of iron-based catalysts in Fischer-Tropsch synthesis, while the active phase of a cobalt catalyst is metallic cobalt, which is highly active in hydrogenation of CO_2 to produce methane as the main product (Khodakov et al. 2007; Smit and Weckhuysen 2008). As the reverse water-gas shift reaction (1) is endothermic, the reaction temperature should be high to shift the reverse water-gas shift reaction toward the CO production. The reverse water-gas shift reaction is nearly at the equilibrium composition when the temperature is high (e.g., 330–350 °C). However, at a high temperature, cobalt catalysts work for methanation (Reaction 1.3) (Ishida et al. 2013). A cobalt catalyst is highly active and selective toward the formation of methane in hydrogenation of CO_2 instead of CO through Reaction (1.3) due to the weaker adsorption of CO_2 than CO on the cobalt surface, leading to lower C/H ratio for the formation of methane instead of higher hydrocarbons via the Fischer-Tropsch reaction (1.2) (Aziz et al. 2015; Visconti et al. 2009; Diez-Ramirez et al. 2017). Therefore, a high concentration of CO_2 in feed gas would decrease the C_{5+} selectivity of a cobalt catalyst (Yao et al. 2010; Gnanamani et al. 2011).

Research has been conducted to increase the CO_2 hydrogenation over a cobalt-based catalyst. An active component for the RWSG reaction such as iron and copper can be incorporated to a cobalt-based catalyst. Proper catalyst supports, structures, promoters, and pretreatment can further improve the performance of cobalt-based catalysts for the simultaneous reverse water-gas shift reaction and Fischer-Tropsch synthesis during CO_2 hydrogenation.

1.2.2 Bimetallic Cobalt and Iron Catalysts

Both iron and cobalt have some advantages and disadvantages in hydrogenation of CO_2 into liquid fuels. Bimetallic cobalt-iron catalysts were studied. Figure 1.1 shows the effect of various iron fractions in a bimetallic iron-cobalt catalyst on the distribution of alcohols during CO_2 hydrogenation (Gnanamani et al. 2016). The selectivity of methanol increased, while the selectivity of ethanol decreased when the iron content in the bimetallic catalyst increased. A pure cobalt catalyst was more active in hydrogenation of CO_2, but the addition of iron to cobalt increased the production of other hydrocarbons besides methane. A combination of 50% iron and 50% cobalt showed the maxim hydrogenation of CO_2 to long-chain hydrocarbons. From Mössbauer spectroscopy analyses, it can be concluded that the iron carbide ratio increases with the increase of the iron fraction in bimetallic iron-cobalt catalysts.

Boreriboon et al. (2018) conducted a comparative study on iron, cobalt, and copper catalysts supported on titanium for CO_2 hydrogenation. Their results showed that cobalt exhibited the highest CO_2 conversion but had a high selectivity exclusively toward methane, while iron and copper showed high selectivity for higher hydrocarbons and carbon monoxide, respectively. The superior selectivity of an iron catalyst to higher hydrocarbons is because of the presence of iron carbide as the active site in the Fischer-Tropsch synthesis, which is essential for carbon-carbon

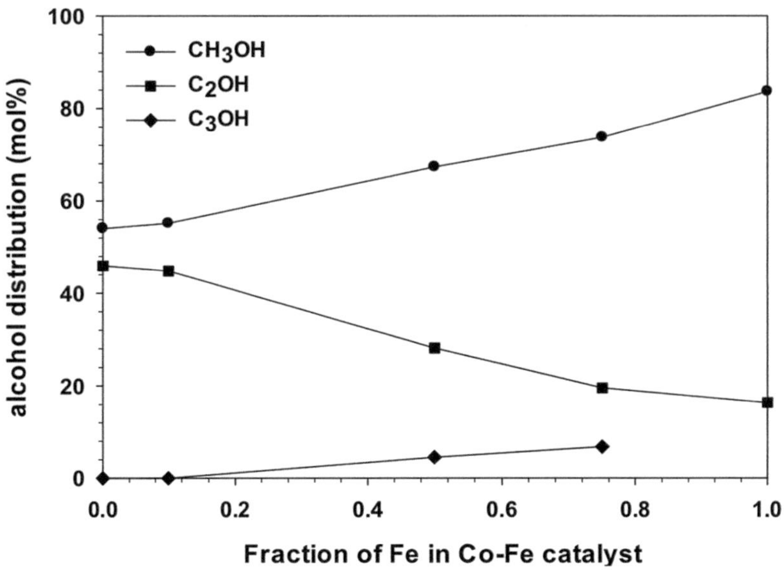

Fig. 1.1 Effect of iron content in a bimetallic iron-cobalt catalyst on the distribution of various alcohols produced by CO_2 hydrogenation (the addition of iron to cobalt increases the production of other hydrocarbons besides methane. With increase in iron content in the bimetallic catalyst, the selectivity of methanol increase, and the selectivity of ethanol decreases). (Reprinted with permission of 2015 American Chemical Society from Gnanamani et al. 2016)

interaction to produce long-chain hydrocarbons. They also found that the addition of a small amount of cobalt on iron could enhance CO conversion and the space-time yields of hydrocarbon products.

Satthawong et al. (2013) examined iron-cobalt bimetallic catalysts that are active and selective for synthesis of olefin-rich, C_2^+ hydrocarbons during CO_2 hydrogenation. The combination of iron and a small amount of cobalt to form a bimetallic catalyst of Fe-Co/Al_2O_3 with 15 wt% total metal loading could significantly promote the synthesis of C_2^+ hydrocarbons during CO_2 hydrogenation. They explored the space-time yields of methane and higher hydrocarbons over iron-cobalt catalysts supported on alumina with different cobalt atomic ratios. The maximum space-time yield of methane that was obtained using the bimetallic catalyst with a cobalt atomic ratio of 0.17 was significantly higher than that obtained using a monometallic iron catalyst. This study concluded that the addition of a small quantity of cobalt to iron could result in substantial promotion of bimetallic effect on higher hydrocarbons production in CO_2 hydrogenation. The higher activity and selectivity of the iron-cobalt bimetallic catalyst indicate that the combination of two metals of iron and cobalt alters the adsorption behavior of the catalyst in such a way that expedites the evolution of carbon chains. It was also evident that the increasing rates of space-time yields for both methane and higher hydrocarbons were similar when the cobalt atomic ratio was lower than 0.17, which explains the bimetallic promotion of higher hydrocarbons without undesirable influence on carbon-carbon interactions to facilitate carbon chain growth for higher hydrocarbons.

1.2.3 Promoters on Cobalt-Based Catalysts for CO_2 Hydrogenation

For catalyzing CO_2 hydrogenation, promoters such as copper, molybdenum, and sodium are typically added to cobalt-based catalysts to improve the reverse water-gas shift reaction and the selectivity of long-chain hydrocarbons in a Fischer-Tropsch synthesis process (Owen et al. 2013; Shi et al. 2018a).

Copper-based catalysts have been used widely to catalyze the reverse water-gas shift reaction. Therefore, the addition of copper metal to a cobalt-based catalyst could improve the efficacy of CO_2 conversion to hydrocarbons in Fischer-Tropsch synthesis. Titania-supported cobalt-copper catalysts and the effect of potassium promoter on CO_2 hydrogenation to long-chain hydrocarbons were studied (Shi et al. 2018a). The homogeneous deposition-precipitation method was used to achieve high loading and dispersion of metal oxides on the support with strong interaction to the support, preventing them from sintering. Although the addition of potassium up to 2.5 wt% caused the decrease in the CO_2 conversion, it increased the selectivity to higher hydrocarbons (C_5^+) with the maximum amount of 35.1 C mol% due to the suppression of methanation. It was also found that the increase of potassium content increased the C_2–C_4 olefin/paraffin production as well. The increase in the selectivity of C_5^+ and in chain growth probability could be attributed to the enhancement of surface basicity, which increases CO_2 uptake and decreases H_2 uptake due to the reduction of electrons from adsorption hydrogen on the surface of metallic cobalt and copper site in presence of potassium (Shi et al. 2018a).

The effect of alkali metal promoters such as sodium, lithium, potassium, rubidium, and cesium on catalytic performance of copper-cobalt catalyst supported on titanium in CO_2 hydrogenation was studied (Shi et al. 2018b). The X-ray powder diffraction patterns and transmission electron microscopy images of calcined and spent catalysts with 2.5 wt% loading of alkali metals obtained by the deposition-precipitation technique showed the high stability and dispersion of active metallic sites on the support. The results from catalytic performance tests showed that the addition of alkali metals except lithium that is the alkali metal with the lowest atomic number decreased CO_2 hydrogenation and increased the selectivity and space-time yield of higher hydrocarbons compared to a copper-cobalt catalyst. The alkali promoters were observed to increase CO_2 chemisorption and decrease H_2 adsorption, leading to reduction in H_2/CO_2 ratio at the catalyst surface to produce CO as an immediate for subsequent Fischer-Tropsch synthesis. Therefore, the CO_2 conversion decreased, the methanation dropped, and the selectivity of long-chain hydrocarbons (C_5^+) increased significantly. The increase in CO_2 conversion over copper-cobalt promoted with lithium compared to alkali copper-cobalt could be explained by the geometric and electronic influence of alkali metals on the catalytic performance. Alkali metal ions with larger atomic numbers might prevent the adsorption of reactants on the catalyst surface. Furthermore, the enriched electron density between lithium and cobalt, attributed to the highest polarization of lithium, blocks the adsorption of reactants and increases the activity of lithium promoted catalyst in

CO_2 hydrogenation. The effect of the addition of alkali metals on the suppression of methanation and the increase of higher hydrocarbon yields depends on their basicity on the surface of the catalyst and their atomic weight. Among the studied alkali metals, sodium exhibited the highest hydrocarbon selectivity at maximum of 42.1% with a CO_2 conversion of 18.4%, which is due to its lower atomic weight leading to higher aspect ratio of sodium atoms to Co or Cu and also due to the strongest catalyst surface basicity exhibited by Na-Co-Cu/TiO_2 catalyst (Shi et al. 2018b).

Cobalt molybdate catalysts were found to have great activity for the hydrogenation of CO_2 into alcohols. Tetrahedrally coordinated Mo^{6+} ions increase the selectivity of alcohols during CO_2 hydrogenation over a cobalt-based catalyst.

1.2.4 Effect of the Supports and Structure of Cobalt-Based Catalysts

Al_2O_3 supports are widely used in industrial Fischer-Tropsch processes. However, studies showed that some other inorganic oxide support such as TiO_2 had better performance than Al_2O_3 in Fischer-Tropsch synthesis with syngas (Jacobs et al. 2002). The supported cobalt catalysts are active for the CO_2 hydrogenation, but the conversion depends on the reduction temperature, loading, and the type of the supports. Das and Deo (2011) found that the reduction of calcined Co_3O_4 on an alumina support could be complete at 550 °C, but the reduction of Co_3O_4 on the silica support was incomplete at 550 °C. Several cobalt-based catalysts on various supports including Al_2O_3, CeO_2, ZrO_2, MgO, SiO_2, and ZSM-5 with sodium (Na) and molybdenum (Mo) as dopants have been studied for simultaneous catalysis of reverse water-gas shift reaction and Fischer-Tropsch synthesis during CO_2 hydrogenation (Owen et al. 2016). It was found that the surface area of the support and metal-support interaction as shown in Table 1.1 had significant effects on the size of cobalt crystallite, the catalytic activity, and the probability of chain growth during CO_2 hydrogenation (Owen et al. 2016).

Among those inorganic oxide supports, the cobalt-based catalyst supported on SiO_2 and ZSM-5 with smaller cobalt particle size and highest surface area exhibited the highest CO_2 conversion with similar selectivity for both CO and hydrocarbons as shown in Table 1.2 (Owen et al. 2016). It was found that cobalt particles with sizes smaller than 2 nm supported on MgO had low activity for the reverse water-gas shift reaction and negligible Fischer-Tropsch synthetic activity. The increase of cobalt particle size to 15 nm supported on SiO_2 and ZSM-5 substantially increased the CO_2 conversion and selectivity of hydrocarbons compared to CO.

But ZSM-5 showed the highest methane selectivity. Therefore, in this case, metal-support interaction is the determining factor for further distribution of higher hydrocarbons. Through the temperature programmed reduction analysis, it was observed that the cobalt supported on ZSM-5 had the lowest reduction temperature

Table 1.1 Surface area and Co_3O_4 crystallite size of supported 20 wt% Co-1 wt%Na-1 wt% Mo catalysts

Catalyst support	Surface area (m^2/g)	Co_2O_4 crystallite size (nm)
SiO_2	56	15
CeO_2	7.0	34
TiO_2	14	23
Al_2O_3	6.1	30
MgO	54	–
ZrO_2	6.6	24
ZSM-5	180	15

Modified after Owen et al. (2016)
The surface area of the catalyst support has a substantial effect on the cobalt crystallite size, which affects catalytic activity and probability of chain growth during CO_2 hydrogenation. ZSM-5 is an aluminosilicate zeolite containing several pentasil units, which consist of 85-membered rings

Table 1.2 Catalytic activity of 20 wt% Co-1 wt%Na-1 wt% Mo catalysts on various supports

Catalyst support	CO2 conversion (%)	CO selectivity (%)	Hydrocarbon selectivity (%)	C_1 (%)	C_{2+} (%)
SiO_2	30.0	21.5	78.5	57.1	42.9
CeO_2	15.1	70.2	29.8	22.1	77.9
TiO_2	13.5	66.7	33.3	23.1	76.9
Al_2O_3	15.4	57.3	42.7	29.3	70.7
MgO	8.7	100	–	0.0	0.0
ZrO_2	14.4	16.7	83.3	26.8	73.2
ZSM-5	29.1	20.6	79.4	95.7	4.3

Modified after Owen et al. (2016)
The highest CO_2 conversion was achieved by smaller cobalt particle size and higher surface area. ZSM-5 shows the highest methane selectivity
Reaction conditions 200 °C, 1 atm, 3:1 H_2/CO_2, 0.35 h^{-1} WHSV

of Co_3O_4 to metallic cobalt, which indicates the weakest metal-support interaction resulting in the formation of shorter hydrocarbon chains. However, the TiO_2 support obtained the highest probability of chain growth, low methane selectivity, and highest C_2–C_4 unsaturated products. It was also found that large cobalt particle size (>20 nm) on other supports such as TiO_2 favored the formation of long-chain hydrocarbons (Owen et al. 2016).

Because of the highest selectivity of TiO_2 and highest conversion of SiO_2 as support of a cobalt catalyst, a mixture of these two catalyst supports could improve the yield of a Fischer-Tropsch process using a cobalt-based catalyst, but the synergetic effect of TiO_2 and SiO_2 supports for both high selectivity and high CO_2 conversion might not be achieved, although the conversion and selectivity could be optimized by changing the proportions of those two supports in the catalyst (Owen et al. 2016).

The activity of a mesoporous cobalt oxide catalyst promoted with platinum (Pt/Co_3O_4-mesoporous) in CO_2 hydrogenation for higher alcohols production was studied and compared to that of Co_3O_4 (Liu et al. 2018). The content of metallic cobalt sites on the surface of the catalyst resulting from the partial reduction of Co_3O_4 is very critical on CO_2 conversion. The highest CO_2 conversion efficiency was achieved using the Pt/Co_3O_4 catalyst reduced at 300 °C due to its highest amount of cobalt metallic sites. However, methane was the predominant product from the CO_2 hydrogenation using the Pt/Co_3O_4 catalyst. The space-time yields of total alcohols and higher alcohols over Co_3O_4 mesoporous supported structure were 5 and 40 times of those obtained over Co_3O_4 nanoparticles, respectively. Since the studied mesoporous and nanoparticle Co_3O_4 had similar contents of metallic cobalt, the difference in yields should be related to the mesoporous structure of Co_3O_4 which was more favorable for carbon chains growth. CH_2 and CH_3 species generated from hydrogenated CO_2 have higher affinity to interact with each other in the mesoporous network of Co_3O_4 than Co_3O_4 nanoparticle, resulting in the formation of C_2–C_4 chains and suppressing the tendency of to form methane. Therefore, the Co_3O_4 mesoporous structure is superior for the production of higher alcohols and higher alcohols. Furthermore, the mesoporous structure showed higher stability during the reaction which is attributed to its mesoporosity and its high surface area, while the stability of Co_3O_4 nanoparticles decreased with time. With the addition of platinum promoter to the mesoporous Co_3O_4, carbon monoxide and total alcohol selectivity increased, and the methane selectivity decreased. However, Owen et al. (2013) found that there was a sharp reduction in the hydrogenation efficiency of CO_2 using a cobalt-based catalyst with the addition of noble metals including palladium, platinum, and rubidium. The possible reason was the weak activity of noble metals in CO_2 and CO hydrogenation.

Molecular sieves have been widely used to develop nanostructured catalysts due to their well-controlled surface and porosity, leading to the formation of specific phases nearby the active sites of catalysts. Metal clusters with organic linkers have been studied to develop a nano-dimensional porous heterogeneous catalyst, which has highly dispersed active sites, well-defined crystal structure, adjustable porosity, and other physiochemical properties by the modification of metal clusters of organic ligands. Cobalt catalysts obtained by different types of metal-organic framework, including microwave-assisted synthesizing nano-crystallites and micro-crystallites produced by conventional solvothermal technique have found to have appropriate activity and selectivity in Fischer-Tropsch synthesis due to their well-controlled morphology and dispersion of cobalt nanoparticles. Those catalysts have a higher percentage of reduced cobalt sites to metallic cobalt nanoparticles as well as higher density and better dispersion of cobalt sites than cobalt catalysts with conventional supports. The only cobalt phase present in these conventional supports is face-centered cubic cobalt, which shows the poorer interaction with carbon monoxide than those hexagonal close-packed cobalt sites. Hexagonal close-packed cobalt phases are desirable because of their suitable activity for Fischer-Tropsch synthesis. However, the addition of promoters such as Si into a Fischer-Tropsch catalyst was able to transform face-centered cube cobalt nanoparticles to hexagonal close-packed cobalt particles through cobalt carbide intermediate, Co_2C, leading to the density expansion of cobalt sites, suitable F-T activity, and high selectivity toward higher hydrocarbons, C_5^+, generating diesel as a main product (Pei et al. 2017).

1.2.5 *Pretreatment of Cobalt Catalysts*

The activation of a sodium-promoted cobalt catalyst on a silica support with the pretreatment using various gases including syngas, carbon monoxide, and hydrogen showed that the presence of partially reduced active phase of cobalt but not totally reduced to metallic cobalt under syngas pretreatment had lower activity for the hydrogenation of CO and CO_2 (Gnamani et al. 2015). As shown in Fig. 1.2, the cobalt catalysts pretreated with CO had the highest alcohol selectivity and lowest methanation activity because the partially reduced catalyst had an appropriate ratio of oxidized cobalt to metallic cobalt and carburized cobalt. The effect of alkali metal promoter is to retain the carburized form of cobalt by preventing it from rapid conversion to metallic cobalt, which can subsequently suppress its hydrogenation activity and convert carbon monoxide to primary alcohols through the pathway shown in Fig. 1.3. The role of alkali promoter in retaining the carburized form of cobalt is also demonstrated in X-ray powder diffraction profiles of fresh and used, undoped, and doped catalysts with sodium as the promoter as shown in Fig. 1.4 (Gnamani et al. 2015).

1.2.6 *Effect of Pressure and Ratio of the Feed Gas*

Dorner et al. (2009) investigated the composition of the feed gas and reaction pressure on the catalytic conversion of CO_2 on a cobalt-platinum catalyst supported on alumina. They examined various gas ratios of H_2/CO_2 including 3:1, 2:1, and 1:1. Their results showed that as the H_2/CO_2 ratio decreased, the yield of higher hydrocarbon increased, while the methane yield decreased. At a constant H_2/CO_2 ratio, the rate of CO_2 hydrogenation drops with the decrease of pressure, while the methanation suppresses, and the selectivity toward higher hydrocarbon increases.

Figure 1.5 shows the distribution of long-chain hydrocarbons, C_7–C_{20}, produced from the Fischer-Tropsch synthesis over a cobalt-iron catalyst supported on alumina at various compositions of feed gas (Diaz et al. 2014).

It was found that for a desired constituent of produced hydrocarbons, a specific ratio of hydrogen, carbon monoxide, and carbon dioxide should be exploited (Diaz et al. 2014). The Fischer-Tropsch synthesis with a low fraction of CO_2 in the feed gas produces kerosene as the main product. The increase of CO_2 content in the feed gas shifts the Fischer-Tropsch synthesis toward the production of kerosene and gasoline fractions.

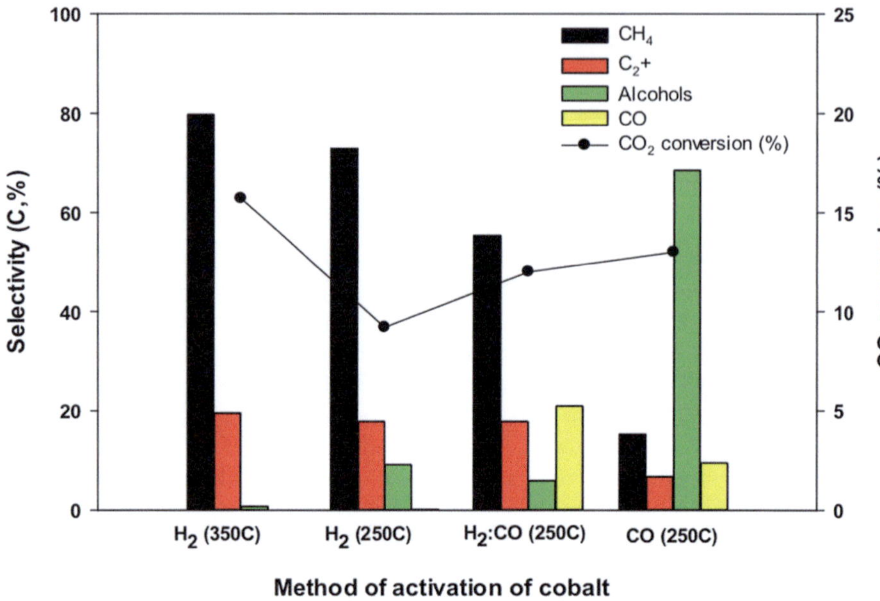

Fig. 1.2 Effect of pretreatment condition of silica supported cobalt catalysts promoted with sodium (cobalt catalysts pretreated with CO shows the highest alcohol selectivity and lowest methanation activity. A cobalt catalysts pretreated with H_2 at 350 °C exhibited the lowest alcohol selectivity and the highest methanation activity). (Reprinted with permission of 2015 Elsevier B.V. from Gnanamani et al. 2015)

1.3 Hydrogenation of CO_2 over a Cobalt Catalyst in a Solution

CO_2 can also be hydrogenated over a metallic cobalt catalyst in a basic solution to produce formate (Patel et al. 2018). Cobalt nanoparticles supported on nitrogen-rich graphitic carbon was found to be highly active for the hydrogenation of CO_2 in 1 M KOH solution at 120 °C at 1:1 partial pressure of CO_2/H_2 (Patel et al. 2018).

1.4 CO_2 Reforming of CH_4 over a Cobalt Catalyst

The dry reforming can reduce CO_2 emissions and convert CO_2 and CH_4 into valuable syngas. The combination of dry reforming (Reaction 1.4) and steam reforming (Reaction 1.5) can produce syngas with the ideal ratio of H_2/CO for the subsequent Fischer-Tropsch synthesis (Zhang et al. 2007):

$$CO_2 + CH_4 \rightarrow 2CO + 2H_2 \quad \Delta H_{298K} = -172\ \text{kJ mol}^{-1} \tag{1.4}$$

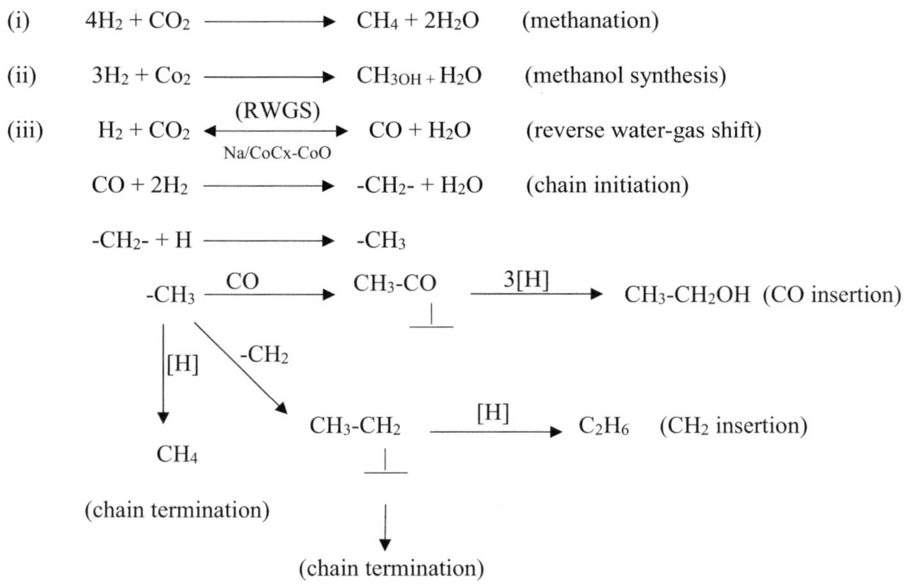

Fig. 1.3 Mechanism of hydrogenation of CO_2 into alcohols over a carburized cobalt catalyst (alkali metal promoter retains the carburized form of cobalt by preventing it from rapid conversion to metallic cobalt. Carburized form of cobalt suppresses its hydrogenation activity and converts carbon monoxide to primary alcohols). (Modified after Gnanamani et al. 2015)

$$H_2O + CH_4 \rightarrow CO + 3H_2 \quad \Delta H_{298K} = 206\ \text{kJ mol}^{-1} \tag{1.5}$$

Coke formation from methane decomposition and CO disproportionation that deactivates catalysts is the greatest challenge for CO_2 reforming of CH_4 (Mirzaei et al. 2015):

$$2CO \rightarrow CO_2 + C \quad \Delta H_{298K} = -172\ \text{kJ mol}^{-1} \tag{1.6}$$

$$CH_4 \rightarrow 2H_2 + C \quad \Delta H_{298K} = 75\ \text{kJ mol}^{-1} \tag{1.7}$$

Noble metals such as Rh, Ru, Au, Pd, and Pt are very active and have high resistance to coking, but they are expensive. There is an increasing interest to develop novel ferrous metal catalysts such as Fe-, Co-, and Ni-based catalysts for CO_2 reforming of CH_4 (Ferreira-Aparicio et al. 1998; Ji et al. 2001). Several methods including the modification of the catalyst supports, the active metal additives, and the promoters have been studied to develop a Ni-based catalyst for CO_2 reforming of CH_4. The synergy between Ni and Co can significantly improve catalyst performance and reduce carbon formation during dry reforming. The high activity and excellent stability of Ni-Co catalyst was closely related to its high metal dispersion, strong metal-support interaction, and formation of stable solid solutions (Zhang et al. 2007). It was reported that the interaction between Ni and Co and

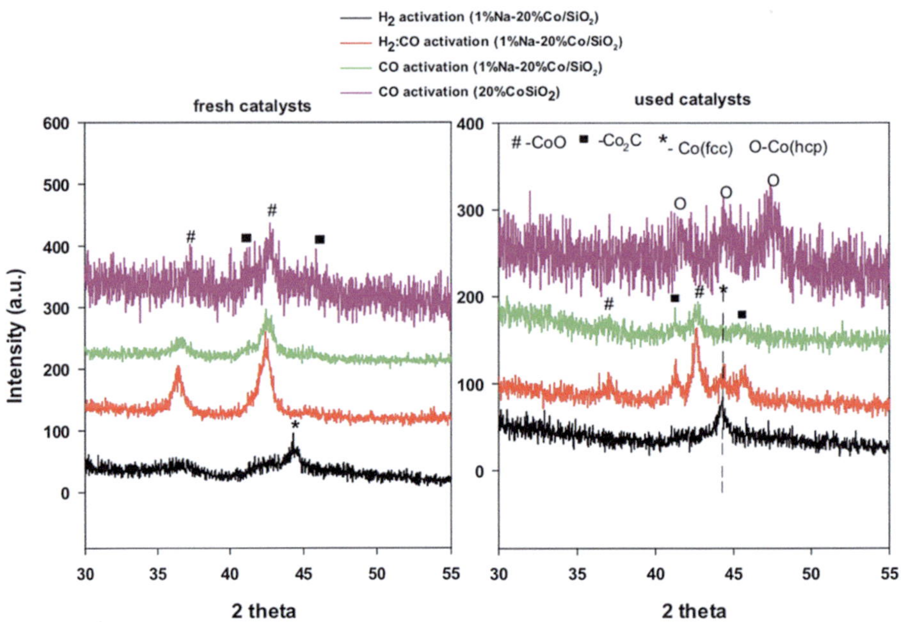

Fig. 1.4 X-ray powder diffraction of fresh and spent cobalt catalysts with different activation conditions (the pretreatment using syngas or carbon monoxide of a sodium-promoted cobalt catalyst on a silica support was not totally reduced to metallic cobalt, compared to the pretreatment using hydrogen gas). (Reprinted with permission of 2015 Elsevier B.V. from Gnanamani et al. 2015)

among the metals and support structures in the bimetallic catalysts slowed down the reduction, stabilized the catalysts, and resulted in small sizes of metal nanoparticles that are uniformly dispersed on catalyst surfaces (Wang et al. 2012). The monometallic Co/TiO_2 and Co/ZrO_2 catalysts were found to deactivate rapidly because of the oxidation of metal (Takanabe et al. 2005; Park et al. 2018). Bimetallic Co-Ni/TiO_2 catalysts with an appropriate Co/Ni ratio showed highly stable activities without carbon deposition. The bimetallic Co-Ni/TIO_2 catalyst with 10 mol% substitution of nickel for cobalt suppressed the oxidation of metal, providing a high catalytic stability. However, the catalysts with excess nickel content (>80 mol%) underwent carbon formation (Takanabe et al. 2005).

Some promoters such as Zn were added to the Co-based catalysts for dry reforming of CH_4. It was found that the addition of 1–2 wt% Zn to Co/ZrO_2 could suppress the oxidation of metallic Co and improve the catalyst stability (Park et al. 2018). Co-Zr bimetallic catalyst supported on activated carbon was studied for CO_2 reforming of methane (Zhang et al. 2015a, b). It was found that Zr promoter reduced the interaction between Co and support and thus increases the reducibility and dispersion of Co. They obtained 90.6% conversion of CH_4 and 88.5% conversion of CO_2 at the 88.6% H_2 selectivity and 87.5% CO selectivity in the dry reforming of CH_4 over the Co-Zr bimetallic catalyst.

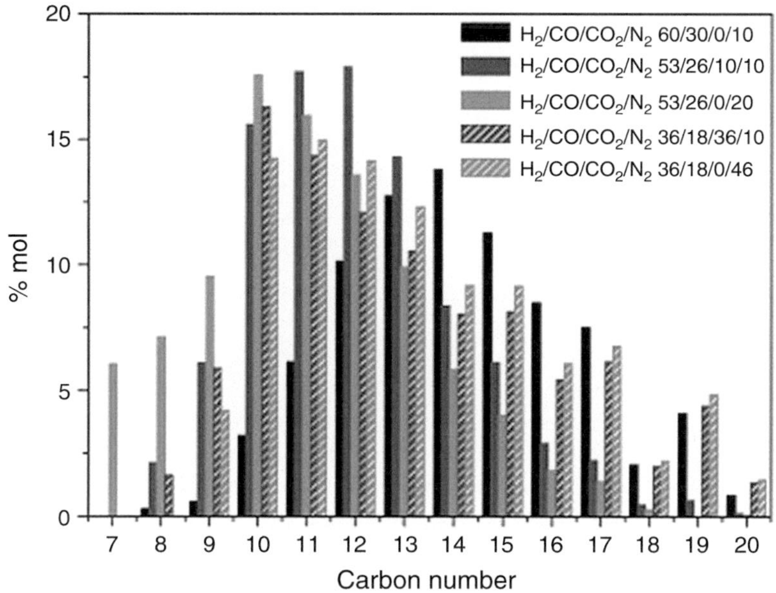

Fig. 1.5 Distribution of long-chain hydrocarbons, C_7–C_{20}, produced from CO_2 hydrogenation at variable concentrations of CO_2 in the feed gas (the higher content of CO_2 in the feed gas is favorable toward the production of gasoline fractions. The lower content of CO_2 in feed gas results in production of kerosene in Fischer-Tropsch). (Reprinted with permission of 2013 Elsevier B.V. from Diaz et al. 2014)

1.5 Electrochemical Reduction of CO_2 over a Cobalt Catalyst

1.5.1 Electrochemical Reduction of CO_2

During electrochemical reduction, the cathode and anode of an electrochemical reactor are fed by humidified CO_2 and water, and CO_2 is reduced to formate, formic acid, alcohols, and hydrocarbons. The selectivity of products depends on many process conditions such as electrocatalysts, electrolyte, and voltages (Yadav and Purkait 2016). The anion derivation of formic acid, $HCOO^-$, is usually generated in a basic condition, and the complement acid is formed in a neutral or acidic environment. It is needed to develop low-cost catalysts with a high selectivity and low operating potential during electroreduction while delivering a high current density for HCOOH and $HCOO^-$ generation from CO_2 reduction (Daiyan et al. 2017).

In formate formation via electrochemical reduction of CO_2, the radicals of CO_2 produced through electron transfer to CO_2 bind weakly to an electrocatalyst which is usually metals with empty p orbitals, like Sn, Pb, Hg, and In. As this step needs large activation energy, it is the rate-determining stage in CO_2 electrochemical reduction

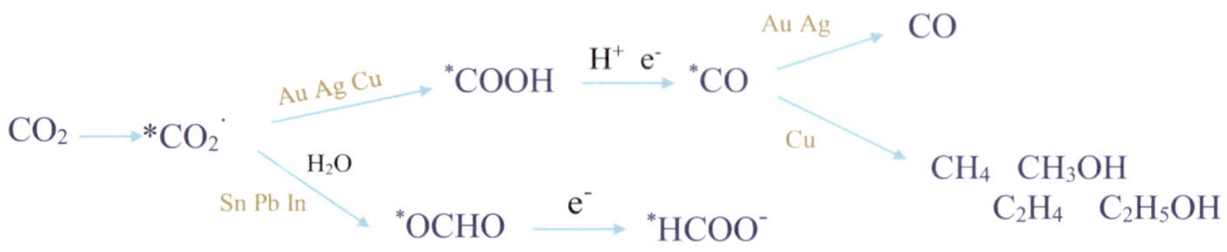

Fig. 1.6 Mechanism of electrochemical reduction of CO_2 on metal-based catalysts (the ability of the catalyst to bind the intermediate radicals and the strength of these bindings are important factors determining the rate and production of the electrochemical reduction. Cu (Copper) catalyst binds strongly to $^{*}CO$ intermediates). (Reprinted with permission of 2017 Wiley from Daiyan et al. 2017)

to formate. Then through the reaction of water and adsorbed CO_2 radicals on the electrocatalyst, the formate is generated. The production of strong binding between CO_2 radicals and electrocatalysts of Au, Ag, and Zn will generate only CO (Fig. 1.6).

Selection of an active and selective electrocatalyst for the electroreduction of CO_2 into hydrocarbons still has main technical obstacles. Metal electrocatalysts such as gold, copper, rubidium, and molybdenum have the issues of positing and over potentiality. Transition metal electrocatalysts such as Co-phthalocyanine (Lieber and Lewis 1984), Fe-porphyrins (Hammouche et al. 1991), and Ru-bipyridines (Ishida et al. 1987) have poor performance.

1.5.2 Cobalt-Based Electrocatalysts for Reduction of CO_2 into Formate

Copper is the most interesting and broadly studied catalyst for the electrochemical reduction of CO_2 because of its ability to transform CO_2 into different products including CO, HCOOH, CH_4, and C_2–C_3 hydrocarbons and alcohols. Several approaches such as bimetallic Cu-Co catalysts have been studied to address the challenges of low energy efficiency, low selectivity, high overpotential, competition with H_2 evolution that a Cu electrode faces (Yadav and Purkait 2016; Bernal et al. 2018). Furthermore, the electroreduction of CO_2 with complex cobalt-based electrocatalysts mainly produces CO.

Altering the atomic-scale structure of cobalt-based electrocatalysts conjugated with partially oxidation would increase their activity and selectivity for the production of formate (Yang et al. 2018). Synthesizing atomic layers of cobalt and their partial oxidization are very challenging. It was found that partially oxidized 4-atom-thick layers of cobalt, as shown in Fig. 1.7, had a faradaic efficacy at 90.1% and current density of 10.59 mA cm^{-2} at lower potential of -0.25 V than the bulk cobalt (Gao et al. 2016a, b, 2017). Table 1.3 provides the performance data of cobalt-based electro catalysts for CO_2 reduction in literature.

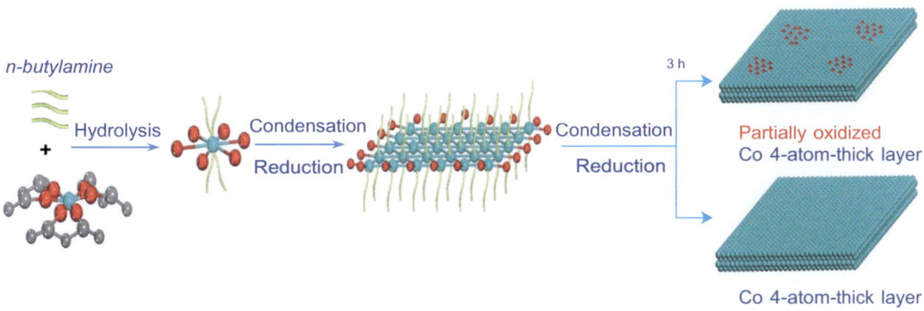

Fig. 1.7 Schematic production procedure of partially oxidized atomic cobalt layer (dimethylformamide reduces the cobalt ions in a controlled way to produce partially oxidized and pure Co atomic layers. Agglomeration is prevented by n-butylamine through its surface energy reduction effect). (Reprinted with permission of 2016 Wiley from Gao et al. 2016a, b)

Table 1.3 Performance of cobalt-based electrocatalyst in carbon dioxide reduction

Catalyst	Electrolyte solution	Potential	CO [%]	$HCOO^-$ [%]	Stability
Ag-co bimetallic catalyst	(1) 0.5 M $KHCO_3$ (2) $TBAPF_6$ in MDF	−2 V vs SHE	7.8	N.A	N.A
Molecular co complexes	0.1 M NBu_4BF_4 in MDF + various w% H_2O	−2 V vs $F_c^{+/o}$	< 1	90	~ 1 h
Co_3O_4 single-unit-cell layer	0.1 M $KHCO_3$	−0.87 V vs SCE	< 10	>85	40 h
Cobalt protoporphyrin	0.1 M $HCLO_4$	−0.6 V vs RHE	60	N.A	1 h
Co + w% pd. + w% K	N.A	N.A	< 11	N.A	N.A
Atomic cobalt layers	0.1 M Na_2SO_4	−0.85 V vs SCE	N.a	~ 90	60 h
Ultrathin Co_3O_4	0.1 M $KHCO_3$	−0.88 V vs SCE	N.A	60	20 h
Co_3O_4 nanofibers	0.1 M $TBAPF_6$ in ACN + $1\%_{vol}$ H_2O	−1.5 V vs NHE	65	27	8 h

Modified after Aljabour et al. (2018)
DMF dimethylformamide, *TBAPF6* tetrabutylammonium hexafluorophosphate, *CAN* acetonitrile

1.5.3 Cobalt Phthalocyanines and Cobalt Porphyrins for CO_2 Reduction

Transition metal in phthalocyanines and porphyrins has attracted much attention of research in electrochemical reduction of CO_2 to hydrocarbon fuels. Their structure is comprised of transition metals with adjustable oxidation levels in the center of macrocyclic ligand matrices. The studies of transition metal phthalocyanines and

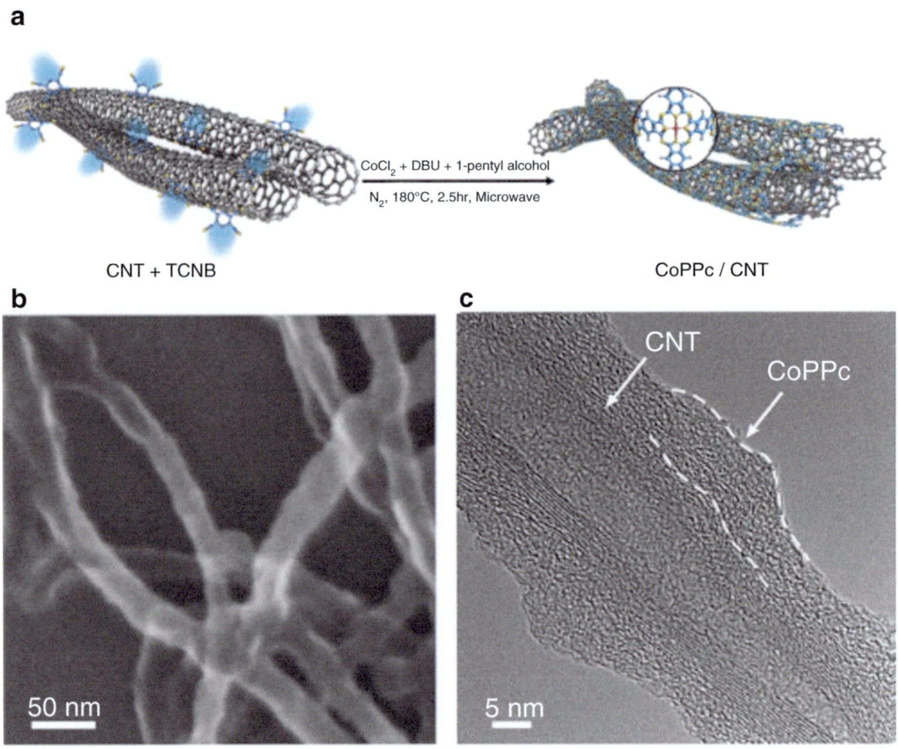

Fig. 1.8 (**a**) Schematic of cobalt polyphthalocyanine synthesis, (**b**) scanning electron microscopy, and (**c**) transmission electron microscopy of cobalt polyphthalocyanine/CNT (in situ polymerization of cobalt phthalocyanine produces a directed polymer layer around carbon nanotubes. An increase in the selectivity of carbon dioxide reduction to carbon monoxide with a significant faradaic efficiency can be achieved through promoting the physical and chemical robustness of cobalt polyphthalocyanine. CNT, carbon nanotube; TCNB, 1,2,4,5-tetracyanobenzene; CoPPc, cobalt polyphthalocyanine). (Reprinted with permission of 2017 Wiley from Han et al. 2017)

porphyrins and their derived compounds have continued for more than four decades. However, there are still a set of drawbacks in their applications such as low faradic efficiency, low reduction current density, elevated overpotential, and lack of stability. Therefore, to address these problems, modification of their architecture could be an attractive route to increase their exposed active sites and tune their electron transmission kinetics. An effort to in situ polymerization of cobalt phthalocyanine produces a directed polymer layer around carbon nanotubes as shown in Fig. 1.8, promoting the physical and chemical robustness of cobalt polyphthalocyanine, rising electrochemically active surface area on carbon nanotubes, and inhibition of organic molecules aggregation, subsequently leading to increase the selectivity of carbon dioxide reduction to carbon monoxide with a significant faradaic efficiency (Han et al. 2017).

1.6 Photocatalytic Reduction of CO_2 over a Cobalt Catalyst

CO_2 can be photocatalytically reduced to produce formic acid, methanol, and formaldehyde in an aqueous solution. TiO_2 is the most widely studied photocatalyst for photocatalytic reduction of CO_2. It was found that the incorporation of cobalt into TiO_2/ZSM-5 by a sol-gel method could increase the photocatalytic reduction efficiency from CO_2 to CH_3OH (Wang et al. 2015). Phthalocyanines (Pc) which have unique properties including excellent semi-conductivity, photoconductivity, chemical stability, and optical absorption in the UV-vis region has been used to develop photocatalysts. Zhao et al. (2009) reported that the total organic carbon yield from the photocatalytic reduction of CO_2 using an in situ CoPc/TiO_2 catalyst was much higher than those of TiO_2 and physical absorbed CoPc/TiO_2 (Zhao et al. 2009).

1.7 Conclusions

Cobalt-based catalysts are widely used in Fischer-Tropsch synthesis of liquid fuels from syngas as they have high performance for the hydrogenation of CO to long-chain hydrocarbons due to their high activity and good selectivity and superior stability. Cobalt and iron Fischer-Tropsch catalysts behave differently in the presence of CO_2. Iron-based catalysts have proved to be the most successful to catalyze reverse water-gas shift reaction for the conversion of CO_2 to CO and further catalytically convert CO to hydrocarbons. If H_2 and CO_2 are used as feeding gases instead of H_2 and CO, a cobalt catalyst becomes selective toward methanation because a cobalt catalyst lacks the ability for catalyzing reverse water-gas shift reaction. An active component for the RWSG reaction such as iron and copper can be incorporated to a cobalt-based catalyst for CO_2 hydrogenation. Proper catalyst supports and structures can further improve the performance of cobalt-based catalysts for CO_2 hydrogenation. The pressure and ratio of the feed gas was found to have significant effects on the distribution of Fischer-Tropsch synthesized hydrocarbons. There are several other approaches to produce chemicals and fuels from CO_2 in chemical processes over a cobalt-based catalyst, which include hydrogenation of CO_2 in a basic solution, the use of CO_2 to reform CH_4 to produce syngas for the subsequent synthesis of Fischer-Tropsch liquids, and electrochemical or photochemical reduction of CO_2. Several methods including the modification of catalyst supports and the use of promoters have been studied to design proper cobalt-based catalysts for those processes.

Acknowledgments A contribution of North Carolina Agricultural and Technical State University, supported by funds provided by US National Scientific Foundation (Grant #: HRD-1736173). Mention of a trade name, proprietary products, or company name is for presentation clarity and does not imply endorsement by the authors or the university.

References

Aga H, Aramata A, Hisaeda Y (1997) The electroreduction of carbon dioxide by macrocyclic cobalt complexes chemically modified on a glassy carbon electrode. J Electroanal Chem 437:111–118. https://doi.org/10.1016/S0022-0728(97)00386-0

Aljabour A, Coskun H, Apaydin DH, Ozel F, Hassel AW, Stadler P, Sariciftci NS, Kus M (2018) Nanofibrous cobalt oxide for electrocatalysis of CO_2 reduction to carbon monoxide and formate in an acetonitrile-water electrolyte solution. Appl Catal B Environ 229:163–170. https://doi.org/10.1016/j.apcatb.2018.02.017

Aziz MAA, Jali AA, Triwahvono S, Ahmad A (2015) CO_2 methanation over heterogeneous catalysts: recent progress and future prospects. Green Chem 17:2647–2663. https://doi.org/10.1039/C5GC00119F

Bernal M, Bagger A, Scholten F, Sinew I, Bergmann A, Ahmadi M, Rossmeisl CBR (2018) CO_2 electroreduction on copper-cobalt nanoparticles: size and composition effect. Nano Energy 53:27–36. https://doi.org/10.1016/j.nanoen.2018.08.027

Bonin J, Maurin A, Robert M (2017) Molecular catalysis of the electrochemical and photochemical reduction of CO_2 with Fe and Co metal based complexes. Recent advances. Coord Chem Rev 334:184–198. https://doi.org/10.1016/j.ccr.2016.09.005

Boreriboon N, Jiang X, Song C, Prasassarakich P (2018) Fe-based bimetallic catalysts supported on TiO2 for selective CO_2 hydrogenation to hydrocarbons. J CO2 Util 25:330–337. https://doi.org/10.1016/j.jcou.2018.02.014

Daiyan R, Lu X, Ng YH, Amal R (2017) Liquid hydrocarbon production from CO_2: recent development in metal-based electrocatalysis. ChemSusChem 10(22):4342–4358. https://doi.org/10.1002/slct.201601980

Das T, Deo G (2011) Synthesis, characterization and in situ DRIFTS during the CO_2 hydrogenation reaction over supported cobalt catalysts. J Mol Catal A Chem 350:75–82. https://doi.org/10.1016/j.molcata.2011.09.008

Díaz JA, Osa AR, Sánchez P, Romero A, Valverde JL (2014) Influence of CO_2 co-feeding on Fischer–Tropsch fuels production over carbon nanofibers supported cobalt catalyst. Catal Commun 44:57–61. https://doi.org/10.1016/j.catcom.2013.07.033

Diez-Ramirez J, Sanchez P, Kvriakou V, Zafeiratos S, Marnellos GE, Konsolakis M, Dorado F (2017) Effect of support nature on the cobalt-catalyzed CO_2 hydrogenation. J CO2 Util 21:562–571. https://doi.org/10.1016/j.jcou.2017.08.019

Dorner RW, Hardy DR, Williams FW, Davis BH, Willauer HD (2009) Influence of gas feed composition and pressure on the catalytic conversion of CO_2 to hydrocarbons using a traditional cobalt-based Fischer–Tropsch catalyst. Energy Fuel 23(8):4190–4195. https://doi.org/10.1021/ef900275m

Dry ME (1996) Practical and theoretical aspects of the catalytic Fischer-Tropsch process. Appl Catal A Gen 138:319–344. https://doi.org/10.1016/0926-860X(95)00306-1

Dry ME (2002) The Fischer-Tropsch process. Catal Today 71:227–241. https://doi.org/10.1016/S0920-5861(01)00453-9

Ferreira-Aparicio P, Guerrero-Ruiz A, Rodriguez-Ramos I (1998) Comparative study at low and medium reaction temperatures of syngas production by methane reforming with carbon dioxide over silica and alumina supported catalysts. Appl Catal A Gen 170:177–187. https://doi.org/10.1016/S0926-860X(98)00048-9

Gao S, Jiao X, Sun Z, Zhang W, Sun Y, Wang C, Hu Q, Zu X, Yang F, Yang S, Liang L, Wu J, Xie Y (2016a) Ultrathin Co_3O_4 layers realizing optimized CO_2 electroreduction to formate. Angew Chem Int Ed 55(2):698–702. https://doi.org/10.1002/anie.201509800

Gao S, Lin Y, Jiao X, Sun Y, Luo Q, Zhang W, Li D, Yang J, Xie Y (2016b) Partially oxidized atomic cobalt layers for carbon dioxide electroreduction to liquid fuel. Nature 529:68–71. https://doi.org/10.1038/nature16455

Gao S, Sun Z, Liu W, Jiao X, Zu X, Hu Q, Sun Y, Yao T, Zhang W, Wei S, Xie Y (2017) Atomic layer confined vacancies for atomic-level insights into carbon dioxide electroreduction. Nat Commun 8:14503. https://doi.org/10.1038/ncomms14503

Gnanamani MK, Shafer WD, Sparks DE, Davis BH (2011) Fischer–Tropsch synthesis: effect of CO_2 containing syngas over Pt promoted Co/γ-Al_2O_3 and K-promoted Fe catalysts. Catal Commun 12:936–939. https://doi.org/10.1016/j.catcom.2011.03.002

Gnanamani MK, Jacobs G, Keogh RA, Shafer WD, Sparks DE, Hopps SD, Thomas GA, Davis BH (2015) Fischer-Tropsch synthesis: effect of pretreatment conditions of cobalt on activity and selectivity for hydrogenation of carbon dioxide. Appl Catal A Gen 499:39–46. https://doi.org/10.1016/j.apcata.2015.03.046

Gnanamani MK, Jacobs G, Hamdeh HH, Shafer WD, Liu F, Hopps SD, Thomas GA, Davis BH (2016) Hydrogenation of carbon dioxide over Co–Fe bimetallic catalysts. ACS Catal 6 (2):913–927. https://doi.org/10.1021/acscatal.5b01346

Goodman ED, Schwalbe JA, Cargnello M (2017) Mechanistic understanding and the rational design of sinter-resistant heterogeneous catalysts. ACS Catal 7:7156–7173. https://doi.org/10.1021/acscatal.7b01975

Guettel R, Kunz U, Turek T (2008) Reactors for Fischer-Tropsch synthesis. Chem Eng Technol 31 (5):746–754. https://doi.org/10.1002/ceat.200800023

Hammouche M, Lexa D, Momenteau M, Savean JM (1991) Chemical catalysis of electrochemical reactions. Homogeneous catalysis of the electrochemical reduction of carbon dioxide by iron (O) porphyrins. Role of the addition of magnesium cations. J Am Chem Soc 113:8455–8466. https://doi.org/10.1021/ja00022a038

Han N, Wang Y, Ma L, Wen J, Li J, Zheng H, Nie K, Wang X, Zhao F, Li Y, Fan J, Zhong J, Wu T, Miller DJ, Lu J, Lee ST, Li Y (2017) Supported cobalt polyphthalocyanine for high-performance electrocatalytic CO_2 reduction. Chem 3(4):652–664. https://doi.org/10.1016/j.chempr.2017.08.002

Iglesia E, Reyes SC, Madon RJ, Soled RS (1993) Selectivity control and catalyst design in the Fischer-Tropsch synthesis: sites, pellets, and reactors. Adv Catal 39:221–302. https://doi.org/10.1016/S0360-0564(08)60579-9

Ishida H, Tanaka K, Tanaka T (1987) Electrochemical CO_2 reduction catalyzed by [Ru(bpy)2(CO)2]2+ and [Ru(bpy)2(CO)Cl]+. The effect of pH on the formation of CO and HCOO−. Organometallics 6:181–186. https://doi.org/10.1021/om00144a033

Ishida T, Yanagihara T, Liu X, Ohashi H, Hamasaki A, Honma T, Oji H, Yokoyama T, Tokunaga M (2013) Synthesis of higher alcohols by Fischer–Tropsch synthesis over alkali metal-modified cobalt catalysts. Appl Catal A Gen 458:145–154. https://doi.org/10.1016/j.apcata.2013.03.042

Jacobs G, Das TK, Zhang Y, Li J, Racoillet G, Davis BH (2002) Fischer-Tropsch synthesis: support, loading, and promoter effects on the reducibility of cobalt catalysts. Appl Catal A Gen 233:263–281. https://doi.org/10.1016/S0926-860X(02)00195-3

Ji L, Tang S, Zeng HC, Lin J, Tan KL (2001) CO_2 reforming of methane to synthesis gas over sol–gel-made Co/γ-Al_2O_3 catalysts from organometallic precursors. Appl Catal A Gen 207:247–255. https://doi.org/10.1016/S0926-860X(00)00659-1

Khodakov AY, Chu W, Fongarland P (2007) Advances in the development of novel cobalt Fischer−Tropsch catalysts for synthesis of long-chain hydrocarbons and clean fuels. Chem Rev 107:1692–1744. https://doi.org/10.1021/cr050972v

Lieber CM, Lewis NS (1984) Catalytic reduction of CO_2 at carbon electrodes modified with cobalt phthalocyanine. J Am Chem Soc 106:5033–5034. https://doi.org/10.1021/ja00329a082

Liu B, Ouyang B, Zhang Y, Lv K, Li Q, Ding Y, Li J (2018) Effects of mesoporous structure and Pt promoter on the activity of Co-based catalysts in low-temperature CO_2 hydrogenation for higher alcohol synthesis. J Catal 366:91–97. https://doi.org/10.1016/j.jcat.2018.07.019

Mirzaei F, Rezaei M, Meshkani F, Fattah Z (2015) Carbon dioxide reforming of methane for syngas production over co–MgO mixed oxide nanocatalysts. J Ind Eng Chem 21:662–667. https://doi.org/10.1016/j.jiec.2014.03.034

Olah GA, Goeppert A, Prakash GKS (2009) Chemical recycling of carbon dioxide to methanol and dimethyl ether: from greenhouse gas to renewable, environmentally carbon neutral fuels and synthetic hydrocarbons. J Org Chem 74(2):487–498. https://doi.org/10.1021/jo801260f

Owen RE, O'Byrne JP, Mattia D, Plucinski P, Pascua SI, Jones MD (2013) Cobalt catalysts for the conversion of CO_2 to light hydrocarbons at atmospheric pressure. Chem Commun 49 (99):11683–11685. https://doi.org/10.1039/C3CC46791K

Owen RE, Plucinski P, Mattia D, Torrente-Murciano L, Ting VP, Jones MD (2016) Effect of support of Co-Na-Mo catalysts on the direct conversion of CO_2 to hydrocarbons. J CO2 Util 16:97–103. https://doi.org/10.1016/j.jcou.2016.06.009

Park JH, Yeo S, Kang TJ, Shin HR, Heo I, Chang TS (2018) Effect of Zn promoter on catalytic activity and stability of Co/ZrO2 catalyst for dry reforming of CH_4. J CO2 Util 23:10–19. https://doi.org/10.1016/j.jcou.2017.11.002

Patel P, Nandi S, Maru MS, Kureshy RI, Khan NH (2018) Nitrogen-rich graphitic carbon stabilized cobalt nanoparticles as an effective heterogeneous catalyst for hydrogenation of CO_2 to formate. J CO2 Util 25:310–314. https://doi.org/10.1016/j.jcou.2018.04.015

Pei Y, Li Z, Li Y (2017) Highly active and selective Co-based Fischer–Tropsch catalysts derived from metal–organic frameworks. AIChE J 63(7):2935–2944. https://doi.org/10.1002/aic.15677

Porosoff MD, Yan B, Chen JG (2016) Catalytic reduction of CO_2 by H_2 for synthesis of CO, methanol and hydrocarbons: challenges and opportunities. Energy Environ Sci 9:62–73. https://doi.org/10.1039/C5EE02657A

Porosoff MD, Baldwin JW, Peng X, Mpourmpakis G, Willauer HD (2017) Potassium-promoted molybdenum carbide as a highly active and selective catalyst for CO_2 conversion to CO. ChemSusChem 10:2408–2415. https://doi.org/10.1002/cssc.201700412

Rafati M, Wang L, Shahbazi A (2015) Effect of silica and alumina promoters on co-precipitated Fe-cu-k based catalysts for the enhancement of CO_2 utilization during Fischer-Tropsch synthesis. J CO2 Util 12:34–42. https://doi.org/10.1016/j.jcou.2015.10.002

Rafati M, Wang L, Dayton DC, Schimmel K, Kabadi V, Shahbazi A (2017) Techno-economic analysis of production of Fischer-Tropsch liquids via biomass gasification: the effects of Fischer-Tropsch catalysts and natural gas co-feeding. Energy Convers Manag 133:153–166. https://doi.org/10.1016/j.enconman.2016.11.051

Riduan SN, Zhang Y (2010) Recent developments in carbon dioxide utilization under mild conditions. Dalton Trans 39(14):3347–3357. https://doi.org/10.1039/B920163G

Riedel T, Claeys M, Schulz H, Schaub G, Nam SS, Jun KW, Choi MJ, Kishan G, Lee KW (1999) Comparative study of Fischer–Tropsch synthesis with H_2/CO and H_2/CO_2 syngas using Fe- and Co-based catalysts. Appl Catal A Gen 186:201–213. https://doi.org/10.1016/S0926-860X(99)00173-8

Saeidi S, Amin NAS, Rahimpour MR (2014) Hydrogenation of CO_2 to value-added products-a review and potential future developments. J CO2 Util 5:66–81. https://doi.org/10.1016/j.jcou.2013.12.005

Satthawong R, Koizumi N, Song C, Prasassarakich P (2013) Bimetallic Fe–Co catalysts for CO_2 hydrogenation to higher hydrocarbons. J CO2 Util 3–4:102–106. https://doi.org/10.1016/j.jcou.2013.10.002

Schulz H (1999) Short history and present trends of Fischer–Tropsch synthesis. Appl Catal A Gen 186:3–12. https://doi.org/10.1016/S0926-860X(99)00160-X

Shi Z, Yang H, Gao P, Chen X, Liu H, Zhong L, Wang H, Wei W, Sun Y (2018a) Effect of alkali metals on the performance of $CoCu/TiO_2$ catalysts for CO_2 hydrogenation to long-chain hydrocarbons. Chin J Catal 39(8):1294–1302. https://doi.org/10.1016/S1872-2067(18)63086-4

Shi Z, Yang H, Gao P, Li X, Zhong L, Wang H, Liu H, Wei W, Sun Y (2018b) Direct conversion of CO_2 to long-chain hydrocarbon fuels over K–promoted $CoCu/TiO_2$ catalysts. Catal Today 311:65–73. https://doi.org/10.1016/j.cattod.2017.09.053

Smit E, Weckhuysen BM (2008) The renaissance of iron-based Fischer–Tropsch synthesis: on the multifaceted catalyst deactivation behavior. Chem Soc Rev 37:2758–2781. https://doi.org/10.1039/b805427d

Takanabe K, Nagaoka K, Nariai K, Aika K (2005) Titania-supported cobalt and nickel bimetallic catalysts for carbon dioxide reforming of methane. J Catal 232:268–275. https://doi.org/10.1016/j.jcat.2005.03.011

Vance CK, Bartholomew CH (1983) Hydrogenation of carbon dioxide on group viii metals: III, effects of support on activity/selectivity and adsorption properties of nickel. Appl Catal 7:169–177. https://doi.org/10.1016/0166-9834(83)80005-0

Visconti CG, Lietti L, Tronconi E, Forzatti P, Zennaro R, Finocchio E (2009) Fischer-Tropsch synthesis on a Co/Al_2O_3 catalyst with CO_2 containing syngas. Appl Catal A Gen 355:61–68. https://doi.org/10.1016/j.apcata.2008.11.027

Visconti CG, Martinelli M, Falbo L, Fratalocchi L, Lietti L (2016) CO_2 hydrogenation to hydrocarbons over Co and Fe-based Fischer-Tropsch catalyst. Catal Today 277:161–170. https://doi.org/10.1016/j.cattod.2016.04.0100920-5861/

Visconti CG, Martinelli M, Falbo L, Infantes-Molina A, Lietti L, Forzatti P, Iaquaniello G, Palo E, Picutti B, Brignoli F (2017) CO_2 hydrogenation to lower olefins on a high surface area K-promoted bulk Fe-catalyst. Appl Catal B Environ 200:530–542. https://doi.org/10.1016/j.apcatb.2016.07.047

Wang H, Miller JT, Shakouri M, Xi C, Wu T, Zhao H, Akatay MC (2012) XANES and EXAFS studies on metal nanoparticle growth and bimetallic interaction of Ni-based catalysts for CO_2 reforming of CH_4. Catal Today 207:3–12. https://doi.org/10.1016/j.cattod.2012.09.015

Wang J, Jin YH, Ouyang T, Zhang Q, Chang CT (2015) Photocatalytic reduction of CO_2 to energy products using Cu–TiO_2/ZSM-5 and Co–TiO_2/ZSM-5 under low energy irradiation. Catal Commun 59:69–72. https://doi.org/10.1016/j.catcom.2014.09.030

Wei J, Sun J, Wen Z, Fang C, Ge Q, Xu H (2016) New insights into the effect of sodium on Fe3O4-based nanocatalysts for CO_2 hydrogenation to light olefins. Cat Sci Technol 6:4786–4793. https://doi.org/10.1039/C6CY00160B

Yadav VSK, Purkait MK (2016) Effect of copper oxide electrocatalyst on CO_2 reduction using Co_3O_4 as anode. J Sci Adv Mater Devices 1:330–336. https://doi.org/10.1016/j.jsamd.2016.07.006

Yang G, Yu Z, Zhang J, Liang Z (2018) A highly efficient flower-like cobalt catalyst for electroreduction of carbon dioxide. Chin J Catal 39:914–919. https://doi.org/10.1016/S1872-2067(18)63021-9

Yao Y, Hildebrandt D, Glasser D, Liu X (2010) Fischer–Tropsch synthesis using H_2/CO/CO_2 syngas mixtures over a cobalt catalyst. Ind Eng Chem Res 49(21):11061–11066. https://doi.org/10.1021/ie100414y

Zhang Y, Jacobs G, Sparks DE, Dry ME, Davis BH (2002) CO and CO_2 hydrogenation study on supported cobalt Fischer-Tropsch synthesis catalysts. Catal Today 71:411–418. https://doi.org/10.1016/S0920-5861(01)00468-0

Zhang J, Wang H, Dalai AK (2007) Development of stable bimetallic catalysts for carbon dioxide reforming of methane. J Catal 249:300–310. https://doi.org/10.1016/j.jcat.2007.05.004

Zhang G, Hao L, Jia Y, du Y, Zhang Y (2015a) CO_2 reforming of CH_4 over efficient bimetallic Co-Zr/AC catalyst for H_2 production. Int J Hydrog Energy 40:12868–12879. https://doi.org/10.1016/j.ijhydene.2015.07.011

Zhang J, Lu S, Xiaojuan FS, Ma Q, Zhao T (2015b) Selective formation of light olefins from CO^2 hydrogenation over Fe–Zn–K catalysts. J CO2 Util 12:95–100. https://doi.org/10.1016/j.jcou.2015.05.004

Zhao Z, Fan J, Xie M, Wang Z (2009) Photo-catalytic reduction of carbon dioxide with in-situ synthesized CoPc/TiO_2 under visible light irradiation. J Clean Prod 17:1025–1029. https://doi.org/10.1016/j.jclepro.2009.02.016

Chapter 2
Conversion of Carbon Dioxide Using Lead/Composite/Oxide Electrode into Formate/Formic Acid

Xiaowei An, Akihiro Yoshida, Abuliti Abudula, and Guoqing Guan

Contents

2.1 Introduction 27
2.2 Electrode Composition 28
  2.2.1 Lead Metal 28
  2.2.2 Lead-Based Composites 30
  2.2.3 Lead Oxides 32
2.3 Catalytic Mechanism 33
2.4 Reactor and Electrode Type 35
  2.4.1 Traditional Electrode 35
  2.4.2 Gas Diffusion Electrodes 36
  2.4.3 Other Types 36
2.5 Effects of Operation Conditions 36
2.6 Conclusions 38
References 40

Abstract High concentration of carbon dioxide (CO_2) in the atmosphere has caused a series of environmental problems. Electrochemical reduction of carbon dioxide has attracted worldwide attention since it could effectively transfer this global warming gas into high-value-added chemicals. It is an important means of energy storage and conversion due to the use of clean electrical energy as a driving force. The core of

---

X. An
Energy Conversion Engineering Laboratory, Institute of Regional Innovation (IRI), Hirosaki University, Aomori, Japan

Graduate School of Science and Technology, Hirosaki University, Hirosaki, Japan
e-mail: h17ds251@hirosaki-u.ac.jp

A. Yoshida · G. Guan (✉)
Energy Conversion Engineering Laboratory, Institute of Regional Innovation (IRI), Hirosaki University, Aomori, Japan
e-mail: ayoshida@hirosaki-u.ac.jp; guan@hirosaki-u.ac.jp

A. Abudula
Graduate School of Science and Technology, Hirosaki University, Hirosaki, Japan
e-mail: abuliti@hirosaki-u.ac.jp

Inamuddin et al. (eds.), *Conversion of Carbon Dioxide into Hydrocarbons Vol. 1 Catalysis*, Environmental Chemistry for a Sustainable World 40,
https://doi.org/10.1007/978-3-030-28622-4_2

carbon dioxide electrochemical reduction lies in electrocatalysts. Among many electrocatalysts, lead (Pb), a metal with high hydrogen evolution overpotential, is often used as an electrocatalyst to electrochemically reduce carbon dioxide to formate/formic acid ($HCOO^-$/HCOOH).

This chapter summarizes the research results of the lead/composite/oxide electrodes for the carbon dioxide electroreduction since the nineteenth century, which includes the aspects of electrode composition, catalytic mechanism, reactor, and electrode form and the influence of the reaction conditions in the catalytic process. It points out that the current research on lead-based electrocatalysts for electrochemical reduction of carbon dioxide is still insufficient, and more different nanocatalysts with different compositions and surface morphologies should be developed. Moreover, the outlook on the research and development of such lead/composite/oxide electrocatalysts is given based on how to design electrocatalysts with low overpotential, high current density, and long-term stability.

Keywords Carbon dioxide · Lead · Formate · Formic acid · Electrocatalysts · Electroreduction · Faradaic efficiency · Overpotential · Current density · Catalytic mechanism

Abbreviations

Ag	Silver
Au	Gold
C_2H_5OH	Ethanol
Cd	Cadmium
CH_3OH	Methanol
CH_4	Methane
CNT	Carbon nanotubes
CO	Carbon monoxide
CO_2	Carbon dioxide
Cu	Copper
Ga	Gallium
GDE	Gas diffusion electrode
H^+	Proton
$HCOO^-$	Formate
HCOOH	Formic acid
Hg	Mercury
HRSEM	High-resolution scanning electron microscope
In	Indium
K_2CO_3	Potassium carbonate
KCl	Potassium chloride
$KHCO_3$	Potassium bicarbonate
$NaHCO_3$	Sodium bicarbonate
NaOH	Sodium hydroxide

NHE	Normal hydrogen electrode
Ni	Nickel
Pb	Lead
Pb^0	Metal lead
PbO	Lead oxide
Pd	Palladium
Pt	Platinum
RHE	Reversible hydrogen electrode
SEM	Scanning electron microscope
Sn	Tin
SnO_x	Tin oxide species
Ti	Titanium
Zn	Zinc

2.1 Introduction

Since industrialization, environmental problems caused by excessive carbon dioxide (CO_2) emissions seriously damage the earth's ecological environment (Pachauri et al. 2014). It is becoming more and more urgent to reduce carbon dioxide emission and/or capture of the carbon dioxide. On the other hand, carbon dioxide, which is also a natural carbon source on the earth, can be converted into hydrocarbons or other valuable chemicals or enter the biological chain through photosynthesis of plants. Recently, photochemical/electrochemical conversion of carbon dioxide simulated plant photosynthesis is considered to be a promising carbon dioxide utilization method. It is driven by renewable solar energy or surplus electric energy with the advantages of mild and controllable reaction condition, simple reaction device, and convenient operation and short reaction time (Costentin et al. 2013).

By electrochemical method, carbon dioxide can be reduced to formate/formic acid ($HCOO^-$/HCOOH) (Zheng et al. 2017), carbon monoxide (CO) (Li et al. 2018), methane (CH_4) (Liu et al. 2018), methanol (CH_3OH) (Lu et al. 2018), ethanol (C_2H_5OH) (Liu et al. 2017), and other valuable liquid fuels. Among them, formic acid is an important basic organic chemical which can be used as a fuel, a means of hydrogen storage, or a feedstock in the synthesis of other fine chemicals (Zheng et al. 2017). Thusly, electroreduction of carbon dioxide to formic acid is considered as an economically feasible way to utilize carbon dioxide resource (Verma et al. 2016).

In recent years, various metal-based electrocatalysts have been developed for the reduction of carbon dioxide to formic acid (Zhang et al. 2018). Among the metal-based electrocatalysts, the low price of lead (Pb 1.9 $ vs. Sn 19.9 $ and Cd 2.5 $ per kilogram) makes it the most competitive catalyst in the practical process (García et al. 2018). Therefore, despite it being considered as a toxic metal, it still attracted great attention in electrochemical reduction of carbon dioxide.

Lead is the nonradioactive element with the largest atomic weight. Owing to its high hydrogen evolution overpotential, it can use as the electrocatalyst for the electrochemical reduction of carbon dioxide. However, compared to other metals, perhaps because of its toxicity, only few studies can be found on its applications for the carbon dioxide electroreduction.

This chapter will discuss the research progress of carbon dioxide electroreduction with the lead/composite/oxide electrodes in terms of electrode material composition, catalytic mechanism, reactor and electrode form, and the influence of the reaction conditions in the catalytic process and give the outlook on the research and development of such lead/composite/oxides electrocatalysts based on how to design electrocatalysts with low overpotential, high current density, and long-term stability.

2.2 Electrode Composition

2.2.1 Lead Metal

In 1994, Hoi et al. used a series of metal electrodes for the electrochemical reduction of carbon dioxide and found that there are two main groups of metal electrocatalysts. One is used to catalyze the reduction of carbon dioxide to formic acid, which includes lead (Pb), **mercury (Hg), indium (In), tin (Sn), cadmium (Cd), and titanium (Ti),** and the other is to reduce carbon dioxide to carbon monoxide, such as copper (Cu), gold (Au), silver (Ag), zine (Zn), **palladium (Pd)**, gallium(Ga), nickel (Ni), and **platinum (Pt)**. Among them, by using the lead electrode, a faradaic efficiency of 97.4% at −1.63 V (vs. normal hydrogen electrode) can be obtained (Hori et al. 1994). Thereafter, a faradaic efficiency close to 100% at a carbon dioxide gas pressure of 60 atmosphere pressure and a constant current density of 200 mA cm^{-2} was obtained by Todoroki et al. (1995). In 1995, the effect of temperature on the electrochemical reduction of carbon dioxide using lead-based electrode under a high pressure was investigated by Mizuno et al. They found that the highest faradaic efficiency reached 91.2% at 60 °C (Mizuno et al. 1995). These researches on the electrochemical reduction of carbon dioxide by using lead electrodes are primarily focused on the bulk metal lead, mainly for the optimization of operating conditions such as temperature and pressure. The faradic efficiency obtained differs from the lead materials used and the experimental equipments in the different research teams. However, it can be observed in these results that lead metal is an excellent electrocatalyst for the reduction of carbon dioxide to formic acid.

After entering the twentieth century, the researchers found that the surface structure and morphology of the catalyst could greatly affect the catalytic performance. The porous and nanostructured surface morphology of catalyst could enhance the electrochemically active surface area of the electrode, resulting in the increase in the current density. For example, Wang et al. prepared a three-dimensional porous lead electrode on a copper substrate by electrodeposition

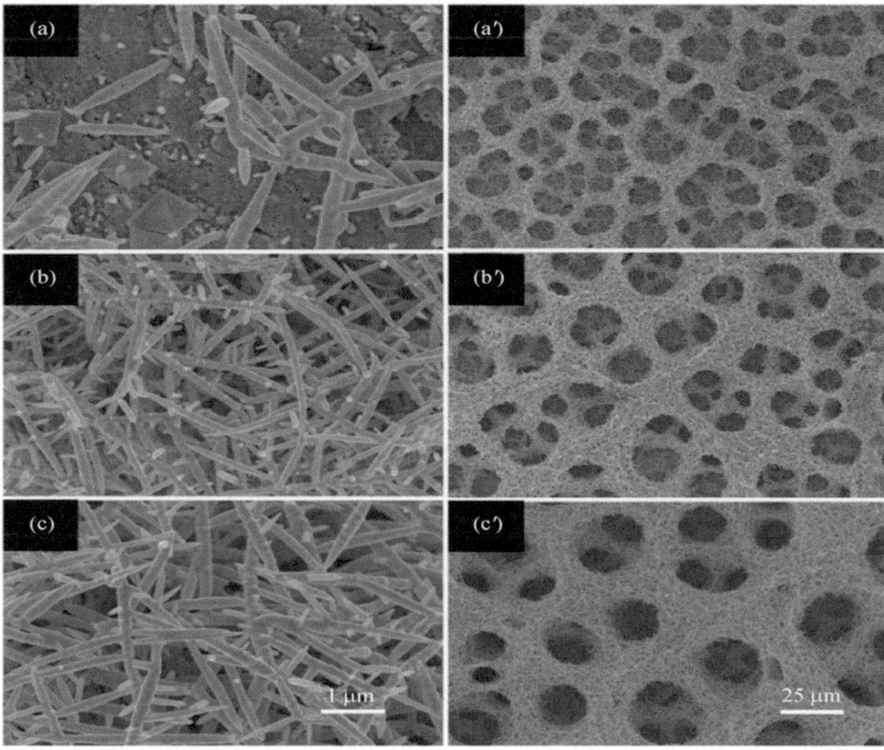

Fig. 2.1 Scanning electron microscope (SEM) images of the porous Pb foams deposited for (a, a′) 3 s, (b, b′) 5 s, and (c, c′) 20 s at -6 A cm^{-2}. The three-dimensional porous lead electrodes were obtained on a copper substrate by electrodeposition method. (Reprinted from Wang et al. 2016 with permission of Springer)

method (Fig. 2.1). At a potential of -1.7 V, a faradaic efficiency of 96.8% was obtained with the current density of about 5 mA cm^{-2}. Compared to the Pb plate, the performance was greatly improved (Wang et al. 2016). Fan et al. prepared a dendrite-like lead film with porous structure by a dynamic hydrogen-bubble templating method for carbon dioxide reduction. As shown in Fig. 2.2, the special structure and morphology make the electrochemical active surface area increased up to 1500 cm^2. As a result, in $KHCO_3$ electrolyte, the faradaic efficiency of formate reached 97% with the current density of 7.5 mA cm^{-2} at the -0.99 V overpotential (vs. reversible hydrogen electrode) after electrolysis for 6 h (Fan et al. 2017). However, for the development of nanostructured lead-based electrocatalysts, the current research work is still not enough for their large-scale applications. Although some works have achieved more than 90% of faradaic efficiency, there still have shortcomings such as high overpotential and low current density. The future researches should focus on the development of nanoscale electrocatalysts with different nanostructures and morphologies for the electroreduction of carbon dioxide.

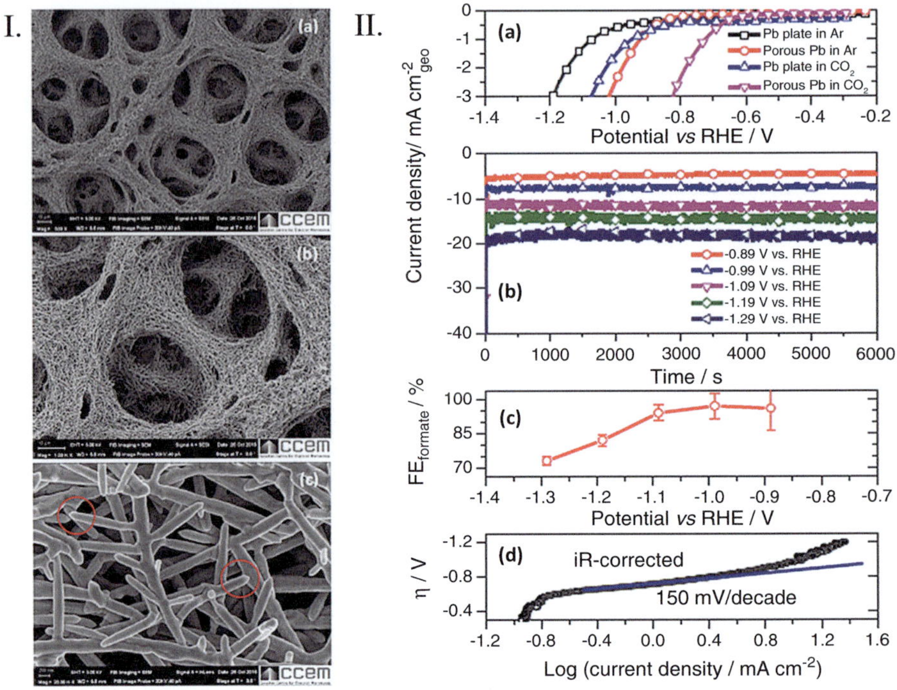

Fig. 2.2 High-resolution scanning electron microscope (HRSEM) images of a porous Pb film deposited at $-4.0\ \text{A cm}^{-2}$ during 40 s. Three different magnifications of the same electrode are shown in (I-**a**), (I-**b**), and (I-**c**). The red circles are representative examples of tips that were used to measure the tip diameter. In (II-**a**), linear sweep voltammeter of a Pb plate and porous Pb films in Ar- and CO_2-saturated 1 M $KHCO_3$. The scan rate was 5 mV s^{-1}. In (II-**b**), variation of the CO_2 electroreduction current density at different electrolysis potentials. In (II-**c**), variation of the formate faradic efficiency with respect to the electrode potential. In (II-**d**), variation of the CO_2 reduction overpotential with respect to the logarithm of the current density. The porous Pb film was deposited at $-4.0\ \text{A cm}^{-2}$ during 40 s. A dendrite-like porous Pb film shows excellent electrocatalytic performance for conversion of carbon dioxide to formate. FE, faradaic efficiency; RHE, reversible hydrogen electrode. (Reprinted from Fan et al. 2017 with permission of Royal Society of Chemistry)

2.2.2 Lead-Based Composites

Since simple lead metal electrocatalysts generally have disadvantages such as high overpotential and low current density, the high temperature and high pressure are always necessary to achieve a high current density. This will make the design of the reactor more complicated in the actual operation. To solve these problems, development of lead composite electrocatalysts has been considered. In 2015, Park et al. prepared tin-lead (Sn-Pb) alloys by the electrodeposition method and used for the electrochemical reduction of carbon dioxide. Their results indicate that the tin and lead combination can hinder the lead oxide (PbO) film formation with low

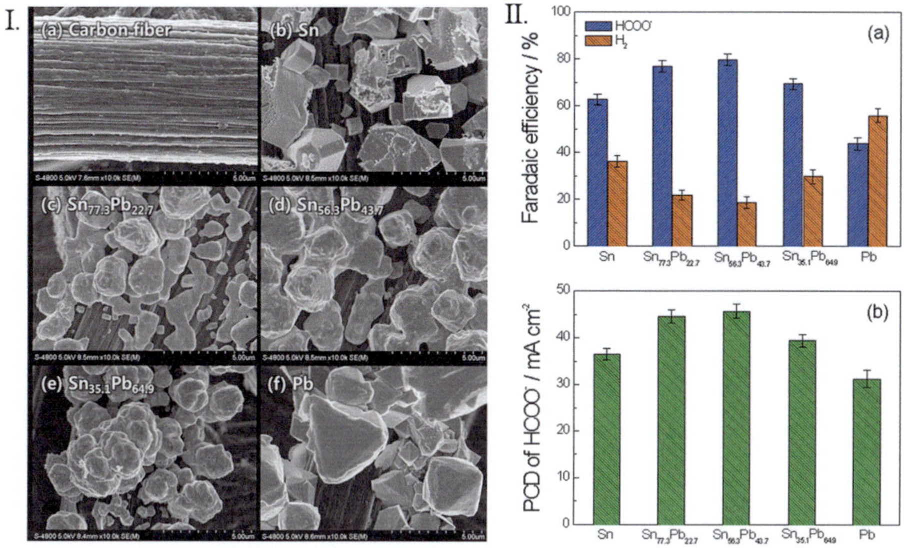

Fig. 2.3 Scanning electron microscope (SEM) images of (I-**a**) pristine carbon paper and deposits of electrodes obtained by electrodeposition with different compositions: (I-**b**) Sn, (I-**c**) $Sn_{77.3}Pb_{22.7}$, (I-**d**) $Sn_{56.3}Pb_{43.7}$, (I-**e**) $Sn_{35.1}Pb_{64.9}$, and (f) Pb. The comparison results of (II-**a**) faradaic efficiencies of $HCOO^-$ and H_2 and (II-**b**) partial current density of $HCOO^-$ for the Sn-Pb alloys with Sn and Pb after 2 h of electrolysis at −2.0 V (vs. Ag/AgCl). The highest faradic efficiency of 79.8% with the highest partial current density of 45.7 mA cm^{-2} was obtained by using $Sn_{56.3}Pb_{43.7}$ alloy catalyst. PCD, partial current density. (Reprinted from Choi et al. 2016 with permission of American Chemical Society)

conductivity on the surface of the alloy but result in the formation of the surface tin oxide species (SnO_x) and metal lead (Pb^0). By using this alloy electrocatalyst, the faradaic efficiency and the partial current density of formate production obtained at −2.0 V (vs. Ag/AgCl) were 16% and 25% higher increases compared to the single metal electrodes. The highest faradic efficiency and partial current density were 79.8% and 45.7 mA cm^{-2}, respectively, by using $Sn_{56.3}Pb_{43.7}$ alloy catalyst (Fig. 2.3) (Choi et al. 2016).

Wang et al. developed a catalyst with three-dimensional copper-lead (Cu-Pb) hierarchical nanostructure by electrodepositing-annealing-electroreduction approach. When it was used for the electrochemical reduction of carbon dioxide, the hydrogen evolution was suppressed by 35.6%, but carbon monoxide and formic acid were enhanced by 29.6% and 9.2%, respectively, as compared with the copper nanowires (Wang et al. 2019). Compared to the copper metal electrode, the selectivity of this material was improved, but compared to the pure lead electrode, it had worse selectivity. Apart from compositing with other metals, compositing of lead with the commonly used carbon-based materials was also considered. For example, García et al. deposited lead nanoparticles (5–10 nm) on carbon nanotubes (CNT) by using supercritical fluids to obtain lead/CNT catalysts and then evaluated its electrocatalytic activity for the reduction of carbon dioxide by using a continuous operation cell in gas phase. Herein, the effects of anolyte concentration, carbon dioxide flow rate, temperature, and current density on the formate formation rate were investigated (García et al. 2018). Compared to pure lead metal, the current density was improved, but the formic acid selectivity was decreased to some extent. Therefore, there is still large exploration space to improve the performance of lead composite electrodes.

2.2.3 *Lead Oxides*

It is found that the surfaces of many metals are easy to be oxidized to the metal oxide films, but these films cannot be completely reduced to metal in the reduction of carbon dioxide. However, this film also plays an important role in the electrochemical reduction of carbon dioxide. Especially, it is confirmed that the tin oxide film could effectively inhibit the evolution of hydrogen (Chen and Kanan 2012; Cui et al. 2016), and similarly, this phenomenon also exists on the lead electrode. Lee and Kanan prepared nanocrystalline lead films by reducing PbO_2 precursors and used for carbon dioxide reduction. Comparing with the polycrystalline lead foil, the nanocrystalline lead film had lower hydrogen ion reduction activity but maintained the excellent carbon dioxide reduction activity (Fig. 2.4). Herein, it is found that the difference of the two electrodes was the coverage of a metastable lead oxide surface layer, which is passivating for H^+ reduction but active for carbon dioxide reduction (Lee and Kanan 2014). Thereafter, Yadav and Purkait synthesized Pb_2O electrocatalyst for the carbon dioxide electroreduction. However, only 60% formic

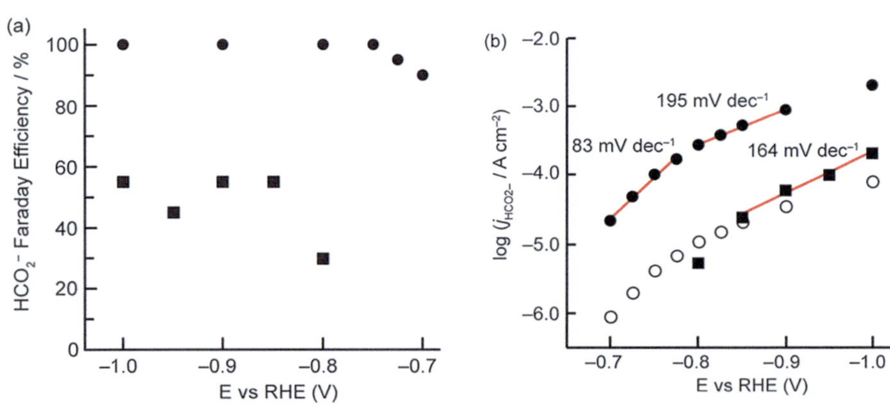

Fig. 2.4 Comparison of CO_2 reduction on Pb foil and oxide-derived Pb in CO_2-saturated 0.5 M $NaHCO_3$. (**a**) Plot of faradaic efficiency vs. potential for Pb foil (■) and oxide-derived Pb (●), (**b**) Tafel plots of the partial current density for HCO_2^- on Pb foil (■) and oxide-derived Pb (●), and partial current density on oxide-derived Pb corrected for roughness factor (○). Obviously, oxide-derived Pb exhibits higher carbon dioxide electroreduction efficiency. (Reprinted from Lee and Kanan 2014 with permission of American Chemical Society)

acid faradaic efficiency was obtained, which was considered to be resulted from the difference in lead surface valence (Yadav and Purkait 2015). In general the meta-stable lead oxide surface layer can inhibit hydrogen evolution. However, it is necessary to understand how the distribution of these surface oxides and the valence of lead can affect the electrocatalytic process. Thusly, in situ experimental confirmation assisted with the quantum chemical calculations is needed in the future study.

2.3 Catalytic Mechanism

It is very important to understand the catalytic mechanisms of various electrode materials. In the early study, it was found that the electrochemical reduction of carbon dioxide to formic acid was mainly attributed to the high hydrogen evolution potential of the metals such as lead (Pb), **mercury (Hg), indium (In), tin (Sn), cadmium (Cd), and titanium (Ti)**. In general, the hydrogen evolution reaction and the carbon dioxide reduction reaction are two competitive reactions in the process of electrochemical reduction of carbon dioxide. As such, when the metal has a high hydrogen evolution overpotential, hydrogen will be difficult to generate; thereby the carbon dioxide reduction can be promoted. About the detailed catalytic mechanism, in the earlier studies, it is speculated that carbon dioxide is firstly adsorbed on the surface of the electrode and then reduced to CO_2^- (Mizuno et al. 1995). However, in some experimental studies, it was observed that the carbon dioxide reduction onset potential is in the range of -1.6 to -1.0 V vs. Ag/AgCl (Hori et al. 1985; Noda et al. 1990), which is more positive than the reduction potential of CO_2 to CO_2^- (-2.0 V vs. Ag/AgCl) (Surdhar et al. 1989; Schwarz and Dodson 1989). This indicates that the electroreduction pathway may not require the process of reduced carbon dioxide to CO_2^-.

In the further studies, it is found that there are two kinds of intermediates in the reduction of carbon dioxide to carbon 1 (C1) compounds. One is ∗COOH, and in this case the surface of the catalyst is combined with the carbon of the intermediate, and the other is ∗OCHO, and in this case, the surface of the catalyst is combined with the O of the intermediate. Feaster et al. experimentally and theoretically investigated the electrochemical reduction of carbon dioxide on the Pb electrode. As shown in Fig. 2.5, ∗COOH is the key intermediate for carbon monoxide production, and meanwhile ∗OCHO is the key intermediate for the carbon dioxide reduced to formate (Feaster et al. 2017).

For the formation of intermediates, a widely accepted one is the proton-assisted mechanism (Fig. 2.6a). However, this mechanism could not give a reasonable explanation for low faradaic efficiency for hydrogen production and the origin of superior selectivity toward formic acid over carbon monoxide on the lead electrode. Back et al. proposed an alternative proton-coupled-electron-transfer mechanism as shown in Fig. 2.6b, which is well consistent with the existing experimental observations. This mechanism can be used to explain the origin of the high selectivity of formic acid production over carbon monoxide as well as hydrogen by the

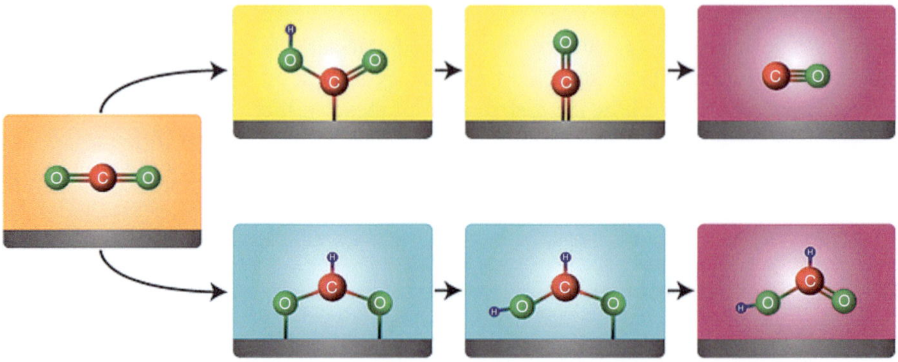

Fig. 2.5 Mechanism that includes pathways for CO and $HCOO^-$ production from CO_2. CO_2 may bind to the electrode surface in an initial electrochemical step via the carbon or the oxygens (resulting in a single adsorption intermediate, ∗COOH, or a bidentate ∗OCHO intermediate, respectively). The second electrochemical step results in the production of CO or $HCOO^-$. For metals that are far from the optimal ∗OCHO binding energy but near the optimal ∗COOH binding energy, it is possible that ∗COOH may be the intermediate for $HCOO^-$ production. (Reprinted from Feaster et al. 2017 with permission of American Chemical Society)

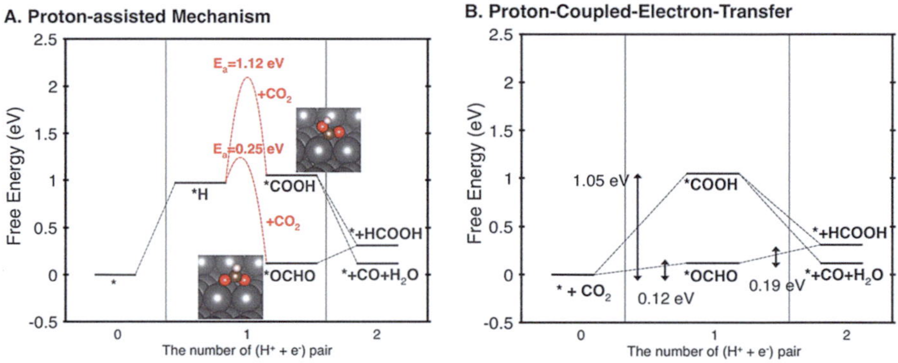

Fig. 2.6 Free energy diagrams for the CO_2 reduction reaction via (**a**) the proton-assisted mechanism and (**b**) the proton-coupled-electron-transfer mechanism. Red lines indicate the activation barriers for the coupling reaction of adsorbed ∗H with CO_2 or co-adsorbed ∗H. The insets show the optimized geometries of the reaction intermediates. (Reprinted from Back et al. 2016 with permission of Royal Society of Chemistry)

O-affinitive characteristics of lead (Back et al. 2016). Nowadays, this proton-coupled-electron-transfer mechanism has been applied to explain the electrochemical reduction of carbon dioxide catalyzed by various metal electrocatalysts and widely accepted by the researchers.

2.4 Reactor and Electrode Type

2.4.1 Traditional Electrode

In the traditional processes of carbon dioxide electroreduction, the H-type electrolytic cell with a proton exchange membrane in the middle is generally used (Fig. 2.7). Herein, the proton exchange membrane is for preventing the reoxidation of the products on the counter electrode (Ju et al. 2017). The working and the reference electrodes are placed in the cathode region, while the counter electrode is placed in the anode region. In an actual operation, carbon dioxide gas is introduced into the electrolyte in the cathode region and then diffused to the surface of the working electrode, where it is reduced. Owing to the very low solubility of carbon dioxide in aqueous solution at the atmospheric pressure (around 30 mM at 1 atm and ambient temperature) (Hori 2008) and low mass transfer ability, the rate of electrochemical reduction of carbon dioxide in this case is always low, which is not suitable for the large-scale practical application.

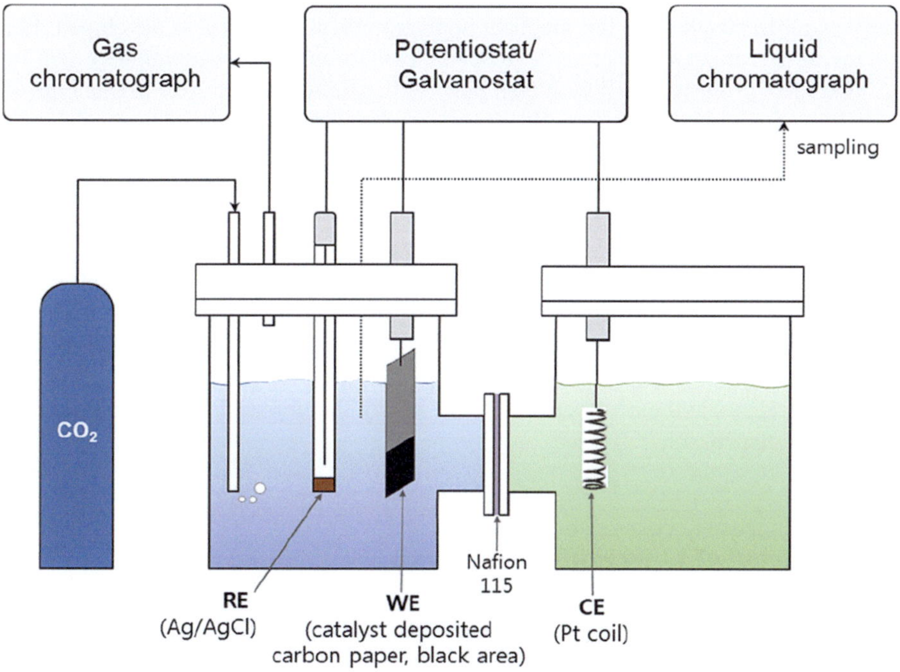

Fig. 2.7 Schematic diagram of the electroreduction of CO_2 in an H-type cell reaction system. *RE* reference electrode, *WE* working electrode, *CE* counter electrode. (Reprinted from Choi et al. 2016 with permission of American Chemical Society)

2.4.2 *Gas Diffusion Electrodes*

In order to solve the problem of slow mass transfer rate of carbon dioxide in aqueous solution, the gas diffusion electrode was developed for the electroreduction of carbon dioxide. The gas diffusion electrode is a special porous membrane electrode, by which a large amount of gas can reach the inside of it and the whole electrolyte can be effectively contact with it. At such an electrode surface, since the gas is directly in contact with the electrode, the catalytic rate can be greatly increased. Machunda et al. obtained a Pb-based gas diffusion electrode with a needlelike structure by electrodeposition, which exhibited a current density as high as 52 mA cm^{-2} at −2 V (vs. Ag/AgCl), much better than other traditional electrodes (Machunda et al. 2010).

2.4.3 *Other Types*

Apart from the above two common types of reactors and electrodes, some other devices more suitable for the practical applications have been also developed. For instances, Alvarez-Guerra et al. developed a continuous electroreduction system by using a filter-press-type cell with lead plate as the cathode (Fig. 2.8). In this type of reactor, since the liquid is in a flowing state, the carbon dioxide dissolved in the solution can be better contacted with the electrode by the flow of the liquid, thereby breaking the current density limit to obtain a higher rate of formic acid production (Alvarez-Guerra et al. 2012). In addition, Köleli et al. developed a fixed-bed reactor using Pb and Sn electrodes, which obtained 94% formic acid faradaic efficiency at high temperature and pressure (Köleli et al. 2003; Köleli and Balun 2004). It provided a useful inspiration for the practical industrial application of electroreduction of carbon dioxide. The electroreduction of carbon dioxide is ultimately to achieve large-scale industrial applications; thusly, the design of various forms of reactors and electrodes which can satisfy the practical application requirements is indispensable.

2.5 Effects of Operation Conditions

In the actual electrochemical reduction of carbon dioxide, the catalyst plays a decisive role in the selectivity of the product. However, the reaction conditions such as temperature, pressure, and electrolyte composition in the catalytic process also play important roles. Mizuno et al. found that the effects of temperature on the electrochemical reduction of carbon dioxide on the In, Sn, and Pb electrodes are

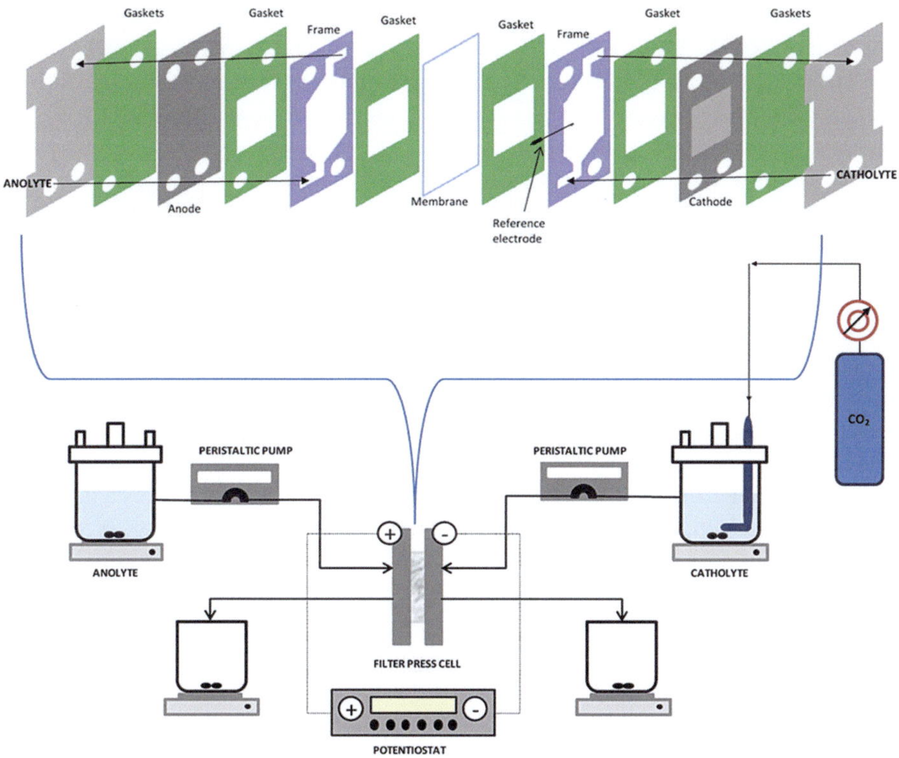

Fig. 2.8 The filter-press electrochemical cell and diagram of the experimental setup. The solution can better contact with the electrode by the flow of the liquid in this system, thereby breaking the current density limit to obtain a higher formate production rate. (Reprinted from Alvarez-Guerra et al. 2012 with permission of Elsevier)

different. For the Pb electrode, in the temperature range of 20–100 °C, the formic acid faradaic efficiency firstly increased and then decreased, whereas the hydrogen faradaic efficiency firstly decreased and then increased, and as a result, at 60 °C, the formic acid faradaic efficiency reached the highest (Mizuno et al. 1995). The influence of temperature on the electrochemical reduction of carbon dioxide system is complicated. It can affect the solubility of carbon dioxide gas in solution, the degree of dissociation of protons in solution, and the rate of hydrogen production or reduction of carbon dioxide. In the future study, the specific impact of temperature on the carbon dioxide reduction mechanism is still needed to explore further.

The effect of pressure on the carbon dioxide electroreduction is relatively simple. Generally, high pressure can increase the solubility of carbon dioxide in the aqueous solution, thus promoting the occurrence of carbon dioxide reduction. Todoroki et al.

found that the faradaic efficiency of formic acid production remained substantially constant in a 0.5 M $KHCO_3$ solution above 30 atm (Todoroki et al. 1995). However, in practical industrial applications, high pressure often consumes a large amount of energy. Thusly, it is expected to perform the electrochemical reduction of carbon dioxide at atmospheric pressure.

In addition, the electrolyte composition also affects the electrochemical reduction of carbon dioxide. Yadav and Purkait tested the electrochemical reduction of carbon dioxide on Pb_2O electrode in $KHCO_3$ and $NaHCO_3$ solutions, respectively, and found that the faradaic efficiency to produce formic acid in $KHCO_3$ was about 10% higher than that in $NaHCO_3$ solution (Yadav and Purkait 2015). Gao et al. explored the electrochemical reduction mechanism during the conversion of carbon dioxide to formic acid on Pb(111) surface using quantum chemical calculation and found that the active intermediates in the electrochemical reduction of carbon dioxide can be affected by the solvation as well as cations. In particular, the efficiency and the selectivity can be influenced by the hydrophilic and hydrophobic characteristics of the cations, which are in agreements with the experimental results (Zhao et al. 2017). Recently, the electrochemical reduction of carbon dioxide in the organic electrolyte was also reported. Unfortunately, the electroreduction of carbon dioxide by lead in an organic electrolyte cannot be found in the literatures. This could be a field to be explored.

2.6 Conclusions

In this chapter, the state of the art on the lead/composite/oxide electrodes used for the carbon dioxide electroreduction was introduced. As can be seen in Table 2.1., the Pb-based materials are excellent electrocatalysts for carbon dioxide reduction. These lead/composite/oxide electrodes exhibit different catalytic properties owing to the different compositions and surface morphologies. However, it is still full of challenges for the developments of high efficient Pb-based electrocatalysts with low overpotentials, high current density, and high selectivity.

Moreover, the proton-coupled-electron-transfer mechanism, which is consistent with the experimental observations, is widely accepted by the researchers. For the reactor and electrode forms, the traditional H-type electrolytic cells and gas diffusion electrodes are mainly used, and meanwhile some new reactors such as the continuous liquid flow and fixed-bed reactor have also been developed. In addition, the temperature, pressure, and electrolyte composition have important influences on the catalytic process. It is expected to develop the electrochemical reduction of carbon dioxide process operated at ambient temperature and pressure conditions with various electrolytes.

Table 2.1. Comparison of electrocatalytic reduction of carbon dioxide to formate/formic acid on different lead/composite/oxide electrodes

Electrocatalyst	Electrolyte	Applied potential (V)	Faradaic efficiency (%)	Current density (mA cm^{-2})	Reference
Pb	0.5 M $KHCO_3$	−1.17 V vs. RHE	~72	5.5	Hori et al. (1985)
Pb	0.1 M $KHCO_3$	−1.63 V vs. NHE	97.4	5	Hori et al. (1994)
Pb	0.5 M $KHCO_3$ (60 atm CO_2)	Constant current density	100	200	Todoroki et al. (1995)
Pb	0.5 M $KHCO_3$ (60 °C)	−1.8 V vs. Ag/AgCl	91.2	–	Mizuno et al. (1995)
Pb-granule (fixed-bed reactor)	0.5 M $KHCO_3$	−1.03 V vs. RHE	74	1.13	Köleli et al. (2003)
Pb-granule (fixed-bed reactor)	0.2 M K_2CO_3 (high press and temperature)	−1.8 V vs. SCE	94	0.5–1.5	Köleli and Balun (2004)
Pb plate	0.5 M NaOH	−0.86 V vs. RHE	65	2.5	Innocent et al. (2009)
Needlelike structure Pb(GDE)	Water vapor	−2 V vs. Ag/AgCl	65	52	Machunda et al. (2010)
Pb plate (filter-press electro-chemical cell)	0.45 M $KHCO_3$ + 0.5 M KCl	–	57	10.5	Alvarez-Guerra et al. (2012)
Porous Pb	0.5 M $KHCO_3$ (5 °C)	−1.7 V vs. SCE	96.8	5	Wang et al. (2016)
Pb dendrites	1 M $KHCO_3$	−0.99 V vs. RHE	97	7.5	Fan et al. (2017)
Sn-Pb alloys	0.5 M $KHCO_3$	−1.8 V vs. Ag/AgCl	79.8	57.3	Choi et al. (2016)
Cu-Pb	0.5 M $KHCO_3$	−0.93 V vs. RHE	20	9.35	Wang et al. (2019)
Pb_2O	0.5 M $KHCO_3$	2 V vs. Co_3O_4	60	~2	Yadav and Purkait (2015)
Oxide-derived Pb	0.5 M $KHCO_3$	−0.8 V vs. RHE	98	0.6	Lee and Kanan (2014)

GDE gas diffusion electrode, *NHE* normal hydrogen electrode, *RHE* reversible hydrogen electrode

Acknowledgments This work is supported by Hirosaki University. X. An greatly acknowledges China Scholarship Council (CSC).

References

Alvarez-Guerra M, Quintanilla S, Irabien A (2012) Conversion of carbon dioxide into formate using a continuous electrochemical reduction process in a lead cathode. Chem Eng J 207-208:278–284. https://doi.org/10.1016/j.cej.2012.06.099

Back S, Kim JH, Kim YT, Jung Y (2016) On the mechanism of high product selectivity for HCOOH using Pb in CO_2 electroreduction. Phys Chem Chem Phys 18(14):9652–9657. https://doi.org/10.1039/c6cp00542j

Chen Y, Kanan MW (2012) Tin oxide dependence of the CO_2 reduction efficiency on tin electrodes and enhanced activity for tin/tin oxide thin-film catalysts. J Am Chem Soc 134(4):1986–1989. https://doi.org/10.1021/ja2108799

Choi SY, Jeong SK, Kim HJ, Baek I-H, Park KT (2016) Electrochemical reduction of carbon dioxide to formate on tin-lead alloys. ACS Sustain Chem Eng 4(3):1311–1318. https://doi.org/10.1021/acssuschemeng.5b01336

Costentin C, Robert M, Savéant J-M (2013) Catalysis of the electrochemical reduction of carbon dioxide. Chem Soc Rev 42(6):2423–2436. https://doi.org/10.1039/C2CS35360A

Cui C, Han J, Zhu X, Liu X, Wang H, Mei D, Ge Q (2016) Promotional effect of surface hydroxyls on electrochemical reduction of CO_2 over SnO_x/Sn electrode. J Catal 343:257–265. https://doi.org/10.1016/j.jcat.2015.12.001

Fan M, Garbarino S, Botton GA, Tavares AC, Guay D (2017) Selective electroreduction of CO_2 to formate on 3D [100] Pb dendrites with nanometer-sized needle-like tips. J Mater Chem A 5 (39):20747–20756. https://doi.org/10.1039/c7ta06528k

Feaster JT, Shi C, Cave ER, Hatsukade T, Abram DN, Kuhl KP, Hahn C, Nørskov JK, Jaramillo TF (2017) Understanding selectivity for the electrochemical reduction of carbon dioxide to formic acid and carbon monoxide on metal electrodes. ACS Catal 7(7):4822–4827. https://doi.org/10.1021/acscatal.7b00687

García J, Jiménez C, Martínez F, Camarillo R, Rincón J (2018) Electrochemical reduction of CO_2 using Pb catalysts synthesized in supercritical medium. J Catal 367:72–80. https://doi.org/10.1016/j.jcat.2018.08.017

Hori Y (2008) Mod. Asp. electrochem. Springer, New York, pp 89–189

Hori Y, Kikuchi K, Suzuki S (1985) Production of CO and CH_4 in electrochemical reduction of CO_2 at metal electrodes in aqueous hydrogencarbonate solution. Chem Lett 14(11):1695–1698. https://doi.org/10.1246/cl.1985.1695

Hori Y, Wakebe H, Tsukamoto T, Koga O (1994) Electrocatalytic process of CO selectivity in electrochemical reduction of CO_2 at metal electrodes in aqueous media. Electrochim Acta 39 (11–12):1833–1839. https://doi.org/10.1016/0013-4686(94)85172-7

Innocent B, Liaigre D, Pasquier D, Ropital F, Léger J-M, Kokoh K (2009) Electro-reduction of carbon dioxide to formate on lead electrode in aqueous medium. J Appl Electrochem 39 (2):227–232. https://doi.org/10.1007/s10800-008-9658-4

Ju W, Bagger A, Hao GP, Varela AS, Sinev I, Bon V, Roldan Cuenya B, Kaskel S, Rossmeisl J, Strasser P (2017) Understanding activity and selectivity of metal-nitrogen-doped carbon catalysts for electrochemical reduction of CO_2. Nat Commun 8(1):944. https://doi.org/10.1038/s41467-017-01035-z

Köleli F, Balun D (2004) Reduction of CO_2 under high pressure and high temperature on Pb-granule electrodes in a fixed-bed reactor in aqueous medium. Appl Catal A Gen 274 (1–2):237–242. https://doi.org/10.1016/j.apcata.2004.07.006

Köleli F, Atilan T, Palamut N, Gizir A, Aydin R, Hamann C (2003) Electrochemical reduction of CO_2 at Pb-and Sn-electrodes in a fixed-bed reactor in aqueous K_2CO_3 and $KHCO_3$ media. J Appl Electrochem 33(5):447–450. https://doi.org/10.1023/A:102447151

Lee CH, Kanan MW (2014) Controlling H^+ vs CO_2 reduction selectivity on pb electrodes. ACS Catal 5(1):465–469. https://doi.org/10.1021/cs5017672

Li J, Chen G, Zhu Y, Liang Z, Pei A, Wu C-L, Wang H, Lee HR, Liu K, Chu S, Cui Y (2018) Efficient electrocatalytic CO_2 reduction on a three-phase interface. Nat Catal 1(8):592–600. https://doi.org/10.1038/s41929-018-0108-3

Liu Y, Zhang Y, Cheng K, Quan X, Fan X, Su Y, Chen S, Zhao H, Zhang Y, Yu H, Hoffmann MR (2017) Selective electrochemical reduction of carbon dioxide to ethanol on a boron- and nitrogen-co-doped nanodiamond. Angew Chem Int Ed Engl 56(49):15607–15611. https://doi.org/10.1002/anie.201706311

Liu X, Yang H, He J, Liu H, Song L, Li L, Luo J (2018) Highly active, durable ultrathin $MoTe_2$ layers for the electroreduction of CO_2 to CH_4. Small 14(16):1704049. https://doi.org/10.1002/smll.201704049

Lu L, Sun XF, Ma J, Yang DX, Wu HH, Zhang BX, Zhang JL, Han BX (2018) Highly efficient electroreduction of CO_2 to methanol on palladium-copper bimetallic aerogels. Angew Chem Int Ed 57(43):14149–14153. https://doi.org/10.1002/anie.201808964

Machunda RL, Lee J, Lee J (2010) Microstructural surface changes of electrodeposited Pb on gas diffusion electrode during electroreduction of gas-phase CO_2. Surf Interface Anal 42 (6–7):564–567. https://doi.org/10.1002/sia.3245

Mizuno T, Ohta K, Sasaki A, Akai T, Hirano M, Kawabe A (1995) Effect of temperature on electrochemical reduction of high-pressure CO_2 with in, Sn, and Pb electrodes. Energy Sources 17(5):503–508. https://doi.org/10.1080/00908319508946098

Noda H, Ikeda S, Oda Y, Imai K, Maeda M, Ito K (1990) Electrochemical reduction of carbon dioxide at various metal electrodes in aqueous potassium hydrogen carbonate solution. Bull Chem Soc Jpn 63(9):2459–2462. https://doi.org/10.1246/bcsj.63.2459

Pachauri RK, Allen MR, Barros VR, Broome J, Cramer W, Christ R, Church JA, Clarke L, Dahe Q, Dasgupta P (2014) Climate change 2014: synthesis report. Contribution of working groups I, II and III to the fifth assessment report of the Intergovernmental Panel on Climate Change. IPCC, p 151

Schwarz H, Dodson R (1989) Reduction potentials of CO_2 and the alcohol radicals. J Phys Chem 93 (1):409–414. https://doi.org/10.1021/j100338a079

Surdhar PS, Mezyk SP, Armstrong DA (1989) Reduction potential of the carboxyl radical anion in aqueous solutions. J Phys Chem 93(8):3360–3363. https://doi.org/10.1021/j100345a094

Todoroki M, Hara K, Kudo A, Sakata T (1995) Electrochemical reduction of high pressure CO_2 at Pb, Hg and In electrodes in an aqueous $KHCO_3$ solution. J Electroanal Chem 394 (1–2):199–203. https://doi.org/10.1016/0022-0728(95)04010-L

Verma S, Kim B, Jhong HRM, Ma S, Kenis PJ (2016) A gross-margin model for defining technoeconomic benchmarks in the electroreduction of CO_2. ChemSusChem 9 (15):1972–1979. https://doi.org/10.1002/cssc.201600394

Wang J, Wang H, Han Z, Han J (2016) Electrodeposited porous Pb electrode with improved electrocatalytic performance for the electroreduction of CO_2 to formic acid. Front Chem Sci Eng 9(1):57–63. https://doi.org/10.1007/s11705-014-1444-8

Wang Y, Hu H, Sun Y, Tang Y, Dai L, Hu Q, Fisher A, Yang XJ (2019) Facile synthesis of nanostructural high-performance cu-Pb electrocatalysts for CO_2 reduction. Adv Mater Interfaces 6(2):1801200. https://doi.org/10.1002/admi.201801200

Yadav VSK, Purkait MK (2015) Synthesis of Pb_2O electrocatalyst and its application in the electrochemical reduction of CO_2 to HCOOH in various electrolytes. RSC Adv 5 (50):40414–40421. https://doi.org/10.1039/c5ra05899f

Zhang W, Hu Y, Ma L, Zhu G, Wang Y, Xue X, Chen R, Yang S, Jin Z (2018) Progress and perspective of electrocatalytic CO_2 reduction for renewable carbonaceous fuels and chemicals. Adv Sci 5(1):1700275. https://doi.org/10.1002/advs.201700275

Zhao CX, Bu YF, Gao W, Jiang Q (2017) CO_2 reduction mechanism on the Pb (111) surface: effect of solvent and cations. J Phys Chem C 121(36):19767–19773. https://doi.org/10.1021/acs.jpcc.7b04375

Zheng X, De Luna P, García de Arquer FP, Zhang B, Becknell N, Ross MB, Li Y, Banis MN, Li Y, Liu M, Voznyy O, Dinh CT, Zhuang T, Stadler P, Cui Y, Du X, Yang P, Sargent EH (2017) Sulfur-modulated tin sites enable highly selective electrochemical reduction of CO_2 to formate. Joule 1(4):794–805. https://doi.org/10.1016/j.joule.2017.09.014

Chapter 3
Thermochemical Conversion of Carbon Dioxide to Carbon Monoxide by Reverse Water-Gas Shift Reaction over the Ceria-Based Catalyst

Joshua Gorimbo and Diane Hildebrandt

Contents

3.1 Introduction ... 44
3.2 Reverse Water-Gas Shift Thermodynamic Considerations ... 46
3.3 Reverse Water-Gas Shift Catalyst ... 50
3.3.1 Supported Metal Catalysts ... 50
3.3.2 Reverse Water-Gas Shift Promoters ... 52
3.4 Chemistry of Cerium During Reduction and Reverse Water-Gas Shift ... 52
3.4.1 CeO_2 Reduction Thermodynamics ... 53
3.4.2 In Situ CeO_2 Reduction ... 53
3.4.3 CeO_2 Reduction Mechanism ... 56
3.5 Conclusion ... 58
References ... 58

Abstract Carbon dioxide (CO_2) is today a point of convergence in the technoscientific research world, due to its contribution to climate change, as levels of CO_2 continue to rise on an increasing trajectory. The reverse water-gas shift reaction (RWGS) is considered one possible way of harnessing atmospheric CO_2, in order to reduce the concentration. In this chapter, attention is given to the catalysis and the thermodynamic of reverse water-gas shift, as it has been indicated that the reaction is feasible at high temperatures (>800 °C). Since the reaction is highly endothermic, energy source alternatives such as the use of concentrated solar energy which virtually provides an unlimited source of cheaper clean energy as well as nonpolluting high-temperature heat are adopted. The reaction is accompanied by several side reactions, such as the Sabatier reaction, methanation, and the Boudouard reaction, which are thermodynamically feasible at a lower temperature than that

---

J. Gorimbo (✉) · D. Hildebrandt
Institute for the Development of Energy for African Sustainability (IDEAS) Research Unit, College of Science, Engineering and Technology (CSET), University of South Africa (UNISA), Johannesburg, South Africa

Inamuddin et al. (eds.), *Conversion of Carbon Dioxide into Hydrocarbons Vol. 1 Catalysis*, Environmental Chemistry for a Sustainable World 40,
https://doi.org/10.1007/978-3-030-28622-4_3

required for the RWGS reaction. Use of a catalyst lowers the reaction temperature remarkably.

Cerium oxide plays a key role as an active support and a catalyst and is a popular topic of ongoing research into reverse water-gas shift reaction, which is relevant to environmental applications and in energy conversion systems. Ceria-based catalyst plays various roles in augmenting catalytic performance, because of their unique ability to accumulate and release oxygen. This attribute results from the interconversion between Ce^{3+} and Ce^{4+} (which are both stable) which allows the oxide to shift between CeO_2 and CeO_{2-x}. The reverse water-gas shift reaction mechanism is discussed, and the main one involves direct reoxidation of the ceria support by the CO_2 via surface carbonates. In this chapter, attention is paid only to catalytic applications of ceria-containing catalysts, with all other reverse water-gas shift reaction catalysts being ignored.

Keywords Reverse water-gas shift · Sabatier reaction · Carbon dioxide · Carbon monoxide · CO_2 hydrogenation · Cerium dioxide · Ceria · CO_2 conversion · Boudouard reaction · Methanation reaction

Abbreviations

NOAA	National Oceanic and Atmospheric Administration
ESRL	Earth System Research Laboratory
CCU	Carbon capture and utilization
FTS	Fischer-Tropsch synthesis
RWGS	Reverse water-gas shift
TPR	Temperature-programmed reduction

3.1 Introduction

Ongoing efforts are being made to improve existing methods to reduce atmospheric carbon dioxide (CO_2). The driving force is a desire to attain reasonable atmospheric carbon dioxide levels at a lower cost. Carbon dioxide is the most abundant carbon source available, and it is known to have a negative effect on the environment. Carbon dioxide is one of the naturally occurring greenhouse gases, but human activities, such as burning fossil fuels to meet energy demands (power plants and car engines), have been a major contributor to increased CO_2 emissions. Nitrous oxide, water vapor, and methane are three of the other greenhouse gases that contribute to the warming of the Earth's atmosphere, through a process of absorbing energy from the sun and directing it back to the Earth's surface. Therefore, an increase in CO_2 means a surplus of greenhouse gases that trap additional heat,

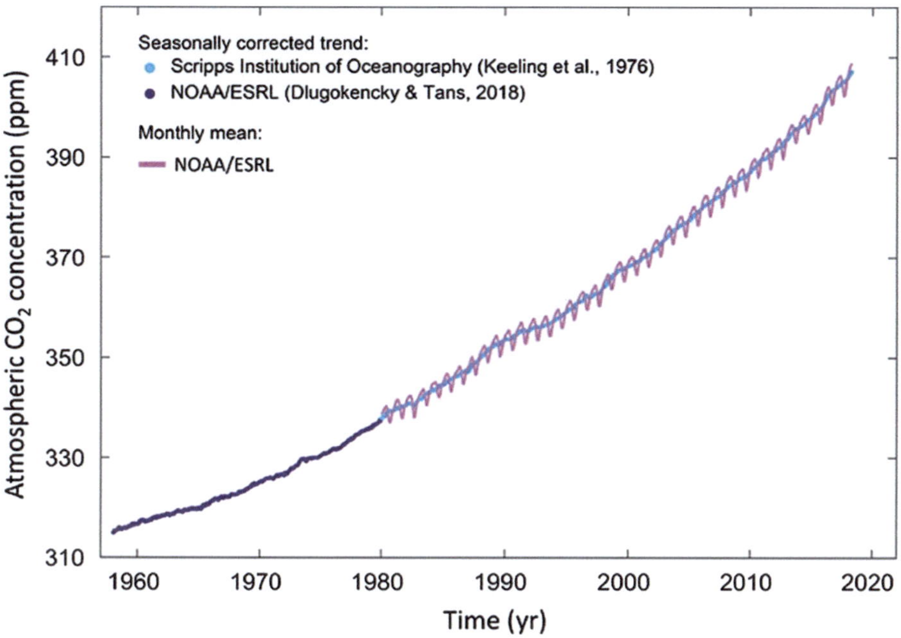

Fig. 3.1 Surface average atmospheric CO_2 concentration (ppm). Values are calculated averages of direct atmospheric CO_2 measurements from different stations (Le Quéré et al. 2018)
NOAA National Oceanic and Atmospheric Administration, *ESRL* Earth System Research Laboratory

which results in melting ice caps and rising ocean levels. This, in turn, results in flooding and global warming.

Surface average atmospheric CO_2 concentrations (ppm) are reported and released annually in the Global Carbon Budget (Le Quéré et al. 2018) publication. These annual reports show that the concentration of CO_2 in the atmosphere has an increasing trajectory. For instance, Fig. 3.1 indicates that an increase was recorded from ca.350 ppm in 1990 to 405.0 ± 0.1 ppm CO_2 in 2017. Earlier studies done at Mauna Loa Observatory, Hawaii, in 1976, also indicated that the annual average CO_2 concentration rose by 3.4% between 1959 and 1971 (Keeling et al. 1976). Therefore, this is a global phenomenon.

The components of the CO_2 budget that are reported annually in the Global Carbon Budget provide disparate estimates for (1) CO_2 released from burning fossil fuel; (2) CO_2 emissions following intentional human undertakings on land, including those that lead to land-use change; (3) their partitioning among the growth rate of atmospheric CO_2 concentration; (4) the uptake of CO_2 (the "CO_2 sinks") in the ocean; and (5) the uptake of CO_2 on land (Le Quéré et al. 2018).

The Budget equation for atmospheric CO_2 are calculated as follows:

$$\Delta CO_2 = F - O - B \tag{3.1}$$

Where:

F = the industrial CO_2 source
O = the net ocean CO_2 sink
B = the net land biospheric CO_2 sink (Keeling and Manning 2014)

Here, B represents the total contributions from both atmospheric or climatic influences and human land-use changes; F incorporates CO_2 released from both cement manufacturing and the burning of fossil fuels.

The Earth System Research Laboratory (ESRL) was established to pursue an extensive and thorough understanding of the Earth system by monitoring parameters such as CO_2 levels, temperature changes, etc. To abate the effects of CO_2, researchers and policy-making boards produced a diversity of solutions, e.g., carbon capture and disposal, carbon capture and utilization (CCU), and sequestration. The conversion of CO_2 into useful products such as chemicals, materials, and fuels is of widespread interest to researchers, as evidenced by the number of reviews, papers, and books that focus on CO_2 utilization (Wolf et al. 2016; Chen et al. 2000).

Researchers have found a way of converting CO_2 into usable energy via the reverse water-gas shift (RWGS) reaction. The reverse water-gas shift reaction can, therefore, be defined as the hydrogenation of CO_2 to produce more valuable chemicals, such as CO (which is the main reactant in Fischer-Tropsch synthesis (Gorimbo et al. 2017)), and in some cases by-products, such as CH_4 and methanol, depending on the conditions used (Wang et al. 2017). Despite many published articles on CO_2 utilization, an organized collection of information about the reverse water-gas shift reaction using Ce-based catalyst in CO production is limited. This chapter was designed with the intention of covering this knowledge gap and collecting all the available views on the reverse water-gas shift reaction to yield a building block (CO) used to produce chemicals via other processes such as Fischer-Tropsch synthesis (FTS).

3.2 Reverse Water-Gas Shift Thermodynamic Considerations

The thermochemical reactions between the catalyst and the reactants are an important aspect of successful CO_2 conversion. CO_2 hydrogenation relies on external heating and thermal diffusion to induce molecular bond-breaking. As mentioned earlier, reverse water-gas shift can be used for the synthesis of CO from CO_2; however, the reaction is highly endothermic. The reaction is equilibrium limited and favored at elevated temperatures, as the reaction is endothermic in nature. The reverse water-gas shift reaction is only thermodynamically feasible at high temperatures, i.e., greater than 822.7 °C. The reverse water-gas shift reaction is expressed as follows:

$$CO_2 + H_2 \leftrightarrow CO + H_2O \tag{3.2}$$

Removal and separation of product can result in the equilibrium shifting toward the products. The thermochemical reactions that occur during CO_2 hydrogenation are driven by vapor equilibrium. For this reaction, there is an equilibrium constant of K, which is defined as follows:

$$K_{eq} = \frac{[CO][H_2O]}{[CO_2][H_2]} \tag{3.3}$$

Thus, the equilibrium constant for the CO_2 reduction reaction depends on the partial pressure of the reactant and products. In turn, the equilibrium condition reflects the standard free energy ΔG for the reaction:

$$\Delta G = -RTln\ K \tag{3.4}$$

Where:

R = the gas constant
T = the absolute temperature

The free energy for CO_2 hydrogenation at different temperature levels can be determined from the graph shown in Fig. 3.2. This figure plots the CO_2 hydrogenation to give two separate sets of products in terms of free energy versus temperature.

In order for the reverse water-gas shift reaction to occur at any temperature, the free energy ΔG must be below the blue line (see Fig. 3.2) and at a temperature greater than indicated by line B (greater than 800 °C) for a non-catalyzed reaction. The same can be said for the Sabatier reaction: for reaction to occur, the ΔG must be below the orange line (see Fig. 3.2) and below the ΔG = 0 line (temperature less than indicated by line A). The reverse water-gas shift reaction is endothermic by ΔH = 9.58 and ΔH = 7.70 kcal at 200 and 1000 °C, respectively, and proceeds catalytically. The reaction has been studied in fixed bed reactors using different catalysts, generally at atmospheric pressure and a temperature range of 200–750 °C.

Thermodynamic assessment at atmospheric pressure shows that CO_2 conversion equilibrium increases with temperature (Fig. 3.2). At temperatures below 595 °C, ΔG values for the Sabatier reaction are negative (see Fig. 3.2)—hence the ease by which CH_4 can be formed. A further increase in temperature suppresses the methane formation, and temperatures beyond 822.7 °C favor the hydrogenation of CO_2 to CO and H_2O. Thermodynamic limitations can, therefore, be used to suppress CH_4 formation and, in some cases, catalyst choices; for instance, minimum or no CH_4 could be detected over the CeO_2 catalyst, indicating that only the reverse water-gas

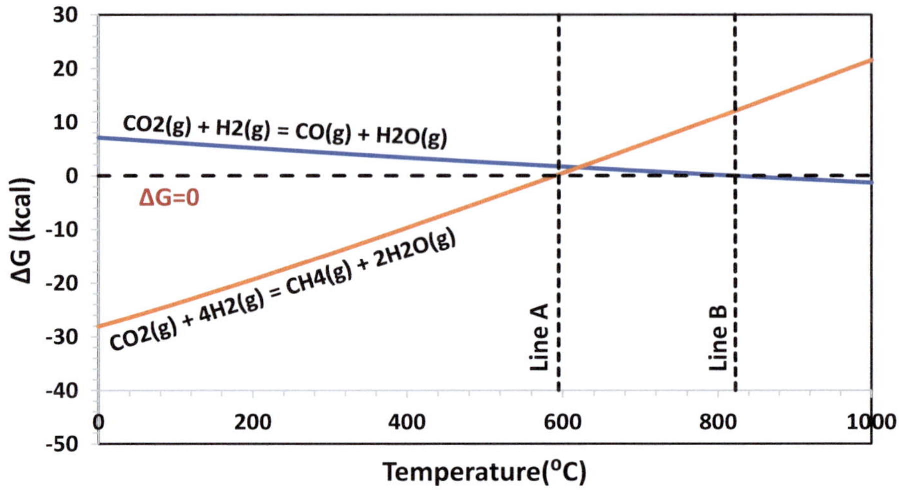

Fig. 3.2 Gibbs free energy calculated at atmospheric pressure versus temperature for reverse water-gas shift and the Sabatier reaction

Table 3.1 ΔH values of reverse water-gas shift and the Sabatier reaction at different temperature levels

	RWGS		Sabatier	
Temperature (°C)	ΔH (kcal)	ΔG (kcal)	ΔH (kcal)	ΔG (kcal)
200	9.58	5.12	−41.36	−19.37
400	9.14	3.32	−43.24	−9.70
600	8.64	1.65	−44.67	0.47
800	8.15	0.10	−45.67	10.92
1000	7.70	−1.33	−46.28	21.53

shift reaction occurs over the catalyst (Dai et al. 2018). Other side reactions are thermodynamically feasible at temperatures below 648 and 700 °C, such as the methanation reaction (Eq. 3.6) and Boudouard reaction (Eq. 3.7), respectively.

$$CO_2 + 4H_2 \leftrightarrow CH_4 + 2H_2O \tag{3.5}$$

$$CO + 3H_2 \leftrightarrow CH_4 + H_2O \tag{3.6}$$

$$2CO \leftrightarrow C + CO_2 \tag{3.7}$$

The Sabatier reaction tends to be exothermic at low temperatures, whereas reversible water-gas shift reaction is endothermic and thermodynamically unfavorable at temperatures less than 1000 °C (see Table 3.1).

The simultaneous spitting of H_2 and CO_2 was shown in experiments to yield syngas with an H_2/CO molar ratio that can be adjusted by altering the feed (H_2/CO_2) molar ratio (Cherkezova-Zheleva and Mitov 2010; Zhu and Wachs 2016). In the

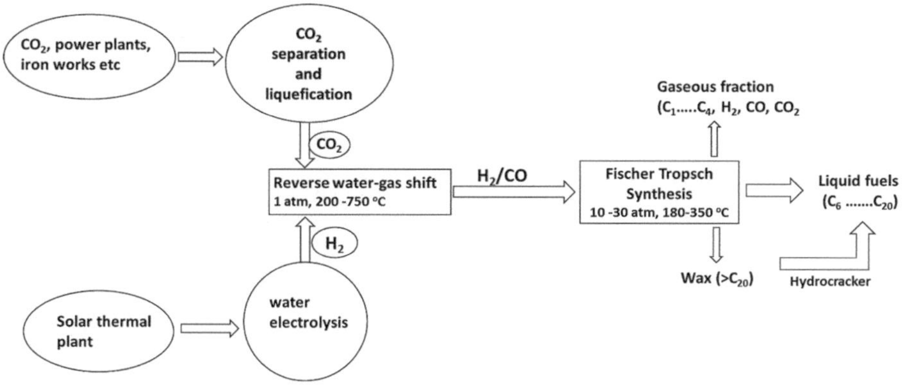

Fig. 3.3 Concept of a process for the production of synfuels from CO_2 by catalytic hydrogenation (Wolf 2016). Solar energy can be used to power water electrolysis that generates hydrogen

case of the RWGS reaction, if excess hydrogen is used in this process and water is removed from the product stream, the resulting gas is a mixture of CO and H_2 (syngas), which can be converted to liquid hydrocarbon fuels (e.g., diesel, kerosene) via the Fischer-Tropsch reaction (Dry 2002; Espinoza et al. 1999). As time progresses, the economic benefits of such processes will further increase due to stricter regulations and legislation in regard to CO_2 emissions (Thiel et al. 2016). This process is most effective when excess electrical energy obtained from renewable resources is used, as in the case of solar energy.

Solar thermochemical processes utilize concentrated solar radiation to drive endothermic reactions (Steinfeld and Palumbo 2004). Concentrated solar radiation can therefore be used as an energy source in the high-temperature process (above 1000 °C) heat as has been used in the thermal production of hydrocarbons (Romero and Steinfeld 2012). If solar energy is utilized in atmospheric reduction of CO_2 concentrations, the economics of the whole process should be more cost-effective than all other alternatives, considering the abundance of solar energy. Solar fuels have been generated from H_2O and CO_2 via thermolysis, thermochemical cycles, and decarbonization processes. All these thermochemical routes are highly endothermic processes that proceeded at high temperature and can be driven by concentrated solar process heat (Furler et al. 2012). Figure 3.3 shows pathways in which hydrocarbons can be produced from carbon dioxide using solar energy.

Therefore, CO_2 is a plausible source of carbon. The reverse water-gas shift reaction generates CO but still requires two moles H_2 if Fischer-Tropsch synthesis route is to be taken to achieve a value-added fuel or chemical. Hydrogen can be produced from water electrolysis (which is an energy intensive process), and electricity needed can also be generated from solar energy (renewable electricity as shown in Fig. 3.3). Instead of storing pure H_2, converting it to liquid hydrocarbons with higher volumetric energy density could prove beneficial as infrastructure already exists for the storage and transportation of liquids like diesel or jet oil (Wolf 2016; Daza and Kuhn 2016).

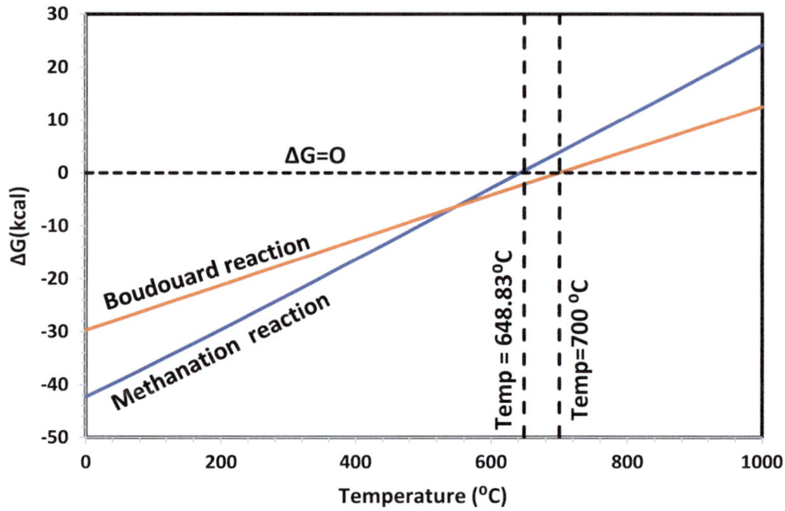

Fig. 3.4 Gibbs free energy calculated at atmospheric pressure versus temperature for the Boudouard reaction and the methanation reaction. Reactions below the ΔG = 0 line are feasible

Control of the CO_2 hydrogenation conditions provides an opportunity to suppress side reactions. The atmosphere is not constant during CO_2 reduction. As shown in Figs. 3.2 and 3.4, the composition gases mainly depend on the temperature, which in turn affect the thermochemical reactions.

The main reaction and the side reactions tend to occur simultaneously at any given instance; as a result, products H_2O, CO, CO_2, CH_4, H_2, and C are found in the reaction medium (Aresta and Dibenedetto 2016). The equilibrium constant of the WGSR reaction increases with increasing temperature. The reaction is thermodynamically favored at elevated temperatures, and since there is no volume change from reactants to products, the reaction is independent of pressure alterations. All the side reactions graphed in Figs. 3.2 and 3.4 indicate the reactions become less feasible with an increase in temperature. An increase in temperature can improve reverse water-gas shift reaction, as per collision theory, as high temperature increases the rate of effective reactant collision. Therefore, CO_2 and H_2 conversions increase with increasing temperature.

3.3 Reverse Water-Gas Shift Catalyst

3.3.1 Supported Metal Catalysts

Various catalysts, including the noble metals Au and Pt, have been tested for reverse water-gas shift reaction, and they were noted to be the most promising catalysts since they are non-pyrophoric and do not need pre-treatment before use (Carrasquillo-flores

Table 3.2 Summary of some experimental catalysts and conditions applied in the hydrogenation of CO_2

	Conditions				
Catalyst used	Temperature (°C)	Pressure	H_2/CO_2	Flowrate	References
MoOx-promoted Au/SiO_2	300	8.1 bar	2:1	15 cm^3 (STP) min^{-1})	Carrasquillo-flores et al. (2015)
Ceria nanocubes	400	Atmospheric	1:1	100 mL/min	Liu et al. (2016)
Fe nanoparticles	600	Atmospheric	1:1	20 mL/min	Kim et al. (2015)
Ni/CeO_2-Al_2O_3	400–750	Atmospheric	4:1	30,000 mLg^{-1} h^{-1}	Yang et al. (2018)
Ni/Ce-Zr-O	550–750	Atmospheric	1:1	50 mL/min	Sun et al. (2015)
Pt/zeolite	200–500	Atmospheric	1:1	50 mL/min	Yang et al. (2017)
CeO_2	300–580	Atmospheric	4:1	50 mL/min	Dai et al. (2018)
Pt/CeO_2	200–500	Atmospheric	1:1	50 mL/min	Chen et al. (2016)
Co-CeO_2	400–600	Atmospheric	1:1	600 mL/min	Wang et al. (2017)
$CuOx/CeO_2$	400	1 bar	1:1	50 mL/min	Ronda-Loret et al. (2018)
Fe_2O_3	250–400	10 and 15 bar	3:1; 6:1	5.7 mL/min	Albrecht et al. (2017)
Cu/ZnO	200–230	Atmospheric	1:1	30 mL/min	Álvarez et al. (2016)

Note the temperature range used in these different experiments when employing different conditions and catalysts

et al. 2015; Chen et al. 2017). These noble metals also possess remarkable thermal stability at elevated temperatures applied in RWGS reactions (Chen et al. 2017; Kim et al. 2013; Pekridis et al. 2007). The reason why their application is limited (even though they are known to be very active for the reverse water-gas shift reaction) is their high cost and restricted natural abundance—hence other alternative materials have been considered as catalysts, e.g., the use of binary oxides as catalyst supports and several transition metals, such as Ni, Co, and Fe, supported on Al_2O_3, CeO_2, SiO_2, TiO_2, and ZrO_2 (Álvarez et al. 2016; Kim et al. 2015; Rafati et al. 2015). Nonetheless, the prolonged stability of these catalysts at a high temperature warrants their application. The RWGS reaction is a reversible endothermic reaction that requires a high operating temperature. With reference to Ni-based catalysts used for the RWGS reaction, high temperatures of up to 750 °C have been used, and still the catalyst showed a high stability for catalytic activity (Sun et al. 2015). Table 3.2 summarizes some of the catalysts and conditions generally used in RWGS.

A plethora of binary metal oxides have been suggested for use in CO_2 hydrogenation; among these, the ceria-based catalyst was selected due to its high oxygen storage ability. Ceria has previously been used in oxidation-reduction (redox) reaction since cerium deftly and reversibly interchanges between Ce^{4+} and Ce^{3+} upon storage and release of oxygen (Trovarelli et al. 2001). In the present work, special attention is paid to the cerium-based catalyst applied in RWGS reaction.

Table 3.1 shows some of the cerium-based catalysts (supported and unsupported) applied to the RWGS reaction over a wide rand of conditions.

The utilization of CeO_2 as a catalyst was reported by Dai et al. (2018), who pointed out that the preparation method greatly affects the structure of the CeO_2 catalyst. What is of importance is that all preparation methods reported resulted in a catalyst with considerable conversion (9.3–15.9%) and 100% selectivity toward CO (Dai et al. 2018). Where CeO_2 was used as a support, such as the reaction conducted over Pt/CeO_2 and compared to pure CeO_2 as catalysts when using temperature ranging from 200 to 500 °C (Chen et al. 2016), activity was always considerably satisfactory. The comparison of the activation energies indicated that the rate of CO produced with Pt/CeO_2 was significantly less than that produced with the pure CeO_2 catalyst. The TPR and XPS results also demonstrate that the addition of platinum enhanced the reducibility of CeO_2 support and significantly increased the amount of Ce^{3+} sites on the catalyst surface (Chen et al. 2016).

3.3.2 *Reverse Water-Gas Shift Promoters*

Fe-based catalysts have been reported as being highly active for RWGS reaction, with activity and selectivity substantially boosted by adding Cs as a promoter. It turns out that the addition of a small percent of another elements as promoters helps to enhance the activity, thermal stability and selectivity of the catalyst.

3.4 Chemistry of Cerium During Reduction and Reverse Water-Gas Shift

The application of cerium, whether promoted or in its purest form, has to do with its ability to interconvert between the $4d^{10}4f^{1}$–Ce(III) and $4d^{10}4f^{0}$–Ce(IV) oxidation states, which is a unique phenomenon in rare-earth metals. The reduction-oxidation chemistry of Ce^{3+}/Ce^{4+} makes the oxides applicable in heterogeneous catalysis through the containment and liberation of oxygen. Studies have reported that the nonstoichiometric cerium oxide system (CeO_{2-x}) possesses an atypically high ion mobility, due to octahedral oxygen vacancy in its lattice (Montini et al. 2016). Cerium oxide with a cubic fluorite structure has received enough attention as a key component in reverse water-gas shift reaction (Dai et al. 2017).

Prior to reverse water-gas shift reaction, cerium-based catalysts undergo in situ reduction, usually with hydrogen (Dai et al. 2018; Chen et al. 2016). Under such reducing atmosphere, cerium dioxide (i.e., CeO_2) liberates oxygen to yield an oxygen-deficient state (Eq. 3.8), while reoxidizing ceria changes it back to CeO_2 (Eq. 3.9).

$$CeO_2 + xH_2 \leftrightarrow CeO_{2-x} + xH_2O \tag{3.8}$$

$$CeO_{2-X} + xCO_2 \leftrightarrow CeO_2 + xCO \tag{3.9}$$

Ceria can store oxygen under oxidizing atmosphere and release the oxygen under reducing conditions, due to the Ce^{4+}/Ce^{3+} redox couple. It is well established that lattice oxygen of ceria participates in these processes. This reversible behavior is feasible only if the cubic fluorite structure is maintained via the redox processes (Zhang et al. 2006). This unique phenomenon attributes to the oxygen storage capacity (OSC) of ceria, which is strengthened by a sizeable surface area and small particle size (Orozco et al. 2017).

Low catalytic activity and reduced surface area are obtained at elevated temperatures, when ceria particles are easily coarsened (Zhang et al. 2006). This phenomenon therefore makes particle size stability a major issue for pure ceria at high temperatures; however, the addition of certain metal oxides can counteract the effect (Mishra 2003; Zhang et al. 2016).

3.4.1 CeO_2 Reduction Thermodynamics

One consequence of what is seen in Fig. 3.5 is the difference in oxidation-reduction processes for cerium oxide. At any temperature below 1000 °C (the temperature region for experimental reverse water-gas shift), cerium exists in the oxidic form. For reduction to occur, energy needs to be fed into the system. Table 3.3 details enthalpy energy at various temperatures to illustrate the concept of preferential oxidation.

The two equations above line $\Delta G = 0$ are a mirror image of the two below the line. Therefore, the position of these equations above or below the G line is dependent on the partial pressure of CO and CO_2 or H_2 and H_2O. The use of oxygen makes the oxidation more feasible. With logic analogous to that indicated in Fig. 3.5 and Table 3.3, the partial pressure of the feed and products needs to be monitored in reverse water-gas shift reactions, as these affect the catalyst phases.

3.4.2 In Situ CeO_2 Reduction

Hydrogen temperature-programmed reduction (H_2-TPR) is mainly conducted separately, so as to establish the reduction temperature of the catalyst. Differently prepared catalysts will have different reduction temperatures. Temperature-programmed reduction (TPR) of $Ce0_2$ shown in Fig. 3.6 shows the presence of two peaks; in the case of Dai et al. (2018), for instance, differently prepared cerium-based catalysts produced two separate peaks upon reduction. The first peak appeared

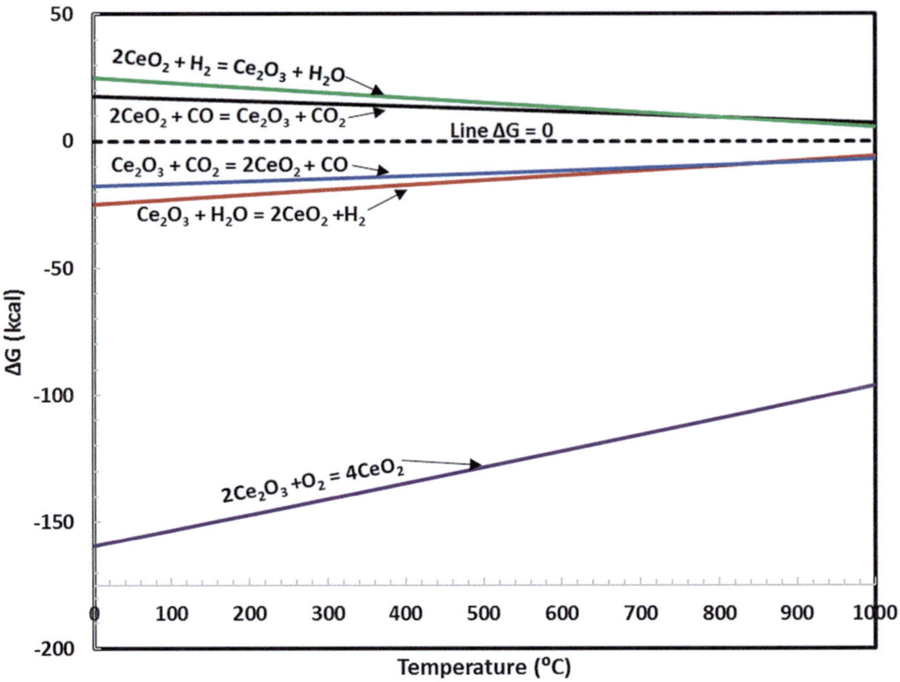

Fig. 3.5 Plot of oxidation-reduction equilibrium for CeO_2/Ce_2O_3 in terms of ΔG versus temperature. All reactions below line $\Delta G = 0$ represent oxidation of cerium; those above the line represent reduction

at a lower temperature, the second one at a higher temperature. In one of the catalysts, the Ce-HT catalyst displayed two reduction peaks at 115–450 °C and 450–650 °C under H_2-TPR (Dai et al. 2018). The first peak could be assigned to the reduction of surface CeO_2 and/or the partial reduction of CeO_2 with the formation of nonstoichiometric Ce oxides. The second peak was attributed to the bulk CeO_2 ---> Ce_2O_3 reduction process.

The presence of another metal reportedly shows no effect on the high-temperature peak but lowers and splits the first peak. Noble metals are known to activate and accelerate the reduction process by hydrogen spillover to the support (Chen et al. 2016). Evidence is available in the literature for the interaction of noble metals supported on CeO_2 after a high-temperature reduction (Goguet et al. 2004). The reducibility of ceria from CeO_2 to $CeO_{2\text{-}X}$ is thermodynamically feasible, and the availability of oxygen vacancies on the surface generally changes the reactant adsorption capacity and the subsequent reactions (Li et al. 1989b).

Table 3.3 ΔH and ΔG values at different temperatures for cerium oxidation and reduction

Temp (°C)	$2CeO_2 + CO(g) = Ce_2O_3 + CO_2(g)$		$Ce_2O_3 + CO_2 = 2CeO_2 + CO$		$2CeO_2 + H_2(g) = Ce_2O_3 + H_2O(g)$		$Ce_2O_3 + H_2O(g) = 2CeO_2 + H_2(g)$		$2Ce_2O_3 + O_2(g) = 4CeO_2$	
	ΔH (kcal)	ΔG (kcal)	ΔH (kcal)	ΔG (kcal)	ΔH (kcal)	ΔG (kcal)	ΔH (kcal	ΔG (kcal)	ΔH (kcal	ΔG (kcal)
200	20.50	15.84	−20.50	−15.84	30.08	20.97	−30.08	−20.97	−176.56	−147.29
400	20.75	13.84	−20.75	−13.84	29.89	17.16	−29.89	−17.16	−177.03	−134.84
600	21.24	11.72	−21.24	−11.72	29.88	13.39	−29.88	−13.39	−177.77	−122.21
800	21.95	9.47	−21.95	−9.47	30.11	9.59	−30.11	−9.59	−178.86	−109.38
1000	22.89	7.06	−22.89	−7.06	30.58	5.73	−30.58	−5.73	−180.31	−96.31

Ce_2O_3 oxidation with either CO_2 or H_2O is exothermic and thermodynamically feasible. The same can be said for oxygen. CeO_2 is endothermic at a temperature less than 1000 °C

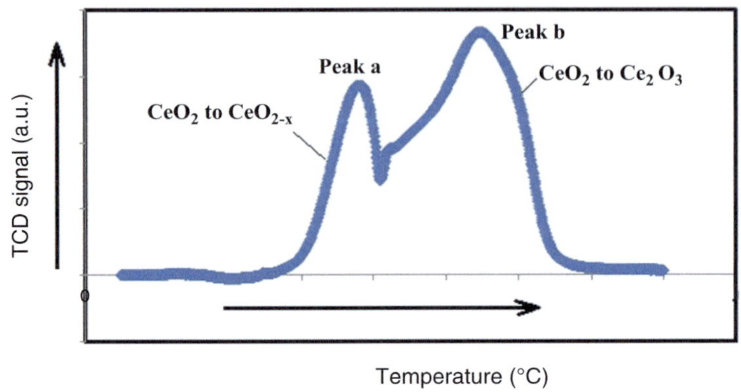

Fig. 3.6 Typical H_2-TPR profile for a cerium-based catalyst showing the characteristic two peaks. The peaks are generally not the same size, i.e., a larger one occurs at a higher temperature (Dai et al. 2018)

3.4.3 CeO_2 Reduction Mechanism

Cerium oxide seems be used often in catalyzing reactions that have to do with oxygen cycling or exchange with the catalyst, such as in the case of CO_2 reduction. The CO_2 reduction mechanism seen in Fig. 3.7 thus shows how the properties and the structure of the catalyst is a good fit for its function.

As might be expected, the presence of oxygen pockets or vacancies on the catalyst surface is the basis of the suggested catalytic mechanism. The adsorption of CO and CO_2 and Pt/CeO_2 was studied by infrared (IR) and X-ray photoelectron spectroscopy (XPS) to elucidate the CO_2 dissociation mechanism.

The reduction of CO_2 was found to proceed via several routes, predominantly the rapid carbonate formation, or carboxylate formation (Appel et al. 1998; Li et al. 1989a, b) and bicarbonate formation (Appel et al. 1998) species on partially reduced $CeO_{2\text{-}x}$. Direct dissociation of CO_2 also happens on the metal-loaded ceria, as depicted in Fig. 3.7 (Bernal et al. 2001; Jin et al. 1987). Lattice oxygen pockets or vacancies in the support, CoO_2, close to the Pt particles (see Fig. 3.7), contribute to the formation of CO from CO_2 (Jin et al. 1987).

Temperature-programmed desorption studies also implicate lattice oxygen at the Pt/CeO_2 interface in the reduction of CO_2 to CO on Pt/CeO_2 (Gorte et al. 1996).

The formate reaction route entails the generation of hydroxyl groups on the CeO_2 surface, which subsequently reacts with CO to form formates. The CO and H_2O are therefore produced from the decomposition of the intermediate species, such as the formates or carbonates adsorbed on the surface of cerium oxide.

The function of the platinum metal is to facilitate the adsorption of CO and to assist in breaking the C-H bond of formate (Goguet et al. 2004). It should be emphasized that multiple reaction pathways could be envisaged for the carbonates,

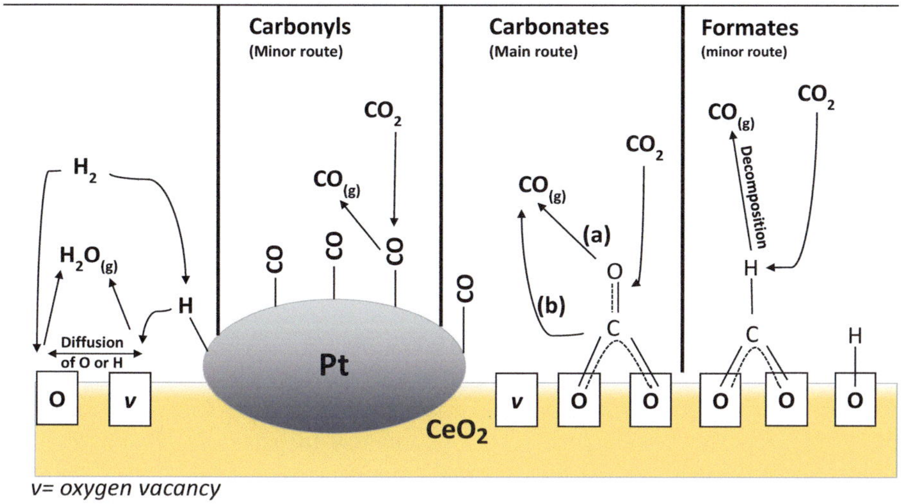

Fig. 3.7 Proposed model for the reaction mechanism of the RWGS reaction over Pt/CeO_2, with various mechanisms shown, such as formates, carbonates, and carbonyls. (Modified from Goguet et al. 2004)

as shown in Fig. 3.5. Another possibility involves carbonates being reacted directly with oxygen vacancies in the ceria, from which CO(g) would be released directly, leaving oxygen in the pockets. This pathway is designated (a) in Fig. 3.5. CO_2 is adsorbed on CeO_2 to form both carbonate and bidentate carbonate (Goguet et al. 2004). The adsorption of CO_2 is enhanced by prereduction of the catalyst, and pre-adsorption of water is reported to have no effect (Jin et al. 1987).

The function of Pt would be to activate H_2, thereby generating oxygen vacancies, and the activated H_2 moves over the cerium oxide surface. Carbon monoxide on metallic Pt is reportedly easier to detach on cerium oxide surface than is CO (Jin et al. 1987). Studies done by Jin et al. (1987) indicate that the CO_2 decomposition actually happens at the Pt/CeO_2 interface such that the CO produced would be adsorbed on Pt and the other oxygen fills the vacancy on the CeO_2. Therefore, CO_2 decomposition is the function of the stoichiometry of the local CeO_2 interface (Jin et al. 1987). The rate-determining step of the CO_2 hydrogenation could safely be given by oxygen vacancy or how the vacancy is moving along the CeO_2 surface and the surface carbonate. The surface migration of the carbonates toward Pt/CeO_2 interface where decomposition via Pt-bound carbonyl intermediate occur could be considered as a reaction pathway and in Fig. 3.5 is designated (b). If this occurs, the rate-determining step would be the desorption of the corresponding Pt-bound CO (ads). The mechanism provides some evidence for interaction of CeO_2 with platinum metals supported thereon (Li et al. 1989b; Goguet et al. 2004).

3.5 Conclusion

Research continues to see developments in distant fields that could be used to improve CO_2 utilization practice. Ceria-based oxides have attracted much attention, as these act as non-innocent support and active co-catalysts for environmental applications and for energy conversion systems. Cerium-based materials are reportedly employed by many researchers in CO_2 conversion, who opt for a low-cost flexible conversion route. The trend amplifies the need for a better understanding of the thermochemical conversion of both CO_2 and the catalyst phases. Researchers have confirmed the unique and incomparable role of ceria and ceria-based catalysts in pollution control, such as the case in RWGS and car converters for exhaust gas. Satisfactory knowledge about the fundamental characteristics of the cerium-based catalyst is available in the literature, and new characterization techniques continue to be developed. These undoubtedly help researchers to gain insight into reaction mechanisms. Ceria, or mixed binary oxides that contain ceria, are critical components in RWGS catalysts, due to their unique oxygen storage capacity.

There is much work ahead as atmospheric CO_2 concentration is on the rise. This calls for further and continuous research in catalyst development and process improvements. Hence, this chapter on thermochemical conversion of CO_2 is just a stepping-stone along the road on an exciting and demanding journey.

Acknowledgments The authors are grateful for the financial support provided by the University of South Africa (UNISA), the National Research Foundation (NRF) of South Africa, and the Institute for the Development of Energy for African Sustainability (IDEAS) research unit at UNISA.

References

Albrecht M, Rodemerck U, Schneider M, Broring M, Baabe D, Kondratenko EV (2017) Unexpectedly efficient CO_2 hydrogenation to higher hydrocarbons over non-doped Fe_2O_3. Appl Catal B Environ 204:119–126. https://doi.org/10.1016/j.apcatb.2016.11.017

Álvarez C, Schumann J, Behrens M, Luis J, Fierro G, Schlögl R, Frei E (2016) Reverse water-gas shift reaction at the Cu/ZnO interface: influence of the Cu/Zn ratio on structure-activity correlations. Appl Catal B Environ 195:104–111. https://doi.org/10.1016/j.apcatb.2016.05.007

Appel LG, Eon JG, Schmal M (1998) The CO_2 – CeO_2 interaction and its role in the CeO_2 reactivity. Catal Lett *56*:199–202. https://doi.org/10.1023/A:1019098121432

Aresta M, Dibenedetto A (2016) Reaction mechanisms in carbon dioxide conversion. Springer, Berlin/Heidelberg. https://doi.org/10.1007/978-3-662-46831-9

Bernal S, Blanco G, Gatica JM, Larese C, Vidal H (2001) Effect of mild re-oxidation treatments with CO_2 on the chemisorption capability of a Pt/CeO_2 catalyst reduced at 500 °C. J Catal 200 (2):411–415. https://doi.org/10.1006/jcat.2001.3210

Carrasquillo-flores R, Ro I, Kumbhalkar MD, Burt S, Carrero CA, Alba-rubio AC et al (2015) Reverse water−gas shift on interfacial sites formed by deposition of oxidized molybdenum moieties onto gold nanoparticles. *J Am Chem Soc* 137(10):317–10325. https://doi.org/10.1021/jacs.5b05945

Chen C, Cheng W, Lin S (2000) Mechanism of CO formation in reverse water–gas shift reaction over Cu/Al_2O_3 catalyst. Catal Lett 68:45–48. https://doi.org/10.1023/A:1019071117449

Chen X, Su X, Liang B, Yang X, Ren X, Duan H et al (2016) Identification of relevant active sites and a mechanism study for reverse water gas shift reaction over Pt /CeO_2 catalysts. J Energy Chem 25:1051–1057

Chen X, Su X, Duan H, Liang B, Huang Y, Zhang T (2017) Catalytic performance of the Pt/TiO2 catalysts in reverse water gas shift reaction: controlled product selectivity and a mechanism study. Catal Today 281:312–318. https://doi.org/10.1016/j.cattod.2016.03.020

Cherkezova-Zheleva Z, Mitov I (2010) In situ Mössbauer investigation of iron oxide catalyst in water gas shift reaction – impact of oxyreduction potential and temperature. J Phys Conf Ser 217. https://doi.org/10.1088/1742-6596/217/1/012044

Dai B, Zhou G, Ge S, Xie H, Jiao Z, Zhang G, Xiong K (2017) CO_2 reverse water-gas shift reaction on mesoporous M-CeO_2 catalysts. *Can J Chem Eng* 95(November 2016):634–642. https://doi.org/10.1002/cjce.22730

Dai B, Cao S, Xie H, Zhou G, Chen S (2018) Reduction of CO_2 to CO via reverse water-gas shift reaction over CeO_2 catalyst. *Korean J Chem Eng 35*(2):421–427. https://doi.org/10.1007/s11814-017-0267-y

Daza YA, Kuhn JN (2016) CO_2 conversion by reverse water gas shift catalysis: comparison of catalysts, mechanisms and their consequences for CO_2 conversion to liquid fuels. RSC Adv 6 (55):49675–49691. https://doi.org/10.1039/c6ra05414e

Dry ME (2002) The Fischer-Tropsch process: 1950–2000. Catal Today 71(3–4):227–241. https://doi.org/10.1016/S0920-5861(01)00453-9

Espinoza RL, Steynberg AP, Jager B, Vosloo AC (1999) Low temperature Fischer–Tropsch synthesis from a Sasol perspective. Appl Catal A Gen 186(1–2):13–26. https://doi.org/10.1016/S0926-860X(99)00161-1

Furler P, Scheffe JR, Steinfeld A (2012) Syngas production by simultaneous splitting of H_2O and CO_2 via ceria redox reactions in a high-temperature solar reactor. Energy Environ Sci 5 (3):6098–6103. https://doi.org/10.1039/c1ee02620h

Goguet A, Meunier FC, Tibiletti D, Breen JP, Burch R (2004) Spectrokinetic investigation of reverse water-gas-shift reaction intermediates over a Pt/CeO_2 catalyst. *J Phys Chem 108*:20240–20246. https://doi.org/10.1021/jp047242w

Gorimbo J, Lu X, Liu X, Yao Y, Hildebrandt D, Glasser D (2017) Low-pressure Fischer–Tropsch synthesis: in situ oxidative regeneration of iron catalysts. Ind Eng Chem Res 56(15):4267–4274. https://doi.org/10.1021/acs.iecr.7b00008

Gorte RJ, Bunluesin T, Cordatos H, Vohs JM, Stubenrauch J (1996) Effect of ceria structure on oxygen migration for Rh/ceria catalysts. J Phys Chem 100(2):785–789. https://doi.org/10.1021/jp952050+

Jin T, Zhou Y, Mains GJ, White JM (1987) Infrared and X-ray photoelectron spectroscopy study of CO and CO, on PtCeO_2. J Phys Chem 91(25):5931–5937. Retrieved from http://pubs.acs.org/doi/pdf/10.1021/j100307a023

Keeling RF, Manning AC (2014) Studies of recent changes in atmospheric O_2 content, Treatise on geochemistry: second edition, vol 5. Elsevier Ltd. https://doi.org/10.1016/B978-0-08-095975-7.00420-4

Keeling CD, Bacastow RB, Bainbridge AE, Ekdahl CA Jr, Guenther PR, Waterman LS, Chin JFS (1976) Atmospheric carbon dioxide variations at Mauna Loa Observatory, Hawaii. Tellus 28 (6):538–551. https://doi.org/10.3402/tellusa.v28i6.11322

Kim SS, Park KH, Hong SC (2013) A study of the selectivity of the reverse water-gas-shift reaction over Pt/TiO_2 catalysts. Fuel Process Technol 108:47–54. https://doi.org/10.1016/j.fuproc.2012.04.003

Kim DH, Han SW, Yoon HS, Kim YD (2015) Reverse water gas shift reaction catalyzed by Fe nanoparticles with high catalytic activity and stability. J Ind Eng Chem 23:67–71. https://doi.org/10.1016/j.jiec.2014.07.043

Le Quéré C, Andrew RM, Friedlingstein P, Sitch S, Hauck J, Pongratz J et al (2018) Global carbon budget 2018. Earth Sci Data 10:2141–2194

Li C, Sakata Y, Arai T, Domen K, Maruya K, Onishi T (1989a) Adsorption of carbon monoxide and carbon dioxide on cerium oxide studied by Fourier-transform infrared spectroscopy. Part 2. Formation of formate species on partially reduced CeO_2 at room temperature. J Chem Soc Faraday Trans 1 85(6):1451–1461. https://doi.org/10.1039/F19898501451

Li C, Sakata Y, Arai T, Domen K, Maruya K, Onishi T (1989b) Carbon monoxide and carbon dioxide adsorption on cerium oxide studied by Fourier-transform infrared spectroscopy. Part 1. Formation of carbonate species on dehydroxylated CeO_2, at room temperature. J Chem Soc Faraday Trans 1 85(4):929–943. https://doi.org/10.1039/F19898500929

Liu Y, Li Z, Xu H, Han Y (2016) Reverse water-gas shift reaction over ceria nanocube synthesized by hydrothermal method. Catal Commun 76(3):1–6. https://doi.org/10.1016/j.catcom.2015.12.011

Mishra BG (2003) Structural, redox and catalytic chemistry of ceria based materials. Bull Catal Soc India 2(2003):122–134

Montini T, Melchionna M, Monai M, Fornasiero P (2016) Fundamentals and catalytic applications of CeO_2-based materials. Chem Rev. https://doi.org/10.1021/acs.chemrev.5b00603

Orozco LM, Renz M, Corma A (2017) Cerium oxide as a catalyst for the ketonization of aldehydes: mechanistic insights and a convenient way to alkanes without the consumption of external hydrogen. Green Chem 19(6):1555–1569. https://doi.org/10.1039/c6gc03511f

Pekridis G, Kalimeri K, Kaklidis N, Vakouftsi E, Iliopoulou EF, Athanasiou C, Marnellos GE (2007) Study of the reverse water gas shift (RWGS) reaction over Pt in a solid oxide fuel cell (SOFC) operating under open and closed-circuit conditions. Catal Today 127:337–346. https://doi.org/10.1016/j.cattod.2007.05.026

Rafati M, Wang L, Shahbazi A (2015) Effect of silica and alumina promoters on co-precipitated Fe–Cu–K based catalysts for the enhancement of CO_2 utilization during Fischer–Tropsch synthesis. *J CO2 Util 12*:34–42. https://doi.org/10.1016/j.jcou.2015.10.002

Romero M, Steinfeld A (2012) Concentrating solar thermal power and thermochemical fuels. Energy Environ Sci 5(11):9234–9245. https://doi.org/10.1039/c2ee21275g

Ronda-Loret M, Rico-Frances S, Sepulveda-Escribano A, Ramos-Fernandez EV (2018) CuO_x/CeO_2 catalyst derived from metal organic framework for reverse water-gas shift reaction. Appl Catal A Gen 562:28–36. https://doi.org/10.1016/japcata.2018.05.024

Steinfeld A, Palumbo R (2004) Solar thermochemical process technology. Encycl Phys Sci Technol 15:237–256. https://doi.org/10.1016/b0-12-227410-5/00698-0

Sun F, Yan C, Wang Z, Guo C (2015) Ni/Ce-Zr-O catalyst for high CO_2 conversion during reverse water gas shift reaction (RWGS). Int J Hydrog Energy *40*:15985–15993. https://doi.org/10.1016/j.ijhydene.2015.10.004

Thiel C, Nijs W, Simoes S, Schmidt J, Van Zyl A, Schmid E (2016) The impact of the EU car CO_2 regulation on the energy system and the role of electro-mobility to achieve transport decarbonisation. Energy Policy 96:153–166. https://doi.org/10.1016/j.enpol.2016.05.043

Trovarelli A, Boaro M, Rocchini E, De Leitenburg C, Dolcetti G (2001) Some recent developments in the characterization of ceria-based catalysts. J Alloys Compd 324:584–591. https://doi.org/10.1016/S0925-8388(01)01181-1

Wang L, Liu H, Chen Y, Yang S (2017) Reverse water–gas shift reaction over co-precipitated Co–CeO_2 catalysts: effect of Co content on selectivity and carbon formation. Int J Hydrog Energy. https://doi.org/10.1016/j.ijhydene.2016.07.048

Wolf A (2016) Syngas production via reverse water-gas shift reaction over a Ni-Al 2 O 3 catalyst: catalyst stability, reaction kinetics, and modeling. Chem Eng Technol 298(6):1040–1048. https://doi.org/10.1002/ceat.201500548

Wolf A, Werner S, Demirel E, Azcan N, Goguet A, Meunier FC et al (2016) Photochemically driven reverse water-gas shift at ambient conditions mediated by a nickel pincer complex. Nat Publ Group 47(2):285–294. https://doi.org/10.1007/s11814-017-0267-y

Yang X, Su X, Chen X, Duan H, Liang B, Liu Q et al (2017) Promotion effects of potassium on the activity and selectivity of Pt/zeolite catalysts for reverse water gas shift reaction. Appl Catal B Environ. https://doi.org/10.1016/j.apcatb.2017.05.067

Yang L, Pastor-Perez L, Gu S, Sepulveda-Escribano A, Reina T (2018) Highly efficient Ni/CeO_2-Al_2O_3 catalysts for CO_2 upgrading via reverse water-gas shift – effect of selected transition metal promoters.pdf. Appl Catal B Environ 232:464–471. https://doi.org/10.1016/j.apcatb.2018.03.091

Zhang F, Chen C, Hanson JC, Robinson RD, Herman IP, Chan S (2006) Phases in ceria-zirconia binary oxide (1-x)$CeO_{2\text{-}x}ZrO_2$ nanoparticles: the effect of particle size. J Am Ceram Soc 1036:1028–1036. https://doi.org/10.1111/j.1551-2916.2005.00788.x

Zhang X, Su H, Zhang Y, Gu X (2016) Effect of CeO_2 promotion on the catalytic performance of Co/ ZrO_2 catalysts for Fischer–Tropsch synthesis. 184:162–168. https://doi.org/10.1016/j.fuel.2016.06.125

Zhu M, Wachs IE (2016) Iron-based catalysts for the high-temperature water-gas shift (HT-WGS) reaction: a review. *ACS Catal* *6*(2):722–732. https://doi.org/10.1021/acscatal.5b02594

Chapter 4
Photocatalytic Systems for Carbon Dioxide Conversion to Hydrocarbons

Amel Boudjemaa and Nabila Cherifi

Contents

4.1 Introduction 65
4.2 Fundamental Aspects for CO_2 Photoconversion 66
4.2.1 Background and General Principles 66
4.2.2 Challenges of CO_2 Photoconversion 66
4.3 Carbon Dioxide Photoreduction over UV-Light Semiconductors 67
4.3.1 Titanium Dioxide (TiO_2) Material 67
4.3.2 TiO_2-Based Photocatalyst 69
4.4 Carbon Dioxide Photoreduction on Visible Light Materials 71
4.4.1 Metal Oxide Photocatalyst 71
4.4.2 Porous Materials 75
4.4.3 Carbon-Based Materials 77
References 80

Abstract CO_2 emissions are a major contributor to the climate change due to the increase of CO_2 concentration in the atmosphere. CO_2 represents 70% of greenhouse gas emissions come essentially from human activities during the combustion of fossil resources such as petrol, natural gas, etc. Many initiatives that allow reducing the CO_2 concentration are developed. One of the most promising approaches is to convert CO_2 into fuels and into valuable chemicals through photocatalysis process. The CO_2 photoreduction into hydrocarbon is a promising method to convert CO_2 taking advantage of the readily available solar energy. To achieve the goal, it is necessary to develop an efficient photocatalyst and enhance both the CO_2 reduction and the efficiency yields. This chapter reviews recent developments, challenges, and novel approaches of CO_2 photoconversion for sustainable fuels. These include providing the photocatalytic properties of UV and visible light photocatalysts applied to CO_2 reduction, as well as the recent advances in the design of photocatalytic systems. At the same time, doping TiO_2 material with various metals

A. Boudjemaa (✉) · N. Cherifi
Centre de Recherche Scientifique et Technique en Analyses Physico-Chimiques (CRAPC), Tipaza, Algeria
e-mail: amel.boudjemaa@crapc.dz; nabila.cherifi@crapc.dz

Inamuddin et al. (eds.), *Conversion of Carbon Dioxide into Hydrocarbons Vol. 1 Catalysis*, Environmental Chemistry for a Sustainable World 40,
https://doi.org/10.1007/978-3-030-28622-4_4

or used as heterojunction allows an improvement photocatalytic activity under visible irradiation. Moreover, the effect of the doping can induce unique properties such as extended light absorption range, charge separation, and also an efficient reactivity. The contribution of carbon-based material and porous materials to the enhanced visible light-driven photocatalysis will be included. So, attractive properties make these materials as a good photocatalysts with immense potential in the elaboration of efficient visible light photocatalysts.

Keywords CO_2 conversion · Photocatalysis · TiO_2 · Carbon-based material · Nanocomposite · Porous materials · Methanol · Hydrocarbons · UV light · Visible irradiation

Abbreviation

UV	ultraviolet
IR	infrared
CO_2	carbon dioxide
CH_4	methane
N_2O	nitrous oxide
SCs	semiconductors
Eg	band gap energy
VB	valence band
CB	conduction band
e^-	electron
h^+	hole
(e^-/h^+)	electron/hole
TiO_2	titanium dioxide
H_2	hydrogen
H^+	proton
HCOOH	formic acid
CH_3OH	methanol
CH_4	methane
1D	one-dimensional
NF	nanofibers
CH_2O	formaldehyde
C_2H_6	ethane
C_3H_8	propane
C_3H_6	propene
C_4H_9	butane
LDHs	layered double hydroxides
HT	hydrotalcite
OH^-	hydroxy group
G	graphene
GO	graphene oxide

rGO	reduced graphene oxide
g-C3N4	graphitic carbon nitride
CNT	carbon nanotubes
CNF	nanofibers

4.1 Introduction

Since the beginning of the twentieth century, the world knows several destructive climatic catastrophes, including hurricanes, drought, floods, fires, and storms. As example, we can easily quote Hugo hurricane recorded in the United States in 1989, catastrophic floods of Haiti in 2003, and southern Europe's canicule in 2007. Recently in 2017, there have been many other environmental damages as Irma hurricane in Florida, deadly heat waves in India, several floods in Southeast Asia, etc. Such extreme weather phenomena presented a broad threat to environment and to human life and caused many diseases and various cases of deaths. It is generally supposed that such environmental catastrophes are in direct relation with elevation in the earth temperature known as global warming which is associated to climate changes occurring throughout the world (Tuckett 2009; Patz et al. 2005). Indeed, as reported in bibliography (Kirk-Davidoff and Keith 2007), over the past 50 years, the temperature of the air has increased by about 0.8 °C. This elevation can reach a value of 5.8 °C by the end of this century, accompanied with a rise in sea level. This situation becomes alarming and has stimulated worldwide research. To better act against these environmental catastrophes, it is necessary to know the reason of such global warming generally related, as explained by many scientists, to greenhouse effects and anthropogenic activities (Xi-Liu and Qing-Xian 2018; Mitchel 1989; Mora et al. 2013; Royer et al. 2007; Perkins et al. 2018). For better understanding, some questions appear essential and require answers. What are greenhouse gases effects and in what their importance reside? What is the relation between such greenhouse gases, global warming, and human activities?

In order to answer these questions, it is important to understand that the earth continually absorbs ultraviolet (UV) and visible radiation from the sun and emits infrared (IR) radiation. The greenhouse gases are gaseous compound in the atmosphere capable to absorb these IR radiations (Barbera et al. 2018). The most important greenhouse gas is carbon dioxide (CO_2), followed by methane (CH_4) and nitrous oxide (N_2O). So, the absorption of the IR radiations by these gases allows ensuring a balance between the incoming and outgoing electromagnetic radiations on the earth, defining the "primary" greenhouse effect. However, with the release of the industrial revolution over the last century, it was noted that the global concentrations of different greenhouse gases have raised significantly. The increase of CO_2 concentration induced the elevation of earth temperature and consequently the problem of global warming and all environmental catastrophes recently observed (McMichael and Costello 2012). Such a situation should not continue, and solutions must be found in urgency to cease these problems. In this chapter we are particularly interested in the conversion of CO_2, as solution to minimize its concentration in the atmosphere. Indeed, as it will be detailed in this contribution, many researchers were interested by converting this greenhouse gas into chemicals values products in both homogenous and heterogeneous medium.

4.2 Fundamental Aspects for CO_2 Photoconversion

4.2.1 Background and General Principles

Photocatalysis is one of the most important fields for renewable and environment-friendly energy. Generally, the catalysts used have semiconductors (SCs) properties with narrow band gap energy (Eg) located between the valence band (VB) and the conduction band (CB). Generally, photocatalysis is based on the light absorption by a catalyst. So, the consumed energy with $h\nu \geq$ Eg excites the electron (e^-) from the VB to the CB, leaving a hole (h^+) in the VB, creating photogenerated electron/hole (e^-/h^+) pair, which are contributed to the oxido-reduction reactions (Yuan et al. 2016; Boudjemaa et al. 2014; Yang and Xu 2016; Hisatomi et al. 2014; Boudjemaa and Trari 2010; Reñones et al. 2016). During the photocatalysis process, (e^-/h^+) are created to transform CO_2 and H_2O into fuels. Many strategies have been developed in order to increase the photocatalytic reactivity which depends on several parameters as the type of the semiconductor where n-type semiconductor electrons are the major charge carrier (the hole for the p-type semiconductor). The second parameter is the irradiation type used to accelerate the recombination of the (e^-/h^+) pair. The pH, the temperature, and the concentration of the electrolyte affect the reactivity via the contact with the targeted molecules, the collision between the semiconductor and the substrate, and the reaction kinetics, respectively. Photocatalysis is successfully used to decompose various kinds of pollutants and enhance the quality of the air, the soil, and the water.

4.2.2 Challenges of CO_2 Photoconversion

To provide appropriate response to the challenge that corresponds to reducing the CO_2 concentration, several techniques have been investigated including CO_2 capture, separation and purification (Xu et al. 2014; Wang et al. 2017; Abbas et al. 2013), electrochemical (Jhong and Ma 2013; Ganesh 2016), thermal catalysis such as hydrogenation (Erdöhelyi et al. 1986; Martin et al. 2017), and biochemical transformation methods (Lee et al. 2012;). On the other hand, the big challenge is related to the ability to transfer CO_2 into fuels without increase in anthropogenic

CO_2 emissions. Therefore, the reduction of CO_2 using solar, wind, nuclear, or geothermal is another challenge estimated to be sustainable alternative to decrease CO_2 emission.

The direct reduction of CO_2 to $CO_2^{\circ-}$ is complex due to its more positive reduction potentials (Halmann and Steinberg 1999). So, for this and compared to the conventional methods, the photoreduction of CO_2 with H_2O using solar energy can consume less energy. The process involves catalysts that are activated under irradiation to ensure effective harnessing of light energy. Therefore, the development of efficient and low-cost effective materials especially those absorbed visible light irradiation is necessary. Another challenge is to increase the reaction yields as enhanced light harvesting based on the material nature, enhancement separation, and transport of the pairs (e^-/h^+) which improved the materials surface and enhanced the reaction activity, selectivity, and stability.

4.3 Carbon Dioxide Photoreduction over UV-Light Semiconductors

Since the discovery of photocatalysis process by Fujishima and Honda on 1972, titanium dioxide (TiO_2) has appeared as an efficient photocatalyst for environmental purification (Fujishima and Honda 1972). In general, the metal d^0 and d^{10} oxide are suitable heterogeneous photocatalytic materials (Kudo and Misekj 2009; Inoue 2009) with appropriate position of CB able to reduce CO_2 as ZrO_2, Ta_2O_5, Ga_2, O_3, etc.

4.3.1 Titanium Dioxide (TiO_2) Material

TiO_2 is present in three crystallographic forms, rutile, anatase, and brookite, with the Eg range from 3.0 eV to 3.2 eV, respectively. TiO_2 photocatalyst is a semiconductor that has been applied for CO_2 reduction to fuels (Tahir and Amin 2013). The first report was published by Inoue et al. in 1979 (Inoue et al. 1979), and since this date, TiO_2 semiconductor has been largely employed in the photocatalysis due to the band edge position, chemical stability, low cost, and low toxicity (Xiong et al. 2015; Pathak et al. 2004; Tian et al. 2009). It is established that the three forms of TiO_2 influence on the efficiency of CO_2 photoreduction. So under UV and visible light, brookite generated a highest yield of CO and CH_4 (Liu et al. 2012), while rutile and anatase phases or a mixed of both produced only CH_4 (Schulte et al. 2010). As a result, the good crystallinity made the material a good photocatalyst because the photogenerated (e^-/h^+) pairs are easy to recombine (Wang et al. 2005). The photoreactivity of the mixed phases was due to the junction effect between rutile and anatase for a good separation of (e^-/h^+) pairs. It is verified that more than 20% of

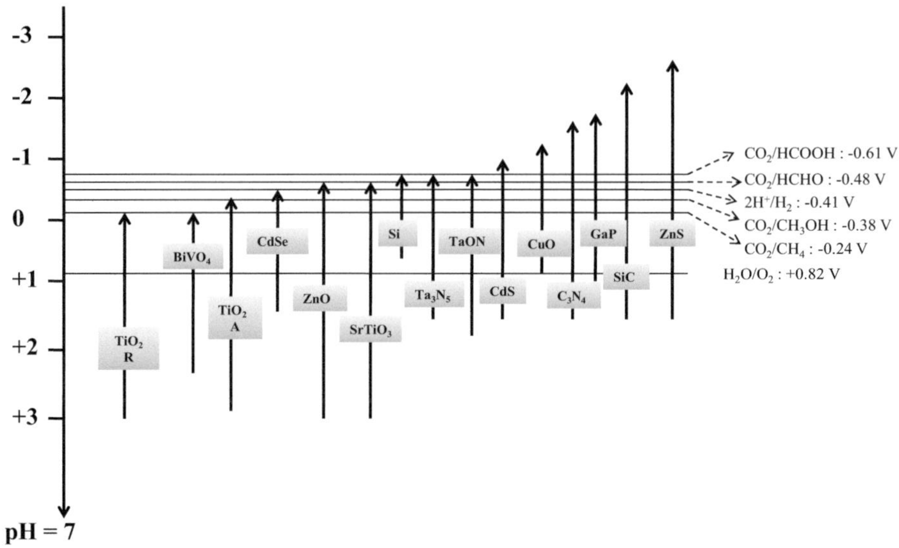

Fig. 4.1 Band bending and the reduction potential of photoconversion of CO_2 neutral electrolyte

rutile exhibited the highest photocatalytic activities (Gao et al. 2014; Wang et al. 2012a, b, c, d). Mainly works have been focused on the utilization of anatase structure (Dhakshinamoorthy et al. 2012).

Moreover, TiO_2 photocatalyst is largely used for CO_2 reduction due to the Eg, CB, and VB values compared to the reduction potential of CO_2 and the oxidation potential of hydrogen (H_2) (Fig. 4.1). The reactions could be occurring between the CB energy and VB energy of TiO_2 where the photocatalyst can reduce CO_2 to formic acid (HCOOH), methanol (CH_3OH), and methane (CH_4). On the other hand, water can be oxidized to form proton (H^+).

The product selectivity is based on the electrons number involved for each reaction:

$$CO_2 + H^+ + 2e^- \rightarrow HCO_2 \tag{4.1}$$

$$CO_2 + 2H^+ + 2e^- \rightarrow CO + H_2O \tag{4.2}$$

$$CO_2 + 4H^+ + 4e^- \rightarrow HCHO + H_2O \tag{4.3}$$

$$CO_2 + 6H^+ + 6e^- \rightarrow CH_3OH + H_2O \tag{4.4}$$

$$CO_2 + 8H^+ + 8e^- \rightarrow CH_42H_2O + H_2O \tag{4.5}$$

$$2CO_2 + 8H^+ + 12e^- \rightarrow C_2H_4 + 2H_2O \tag{4.6}$$

$$2CO_2 + 12H^+ + 12e^- \rightarrow C_2H_5OH + 3H_2O \tag{4.7}$$

$$2CO_2 + 12H^+ + 12e^- \rightarrow C_2H_2 + 4H_2O \tag{4.8}$$

Generally, the highest production rate of carbon monoxide (CO) and hydrogen (H_2) was increased after 6 h of irradiation. When the reaction time increased, CH_4 and CH_3OH were formed and reached the maximum yield after 11 h of irradiation.

To improve the performance of TiO_2, diverse approaches have been made such as band bending engineering, modifying textural, structural, and optoelectronic properties (Yu et al. 2006; Kandiel et al. 2009; Collado et al. 2013). So, several researchers have demonstrated that the combination of one-dimensional (1D) morphologies such as rods, tubes, and fiber structures provide TiO_2 with unique properties favoring the (e^-/h^+) separation (Fan et al. 2012; Liu et al. 2007; Archana et al. 2009). Reñones and coauthors demonstrated that hierarchical mesoporous TiO_2 1-D nanofibers (NF) synthesized by a combination of electrospinning and sol-gel methods improved the photocatalytic reactivity compared to the other bulk TiO_2 (Reñones et al. 2016). The CO_2 conversion was reached with TiO_2NF, and a higher rutile amount (80:20 rutile/anatase) increased the H_2 production 4 times higher (vs. 2.5 times of CO). Another way to increase the photocatalytic reactivity is the utilization of the porous structure which improved the light adsorption capacity due to the important surface area and a multiple interparticle scattering (Yu et al. 2006; Hou et al. 2014; Sun et al. 2015a, b). Mesoporous TiO_2 illustrated improved activity compared to P25 photocatalyst (Joo et al. 2005; Wu et al. 2005; Peng et al. 2005). So, the decrease in the particles size can enhance the redox potential of VB and CB (Martín et al. 2015). Another work demonstrated that TiO_2NF illustrate an improved photoreactivity related to the porous structure. This structure improved by the optoelectronic properties facilitating light absorption by rising the charge separation and interparticle charge transfer along TiO_2 (Moya et al. 2015; Choi et al. 2010; Chuangchote et al. 2008). Also, the photoreactivity improved as TiO_2 particles diameter decreased (Anpo et al. 1995). Many works conclude that the decrease in particle size reduced the recombination of the charge carriers which will be a dominant process. So, e^- and h^+ generated were faster than interfacial charge carrier transfer processes (Zhang et al. 1998; Wang et al. 1997; Chae et al. 2003).

However, TiO_2 with large Eg provide only absorbance in the UV region (Nakata and Fujishima 2012; Pelaez et al. 2012), which absorbs only 4% of solar light. In addition, fast recombination of (e^-/h^+) limited the applicability of TiO_2 as photocatalyst and made it as unfavorable photocatalyst for large-scale production. Many strategies have been adopted to increase the absorption in the visible region (Zhang et al. 2011; Chen and Burda 2008; Etacheri et al. 2010).

4.3.2 *TiO_2-Based Photocatalyst*

The elaboration of TiO_2-based material with narrow Eg band gap energy with appropriate band bending configuration have been investigated. These methods include the modification of the electronic properties (Yang et al. 2009; Wang et al. 2012a, b, c, d), by combining with noble metals (Kong et al. 2017), or with other composite materials (Li et al. 2010). So, these strategies influenced on the photocatalytic activity by reducing the band width and increased the lifetime of separated (e^-/h^+) pairs. For example, the CO_2 photoreduction rate of CH_3OH over TiO_2 was only 0.02 m mol g_{cat}^{-1} h^{-1} (Wu et al. 2005). Also, the effect of doping TiO_2 was investigated by Halmann et al. (Halmann et al. 1983). So, the production rate of HCOOH, formaldehyde (CH_2O), and CH_3OH increased with doping TiO_2.

Transition and Noble Elements

Oluwafunmilola and Maroto-Valer developed M-TiO_2 (M = V, Co, Cr)-nanostructured photocatalyst immobilized onto quartz plates (Ola and Maroto-Valer 2015). The introduction of the metal modified both the crystallinity and the Eg energy. Conversely, high metal concentration has an effect disadvantageous to visible light photocatalytic activity, due to the metal ion species acting as multiple trap sites and thus facilitating (e^-/h^+) recombination (Tian et al. 2018). Consequently, the deposition of nanoparticles like Cu enhanced selectivity toward CH_3OH, whereas Pd or Pt favors the selectivity to CO and CH_4 (Dhakshinamoorthy et al. 2012). Also, Pt, Ru, Rh, Ag, and Ni significantly enhance the rate in several approaches such as charge separation, recombination, and trapping of charge carriers (Zhang et al. 2009; Sasirekha et al. 2006; Kohno et al. 1999; Collado et al. 2013; Fan et al. 2010). The deposition of metal nanoparticles on TiO_2 has as goal the generation of the redox couples and favors the adsorption of CO_2 molecules on metal sites (Tahir et al. 2015).

The effect of silver nanoparticles has been investigated in the gas phase and under UV irradiation (Collado et al. 2013). After 15 h of irradiation, H_2, CO, CH_4, ethane (C_2H_6), propane (C_3H_8), propene (C_3H_6), butane (C_4H_9), and CH_3OH were detected as by-product reaction. So, the deposition of Ag leads to decrease in the yield of CO followed by an increase of CH_4, C_2H_6, and C_3H_7 yields. The enhancement photoconversion was explained by a polymerization mechanism where CO and H_2 were combined to form -CH_2- species (Wojciechowski 1988; de la Peña O'Shea et al. 2007).

Furthermore, the photoconversion of CO_2 into hydrocarbons was also obtained by many authors on numerous modified TiO_2 photocatalysts, for example, on CdS (Bi_2S_3)/TiO_2 (Li et al. 2012a, b), iodine-doped TiO_2 (Zhang et al. 2011), Cu- and cerium-doped TiO_2 (Luo et al. 2011), NiO/$InTaO_4$ (Wang et al. 2010), nanocomposite kaolinite-TiO_2 (Kočí et al. 2011). In homogenous phase, little TiO_2-based catalysts like enzyme-modified TiO_2 (Woolerton et al. 2010), Ti-SBA-15 (Yang et al. 2011), Ag-loaded $ALa_4Ti_4O_{15}$ (A = Ca, Srand Ba) (Iizuka et al. 2011), and Pt-NP loaded TiO_4 (Feng et al. 2011) show enhanced photocatalytic reactivity.

Rare Earth Elements

Using cerium oxide for modified TiO_2 increased the light absorption due to the valences state 3+ and 4+ of Ce (Ce^{4+} and Ce^{3+}). In addition, these valances enhance also the electron transferring in Ce-O. Significant results were obtained due to the insertion of CeO_2 to TiO_2 (Zhou et al. 2014). Some studies about CeO_2 are regrouped in Table 4.1. So, under UV irradiation, CeO_2-TiO_2 demonstrated an efficient reactivity (Zhou et al. 2014). For this, the addition of CeO_2 to TiO_2 decreased the recombination (e^-/h^+) pairs and improved the conversion of CO_2 to CH_3OH and to CH_2O. Matejova et al. and Wang et al. reported that CH_4 can produce under UV light and CO can be formed under solar irradiation (Matejová et al. 2014; Wang et al. 2013a, b, c). The Eg decreased from 3.19 to 2.64 eV after the inclusion of CeO_2 indicating the large application under visible irradiation (Abdullah et al. 2015). The methanol yield was 18.6 $\mu mol/g_{catalyst}$, whereas the yield obtained from TiO_2 alone was 6 $\mu mol/g_{catalyst}$. This result is due to the presence of anatase structure with nanoparticles in agreement with Koci et al. (2009). A general approach for CO_2 photoreduction with H_2O was described by Liu and Li (2014). So, after the photogeneration of the (e^-/h^+) pairs, the intermediate products were formed and converted to the products. Xie et al. proved that when the CO_2 was adsorbed on the photocatalyst surface, the products yield generated increased (Xie et al. 2013a, b). On the other hand, the photoreactivity was improved with the couples material such as the recombination with adsorbed and photocatalyst materials (Liu et al. 2013, 2014).

4.4 Carbon Dioxide Photoreduction on Visible Light Materials

The advanced visible light material with a narrow Eg has taken a great perspective for photocatalytic applications. Researchers in this field are always increased in finding a new photocatalyst with some advanced properties to enhance the photocatalytic performance.

4.4.1 Metal Oxide Photocatalyst

Metal oxide semiconductors are largely used in photocatalysis due to their stability through photooxidation reactions. The photogenerated h^+ in VB has strong oxidizing power, and its ample Eg makes them thermodynamically able to reduce CO_2 (Habisreutinger et al. 2013). The CB of this class of semiconductors has more negative potential compared to (H_2O/H_2) and (CO_2/products) redox couples which makes them a good photocatalysts for hydrogen production and CO_2 reduction

Table 4.1 Summary of CO_2 reduction efficiency with different photocatalysts

Catalysts	Reaction conditions	Selectivity	Remarks	References
Ag_3PO_4/ CeO_2	Under visible light	**Ag_3PO_4/CeO_2**	Ag nanoparticles play a significant role in enhancing the photocatalytic activity	Xiong et al. (2018)
		CH_3OH: 42.5 µmol $gcat^{-1}$		
		C_2H_5OH: 31.5 µmol $gcat^{-1}$ total yield of the alcohols	During the photocatalysis, CeO_2 microspheres provide abundant pores for loading Ag/Ag_3PO_4 nanoparticles, ensuring the stability of the light sensitizers	
		74 µmol $gcat^{-1}$		
		27.2 µmol $gcat^{-1}$	CeO_2 can transform inert linear CO_2 to b-HCO_3^- and b-CO_3^{2-} species to decline reductive potential of CO_2	
		Ag_3PO_4		
		15.1 µmol $gcat^{-1}$		
0.5% Ce/TiO_2	UV light: 9 W hg lamp 0.5 g of catalyst /400 ml NaOH (0.2 M)	**0.5% Ce/TiO_2**	Two forms of Ce4+ ions doped TiO_2; Ce4+ ions incorporated into the lattice of TiO_2 and small CeO_2 nanoparticles	Xiong et al. (2014)
1% Ce/TiO_2		CH_3OH: 150 µmol $gcat^{-1}$		
2% Ce/TiO_2		HCOH: 70 µmol $gcat^{-1}$		
3% Ce/TiO_2		CH_4: 10 µmol $gcat^{-1}$	The products selective can be ascribed to the compromise between charge transfer and thermodynamics	
		1% Ce/TiO_2		
		CH_3OH: 180 µmol $gcat^{-1}$		
		HCOH: 90 µmol $gcat^{-1}$		
		CH_4: 15 µmol gcat-	Ce^{3+}/Ce^{4+} mixture can inhibit the recombination of photogenerated (e^-/h^+)	
		2% Ce/TiO_2		
		CH_3OH: 80 µmol $gcat^{-1}$		
		HCOH: 40 µmol $gcat^{-1}$		
		CH_4: 7 µmol $gcat^{-1}$		
		3% Ce/TiO_2		
		CH_3OH: 30 µmol $gcat^{-1}$		
		HCOH: 20 µmol $gcat^{-1}$		
		CH_4: 2.5 µmol $gcat^{-1}$		

(continued)

Table 4.1 (continued)

Catalysts	Reaction conditions	Selectivity	Remarks	References
Ce/La/ TiO_2	15 W fluorescent lamp, 1 g/L catalyst/200 ml NaOH (0.2 M)	**1.5 wt% Ce/La/TiO_2**	Product rate increased with an increase of Ce concentration	Ahmed Ali et al. (2017)
		CH_3OH: 355.2 μmol/ g^{-}cat		
		CH_4: 289.1 μmol/ g^{-}cat	The product yields for 1.5 wt % Ce/La/TiO_2 were around 1.8, 1.9, and 2.1 times higher than that of La/TiO_2 for CH_3OH, CH_4, and CO, respectively	
		CO: 132.6 μmol/ g^{-}cat		
CeO_2-TiO_2	Visible light irradiation, 500 W xenon lamp 0.4 L/min of CO_2	CH_3OH:		Abdullah et al. (2015)
		18.6 μmol/g (CeO_2-TiO_2)		
		6.0 μmol/g (TiO_2)		
	0.3 g catalyst /300 /mL NaOH (0.1 Mol/L)	CH3OH		
Ce-TiO_2/ SBA-15		CO and CH_4	Dispersion of Ce–TiO_2 on the silica matrix improved both the textural properties and the photocatalytic reactivity compared to pure TiO_2	Zhao et al. (2012)
Ce doped TiO_2	9 W hg lamp, 254 nm Slurry reactor;	CH_3OH: 23.5 μmol/h gcat		Xiong et al. (2015)
	0.5 g/NaOH (0.2 N)			
	CO2 = 5 mL/min			

(Fig. 4.2). However, the positions of CB and VB of various semiconductors must have more negative potential than required for the water oxidation reaction.

Cuprous oxide with Eg ~ 2.0 eV is a p-type semiconductor for visible light hydrogen evolution (Paracchino et al. 2012). Depending on the synthesis process and the reaction conditions, Eg values can be decreased and ranged from 1.3 eV to 1.8 eV (Izaki et al. 2011; Chiang et al. 2012a, b; Chiang et al. 2011). So, the small Eg allows absorbing a large portion of the solar spectrum. The narrow Eg and the suitable position of CB and VB make the material an efficient photocatalyst for CO_2 photoconversion (Tran et al. 2012a, b; Ghadimkhani et al. 2013). Moreover, the high abundance and the low toxicity of CuO make it a more attractive material. In addition, copper-based materials (i.e., Cu, CuO, and Cu_2O) were used as photocathode and show good performances in photoelectrochemical CO_2 reduction (Man et al.

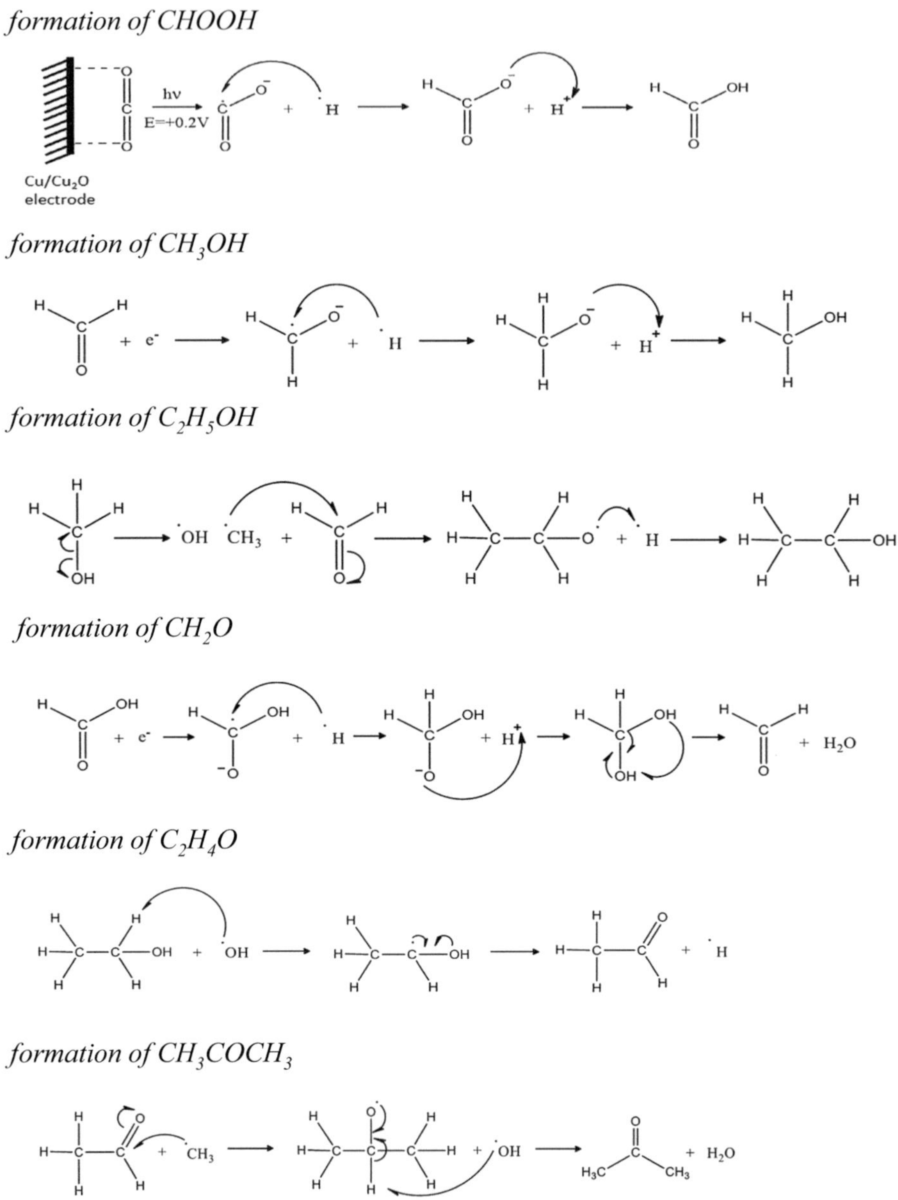

Fig. 4.2 Mechanism of a formation of reaction products through photoelectrochemical reduction of CO_2 (Ferreira de Brito et al. 2015)

2015; Mao et al. 2012; Paracchino et al. 2011). Several works have demonstrated that CuO can serve as a good photocathodic material alone (Izaki et al. 2011; Chiang et al. 2012a, 2012b) or as a modified/heterojunction photoelectrodes with other SCs cited TiO_2 and ZnO (Zhao et al. 2010).

Scheme 4.1 Schematic of CO_2 photocatalytic reduction on a semiconductor photocatalyst (Xie et al. 2016)

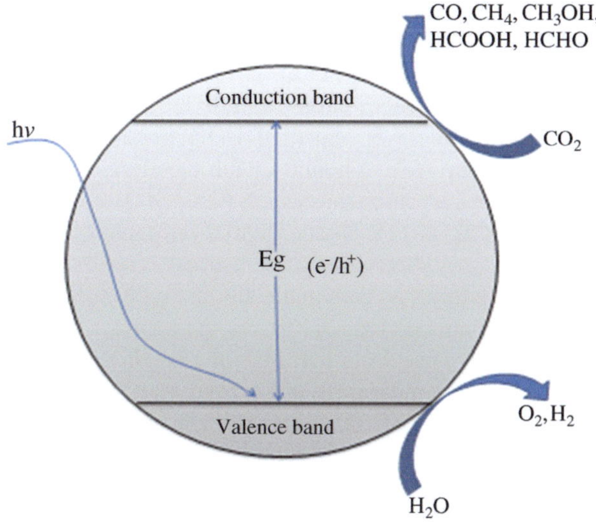

The transition metals Ag, Au, Cd, Cu, Pb, and Sn deposed on CuO/Cu_2O were investigated for controlling the products selectivity in terms of faradaic efficiency (Won et al. 2014). Also, due to the positions of CB and VB, the attractive hopping of (e^-/h^+) through Cu_2O and CuO was provided in the layered structure of CuO/Cu_2O. A higher potential than standard redox potentials of HCOOH and CH_3OH formation from CO_2 was obtained over $Pb/CuO/Cu_2O$. The redox potential of HCOOH, CH_3OH, and CO was ca. -0.61 V, −0.38 V, and − 0.53 V vs. SHE, respectively (Kumar et al. 2012). So, CH_3OH selectivity was narrow ~95% faradaic efficiency (Rajeshwar et al. 2013; Ghadimkhani et al. 2013). Hence, deposited metals facilitated the electron transfer from the material to the electrolyte leading to the reduction of energy loss (Feng et al. 2011). The hydrocarbons selectivity depends on the nature of metal deposed on the materials (Li et al. 2012a, 2012b). The photoreduction of CO_2 to CH_3OH, C_2H_5OH, CH_2O, C_2H_4O, and CH_3COCH_3 related to the presence of OH°, H°, and CH_3° during reaction (Ferreira de Brito et al. 2015) (Scheme 4.1). These radicals depend on both (e^-/h^+) pairs formation during the illumination of Cu/Cu_2O and the pH solution. The sensitivity of the photoreduction selectivity is changed with the pH, the potential, and the electrolyte composition and concentration (Wang et al. 2012a, b, c, d; Ghadimkhani et al. 2013; Li et al. 2014; Liu and Li 2014; Qin et al. 2011; Yuan et al. 2012).

4.4.2 Porous Materials

Layered double hydroxides (LDHs) anionic clays named also hydrotalcite (HT) are usually applied for CO_2 adsorbents (Yong et al. 2001). The photoconversion of CO_2 into fuels was reported for the first time using $[Zn^{II}{}_{1\text{-}x\text{-}y}Cu^{II}{}_{y}M^{III}{}_{x}(OH)_2]^{x+}{}_{2/}$

$_x(CO_3)^{2-}{\cdot}mH_2O$ as photocatalysts (Ahmed et al. 2011). Over Zn-Cu-Al photocatalyst, the CH_3OH selectivity was 26 mol % around 3 times less than Zn-Cu-Ga catalysts (Ahmed et al. 2011). Hybrid TiO_2/LDHs material is also used for CO_2 capture and other photocatalytic reduction processes (Liu et al. 2015). Generally, in the interlayer space of LDH materials, CO_2 adsorbed reacted with the hydroxy group (OH^-) bound of the metal sites and formed an intermediate space named hydrogen carbonate which was converted to HCOOH, CH_2OH, and CH_3OH (Ahmed et al. 2011). Under irradiation, the metal ions in layer facilitated charge separation. Another work demonstrated that the reaction rate was enhanced by replacing interlayer carbonate anions with metal hydroxyl group such as $[Cu(OH)_4]^{2-}$ (Ahmed et al. 2012).

TiO_2 nanoparticles in the layer of CoAl-LDHs prepared with one-pot hydrothermal synthesis (Kumar et al. 2017) reveal a good activity and selectivity to CO where P25 exhibited lower photoactivity which may be due to the fast charge carrier recombination. In this study, the production of hydrocarbon was not observed probably due to both the low adsorption of CO_2 and the potential values of VB and CB of the SC (see Fig. 4.3).

The effect of the preparation conditions on the photoconversion of CO_2 over MgAlTi-LDHs photocatalysts was investigated by Zhao et al. (2016). The results demonstrated that the amorphous TiO_2 present in MgAlTi-LDHs show a good reactivity compared to the commercial TiO_2 (P25). In the same conditions, only CO was produced as a main product with a trace of CH_4, in agreement with (Zhang et al. 2011; Liu et al. 2012; Hong et al. 2014). So, compared to the P25, the photoreactivity over CP was two times higher making MgAlTi-LDH a good photocatalyst appropriate to both the surface area and the particles dispersion. Hong and coauthors investigated the photoreactivity of MgAl-LDHs nanosheets with carbon nitride (C3N4). The carbonate anions enriched in the interlayer of LDH exhibit remarkably high reduction efficiency to CH4 in the presence of a C3N4 used as a photoabsorber (Hong et al. 2014).

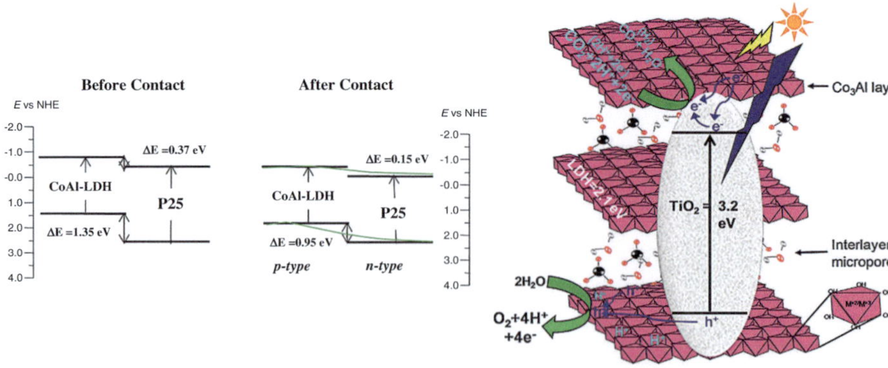

Fig. 4.3 Band bending of nanocomposite (right). (Proposed mechanism of CO_2 photoreduction (left) (Kumar et al. 2017)

4.4.3 Carbon-Based Materials

Recently, it was demonstrated that the carbon-based material improved an enhancement photocatalytic for CO_2 reduction. So, graphene (G), graphene oxide (GO), graphitic carbon nitride (g-C3N4), carbon nanotubes (CNT), and nanofibers (CNF) are promising materials for the photocatalytic reaction due to the chemical stability, large surface area, and electrical conductivity (Putri et al. 2015; Low et al. 2015; Zhang et al. 2014; Benammar et al. 2018). So, the introduction of carbon-based materials can improve CO_2 adsorption due to the presence of π junction on the surface. Carbon materials used as heterosystem or composites act as an electron acceptor.

Graphene and Graphene Oxide

Since the isolation of graphene (GR) by Geim and Novoselov in 2004, the elaboration, characterization, and application of two-dimensional (2D) material have attracted more interest (Lee et al. 2008; Novoselov et al. 2004; Tang et al. 2010, 2015; Tang and Hu 2013). GR with narrow Eg is largely used and improved the visible light photocatalytic efficiency. Therefore, the higher electron mobility is due to the delocalized conjugated π-e^- helpful to improve the photoactivity (Neto et al. 2009; Maka et al. 2010).

GO is an intermediary product involving the solution-phase isolation of GR species from graphite. The 2D-GO nanosheets have hydrophilic properties that make it a stable material due to the presence of organic functions on the surface (Compton and Nguyen 2010). Generally, GO is used as precursor of GR to elaborate GR-based composites. So, compared to GR, GO is an insulator due to the formation of sp^3-hybridized carbon atoms, which disrupt the extended sp^2 conjugated network of the pristine GR nanosheets (Xiang et al. 2015). Generally, GR is modified with metal oxide and result in the elaboration of composite photocatalysts. Numerous studies investigated the reactivity of GO via the CO_2 photoconversion to hydrocarbons like CH_4, CO, H_2, etc. The investigation of the effect of the GR doping on the CO_2 reduction was also reported. So, the conversion rate of CO_2 to CH_3OH on modified GO was sixfold higher than the pure TiO_2 (Hsu et al. 2013). rGO-Cu_2O composite elaborated by a facile one-step microwave-assisted chemical technique exhibited high photocatalytic reactivity compared to Cu_2O (An et al. 2014). Cu_2O deposed on rGO demonstrated an enhanced conversion to CO as the reduction product. Consequently, the rGO can not only act as an ideal electron source to increase and improve the charge recombination but also improve the Cu_2O stability (Tran et al. 2012a, 2012b). rGO-TiO_2 fabricated through solvothermal method exhibited a better photocatalytic activity in the reduction of CO_2 to CH_4 (Tan et al. 2013; An et al. 2014). Liang et al. reported that GR-TiO_2 nanocomposites with low defect densities in the GR were better compared to TiO_2 alone (Liang et al. 2011). Another study used GR-TiO_2 nanosheets in the homogenous medium to investigate

Table 4.2 Comparison of photocatalytic CH_3OH production rates of different photocatalysts

Samples	cm-ZnO	ZnO	G10	g-C3N4
CH_3OH (μmol g^{-1} h^{-1})	0.32	0.37	0.6	0.26

Yu et al. (2015)

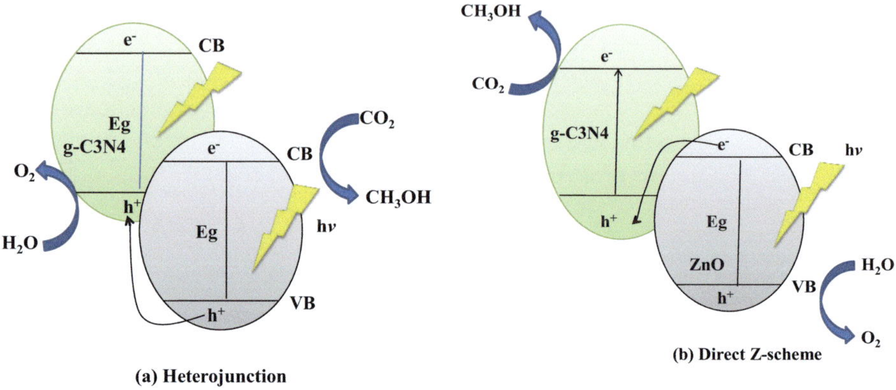

Fig. 4.4 Mechanisms for charge carrier separation: (**a**) conventional heterojunction (**b**) direct Z-scheme mechanisms (Yu et al. 2015)

the photoconversion of CO_2 into CH_4 and C_2H_6 (Liang et al. 2011). Li et al. reported that the composite ZnO-rGO prepared by hydrothermal method exhibited a better photoactivity (Li et al. 2013). So, the loading of rGO reduced the ZnO's size and prevented the self-aggregation of ZnO. Under sunlight irradiation, the rate of CH_3OH production is around 4.5 μmolg^{-1} h^{-1}, whereas Yu et al. reported that the CH_3OH production rate over pure ZnO and g-C3N4 was 0.37 and 0.26 μmol h^{-1} g^{-1}, respectively. It remains relatively low compared to g-C3N4/ZnO composite (0.6 μmol h^{-1} g^{-1}) (Yu et al. 2015). The activity is attributed to the high electron hopping through the heterosystem g-C3N4/ZnO (Table 4.2, Fig. 4.4). GR-WO_3 composites prepared by hydrothermal method were investigated for the photoreduction of CO_2 to CH_4 under visible light irradiation (Wang et al. 2013a, b, c). So, the production rate of CH_4 was around 0.11 μ mol h^{-1}. Over TiO_2 hollow spheres and GR nanosheets, the CO and CH_4 production rates were 8.91 μ mol g^{-1} h^{-1} and 1.14 μ mol g $^{-1}$ h^{-1}, respectively (Tu et al. 2012). Yu et al. reported that under visible light, CdS nanorods and rGO were used as a photocatalyst for the transformation of CO_2 into CH_4 and CH_3OH (Yu et al. 2014a, b).

Graphitic Carbon Nitride (g-C3N4)

g-C3N4 improved their reactivity through energy and environmental applications (Wang et al. 2012a, b, c, d; Cao et al. 2014). The material was successfully elaborated from cyanamide (Wang et al. 2009). Many efforts have been made

g-C3N4 as a photocatalyst (Li et al. 2011; Gao 2012; Xian et al. 2013; Morales-Torres et al. 2013). Due to the semiconductor property, g-C3N4 is largely used in the photocatalytic reduction of CO_2 (Hong et al. 2014; Zhang et al. 2015; Mao et al. 2013).

So, compared to other carbon-based photocatalysts, g-C3N4 exhibits low cost for CO_2 reduction (Zhu et al. 2014). g-C3N4 can be obtained by direct heating of melamine, thiourea, or urea. On the other hand, CB of g-C3N4 is located at −1.23 V where the potential of the reduction of CO_2 to CH_3OH and HCOOH was – 0.38 and V – 0.61 V, respectively (Ohno et al. 2014). Doping g-C3N4 with metals or oxides has been investigated to improve the photocatalytic efficiency compared to g-C3N4 alone (David et al. 2015; Wang et al. 2015). These modifications result in the formation of composites and heterostructures with reduced Eg energy due to the synergistic interaction between g-C3N4 and the metal oxide (Lei et al. 2015; Li et al. 2015). g-C3N4/TiO_2 heterostructure confirmed that the oxidation and reduction reactions are initiated in VB and CB of TiO_2 and g-C3N4, respectively (Li et al. 2015).

Similar results were obtained on a g-C3N4/TiO_2 composite; in order to combine the photocatalytic activity of TiO_2 and the high surface area, (e^-/h^+) pairs transfer the ability of g-C3N4 to form a new efficient catalyst (Yan and Yang 2011; Mitoraj et al. 2010; Lu et al. 2010). Yan et al. reported the photocatalytic TiO_2/g-C3N4 achieved through ball-milling of the two components (Yan and Yang 2011). Therefore, the modification of g-C_3N_4 with Pt improved the photocatalytic performance for the CO_2 conversion into CH_4, CH_3OH, and HCHO (Yu et al. 2014a, 2014b). Furthermore, Au and Ag nanoparticles exhibited enhanced reactivity for photoreduction to CH_3OH, HCOOH, and CH_4 (Ohno et al. 2014). The photoreduction of CO_2 over TiO_2/g-C_3N_4 provides the CH_4 selectivity of 1.53 μ mol g^{-1} h^{-1}, and the selectivity to CO was 10.05 μ mol g^{-1} h^{-1} (Wang et al. 2013a, b, c). Cu-loaded TiO_2 and g-C3N4 show enhancement reactivity for CH_3OH and HCOOH with the yield of 2574 and 5069 m mol/$g_{catalyst}$, respectively (Adekoya et al. 2017). So, doping the photocatalysts with copper reduced the Eg value resulting in reducing the recombination rate of (e^-/h^+) pairs. Under visible irradiation, the addition of the cocatalyst to the nanomaterials improved the photocatalytic reactivity (Chai et al. 2012). The study recorded maximal photoactivity and stability for the composite with a mass ratio of (70:30). g-C3N4-N-TiO_2 illustrated the photoreduction of CO_2 to CO and CH_4 (Zhou et al. 2014) and improvement of the pairs (e^-/h^+) separation. So, the narrow Eg, electronic structure, and the absorption region of g-C3N4 facilitated the charge transfer from Cu to TiO_2. Also, g-C3N4 can be achieved as an excellent photocatalyst (Sun et al. 2015a, b). The study of the photoconversion of CO_2 to the hydrocarbon over ZnO/g-C3N4 indicated the decrease of (e^-/h^+) recombination and improved the photocatalyst reactivity (He et al. 2015). The photoconversion of CO_2 is also obtained using the cobalt complex $Co(bpy)_3Cl_2$-modified g-C_3N_4 (Lin et al. 2013).

References

Abbas Z, Mezher T, Abu-Zahra MRM (2013) Evaluation of CO2 purification requirements and the selection of processes for impurities deep removal from the CO2 product stream. Energy Procedia 37:2389–2396. https://doi.org/10.1016/j.egypro.2013.06.120

Abdullah H, Khan MR, Pudukudy M, Yaakob Z, Ismail NA (2015) CeO2-TiO2 as a visible light active catalyst for the photoreduction of CO2 to methanol. J Rare Earths 33:1155–1161. https://doi.org/10.1016/s1002-0721(14)60540-8

Adekoya DO, Tahir M, Saidina Amin NA (2017). g-C3N4/(Cu/TiO2) nanocomposite for enhanced photoreduction of CO2 to CH3OH and HCOOH under UV/visible light. J CO_2 Util 18:261–274. https://doi.org/10.1016/j.jcou.2017.02.004

Ahmed Ali K, Abdullah AZ, Mohamed AR (2017) Visible light responsive TiO2 nanoparticles modified using Ce and La for photocatalytic reduction of CO2: effect of Ce dopant content. Appl Catal A Gen 537:111. https://doi.org/10.1016/j.apcata.2017.03.022

Ahmed N, Shibata Y, Taniguchi T, Izumi Y (2011) Photocatalytic conversion of carbon dioxide into methanol using zinc–copper–M(III) (M = aluminum, gallium) layered double hydroxides. J Catal 279:123–135. https://doi.org/10.1016/j.jcat.2011.01.004

Ahmed N, Morikawa M, Izumi Y (2012) Photocatalytic conversion of carbon dioxide into methanol using optimized layered double hydroxide catalysts. Catal Today 185:263–269. https://doi.org/10.1016/j.cattod.2011.08.010

An X, Li K, Tang J (2014) Cu2O/reduced graphene oxide composites for the photocatalytic conversion of CO2. ChemSusChem 7(4):1086–1093. https://doi.org/10.1002/cssc.201301194

Anpo M, Yamashita H, Ichihashi Y, Ehara S (1995) Photocatalytic reduction of CO2 with water on various titanium oxide catalysts. J Electroanal Chem 396:21–26. https://doi.org/10.1016/0022-0728(95)04141-a

Archana PS, Jose R, Vijila C, Ramakrishna S (2009) Improved electron diffusion coefficient in electrospun TiO2 nanowires. J Phys Chem C 113:21538. https://doi.org/10.1021/jp908238q

Barbera AC, Vymazal J, Maucieri C (2018) Greenhouse gases formation and emission. In: Reference module in earth systems and environmental science. https://doi.org/10.1016/B978-0-12-409548-9.10895-4

Benammar S, Boudjemaa A, Nezzal G, Gómez-Ruiz S, Meziane D, Bachari K, Lounis A, Coville NJ (2018) Nanoparticles based on copper deposited on carbon spheres. Preparation, characterizations and application for CO2 photo-electrochemical reduction. J Electroanal Chem 809:80–87. https://doi.org/10.1016/j.jelechem.2017.12.007

Boudjemaa A, Trari M (2010) Photo-catalytic hydrogen production over Fe2O3 based catalysts. Int J Hydrog Energy 3:7684. https://doi.org/10.1016/j.ijhydene.2010.05.096

Boudjemaa A, Trari M, Bachari K (2014) Photo-electrochemical properties of an active site working on Fe-SBA-15 catalyst: application to the water reduction under visible light irradiation. Environ Prog Sustain Energy 33:141–146. https://doi.org/10.1002/ep.11766

Cao SW, Yuan YP, Barber J, Loo SCJ, Xue C (2014) Noble-metal-freeg-C3N4/Ni(dmgH)2 composite for efficient photocatalytic hydrogen evolution under visible light irradiation. Appl Surf Sci 319:344–349. https://doi.org/10.1016/j.apsusc.2014.04.094

Chae SY, Park MK, Lee SK, Kim TY, Kim SK, Lee WI (2003) Preparation of size controlled TiO2 nanoparticles and derivation of optically transparent photocatalytic films. Chem Mater 15:3326–3331. https://doi.org/10.1021/cm030171d

Chai B, Peng T, Mao J, Li K, Zan L (2012) Graphitic carbon nitride (g-C3N4)-Pt-TiO2 nanocomposite as an efficient photocatalyst for hydrogen production under visible light irradiation. Phys Chem Chem Phys 14(48):16745–16752. https://doi.org/10.1039/c2cp42484c

Chen X, Burda C (2008) The electronic origin of the visible-light absorption properties of C-, N- and S-doped TiO2 nanomaterials. J Am Chem Soc 130:5018–5019. https://doi.org/10.1021/ja711023z

Chiang C-Y, Aroh K, Franson N, Satsangi VR, Dass S, Ehrman S (2011) Copper oxide nanoparticle made by flame spray pyrolysis for photoelectrochemical water splitting e part

II. Photoelectrochemical study. Int J Hydrog Energy 36(24):15519–15526. https://doi.org/10.1016/j.ijhydene.2011.09.041

Chiang C-Y, Chang M-H, Liu H-S, Tai CY, Ehrman S (2012a) Process intensification in the production of photocatalysts for solar hydrogen generation. Ind Eng Chem Res 51:5207–5215. https://doi.org/10.1021/ie202522g

Chiang C-Y, Shin Y, Aroh K, Ehrman S (2012b) Copper oxide photocathodes prepared by a solutionbased process. Int J Hydrog Energy 37:8232–8239. https://doi.org/10.1016/j.ijhydene.2012.02.049

Choi SK, Kim S, Lim SK, Park H (2010) Photocatalytic comparison of TiO2 nanoparticles and electrospun TiO2 nanofibers: effects of Mesoporosity and Interparticle charge transfer. J Phys Chem C 114:16475–16480. https://doi.org/10.1021/jp104317x

Chuangchote S, Sagawa T, Yoshikawa S (2008) Efficient dye-sensitized solar cells using electrospun TiO2TiO2 nanofibers as a light harvesting layer. Appl Phys Lett 93:033310. https://doi.org/10.1063/1.2958347

Collado L, Jana P, Sierra B, Coronado JM, Pizarro P, Serrano DP, de la Peña O'shea VA (2013) Enhancement of hydrocarbon production via artificial photosynthesis due to synergetic effect of Ag supported on TiO2 and ZnO semiconductors. Chem Eng J 224:128–135. https://doi.org/10.1016/j.cej.2012.12.053

Compton OC, Nguyen ST (2010) Graphene oxide, highly reduced graphene oxide, and graphene: versatile building blocks for carbon-based materials. Small 6:711–723. https://doi.org/10.1002/smll.200901934

David AO, Tahir M, Amin NAS (2015) Copper modified TiO2 and g-C3N4 catalystsfor photoreduction of CO2 to methanol using different reaction mediums, Malays. J Fund Appl Sci 11(3). https://doi.org/10.11113/mjfas.v11n3.376

de la Peña O'Shea VA, Álvarez-Galván MC, Campos-Martín JM, Fierro JLG (2007) Fischer-Tropsch synthesis on mono- and bimetallic Co and Fe catalysts in fixed-bed and slurry reactors. Appl Catal A Gen 326:65–73. https://doi.org/10.1016/j.apcata.2007.03.037

Dhakshinamoorthy A, Navalon S, Corma A, Garcia H (2012) Photocatalytic CO2 reduction by TiO2 and related titanium containing solids. Energy Environ Sci 5(11):9217. https://doi.org/10.1039/c2ee21948d

Erdöhelyi A, Pásztor M, Solymosi F (1986) Catalytic hydrogenation of CO2 over supported palladium. J Catal 98:166–177. https://doi.org/10.1016/0021-9517(86)90306-4

Etacheri V, Seery MK, Hinder SJ, Pillai SC (2010) Highly visible light active TiO2-xNx heterojunction photocatalysts. Chem Mater 22:3843–3853. https://doi.org/10.1021/cm903260f

Fan J, Liu E, Tian L, Hu XY, He Q, Sun T (2010) Study on synergistic effect of N and Ni2+ on Nano Titania in photocatalytic reduction of CO_2. J Environ Eng 137(3). https://doi.org/10.1061/(ASCE)EE.1943-7870.0000311

Fan J, Zhao L, Yu J, Liu G (2012) The effect of calcination temperature on the microstructure and photocatalytic activity of TiO2-based composite nanotubes prepared by an in situ template dissolution method. Nanoscale 4(20):6597. https://doi.org/10.1039/C2NR32048G

Feng X, Sloppy JD, LaTempa TJ, Paulose M, Komarneni S, Bao N, Grimes CA (2011) Synthesis and deposition of ultrafine Pt nanoparticles within high aspect ratio TiO2 nanotube arrays: application to the photocatalytic reduction of carbon dioxide. J Mater Chem 21:13429–13433. https://doi.org/10.1039/C1JM12717A

Ferreira de Brito J, Araujo AR, Rajeshwar K, Zanoni MVB (2015) Photoelectrochemicalreduction of CO2 on Cu/Cu2O films: product distribution and pH effects. Chem Eng J 264:302–309. https://doi.org/10.1016/j.cej.2014.11.081

Fujishima A, Honda K (1972) Electrochemical photolysis of water at semiconductor electrode. Nature 238:37. https://doi.org/10.1038/238037a0

Ganesh I (2016) Electrochemical conversion of carbon dioxide into renewable fuel chemicals – the role of nanomaterials and the commercialization. Renew Sust Energ Rev 59:1269–1297. https://doi.org/10.1016/j.rser.2016.01.026

Gao Z (2012) Influence of environmental humidity on plasma etching polyamide 6 films. Appl Surf Sci 258:5574–5578. https://doi.org/10.1016/j.apsusc.2012.02.004

Gao Y, Wang H, Wu J, Zhao R, Lu Y, Xin B (2014) Controlled facile synthesis and photocatalytic activity of ultrafine high cristallinity TiO2 nanocrystal with tunable anatase/rutil ratios. Appl Surf Sci 294:36–41. https://doi.org/10.1016/j.apsusc.2013.12.107

Ghadimkhani G, Tacconi N, Chanmanee W, Janakyab C, Rajeshwar K (2013) Efficient solar photoelectrosynthesis of methanol from carbon dioxide using hybrid CuO–Cu2O semiconductor nanorod arrays. Chem Commun 49:1297–1299. https://doi.org/10.1039/C2CC38068D

Habisreutinger SN, Schmidt-Mende L, Stolarczyk JK (2013) Photocatalytic reduction of CO2 on TiO2 and other semiconductors. Angew Chem Int Ed 52(29):7372–7408. https://doi.org/10.1002/anie.201207199

Halmann MM, Steinberg M (1999) Greenhouse gas carbon dioxide mitigation science and technology. Lewis Publisher, Boca Raton

Halmann M, Ulman M, Blajeni BA (1983) Photochemical solar collector for the photoassisted reduction of aqueous carbon-dioxide. Sol Energy 31:429–431. https://doi.org/10.1016/0038-092x(83)90145-7

He Y, Wang Y, Zhang L, Teng B, Fan M (2015) High-efficiency conversion of CO2 to fuel over ZnO/g-C3N4 photocatalyst. Appl Catal B 168-169:1–8. https://doi.org/10.1016/j.apcatb.2014.12.017

Hisatomi T, Kubota J, Domen K (2014) Recent advances in semiconductors for photocatalytic and photoelectrochemical water splitting. Chem Soc Rev 43(22):7520–7535. https://doi.org/10.1039/c3cs60378d

Hong J, Zhang W, Wang Y, Zhou T, Xu R (2014) Photocatalytic reduction of carbon dioxide over self-assembled carbon nitride and layered double hydroxide: the role of carbon dioxide enrichment. ChemCatChem 6:2315–2321. https://doi.org/10.1002/cctc.201402195

Hou H, Wang L, Gao F, Wei G, Zheng J, Tang B, Yang W (2014) Fabrication of porous titanium dioxide fibers and their photocatalytic activity for hydrogen evolution. Int J Hydrog Energy 39:6837–6844. https://doi.org/10.1016/j.ijhydene.2014.02.151

Hsu HC, Shown I, Wei HY, Chang YC, Du HY, Lin YG, Tseng CA, Wang CH, Chen LC, Lin YC, Chen KH (2013) Graphene oxide as a promising photocatalyst for CO_2 to methanol conversion. Nanoscale 5:262–268. https://doi.org/10.1039/C2NR31718D

Iizuka K, Wato T, Miseki Y, Saito K, Kudo A (2011) Photocatalytic reduction of carbon dioxide over ag Cocatalyst-loaded ALa4Ti4O15 (A = Ca, Sr, and Ba) using water as a reducing reagent. J Am Chem Soc 133(51):20863–20868. https://doi.org/10.1021/ja207586e

Inoue Y (2009) Photocatalytic water splitting by RuO2-loaded metal oxides and nitrides with d0 and d10 related electronic configuration. Energy Environ Sci 2:364. https://doi.org/10.1039/B816677N

Inoue T, Fujishima A, Konishi S, Honda K (1979) Photoelectrocatalytic reduction of carbon dioxide in aqueous suspensions of semiconductor powders. Nature 277:637–638. https://doi.org/10.1038/277637a0

Izaki M, Nagai M, Maeda K, Mohamad FB, Motomur AK, Sasano J, Shinagawa T, Watase S (2011) Electrodeposition of 1.4-e V-bandgap p-copper (II) oxide film with excellent photoactivity. J Electrochem Soc 158:D578. https://doi.org/10.1149/1.3614408

Jhong H-R"M", Ma S, JA KP (2013) Electrochemical conversion of CO2 to useful chemicals: current status, remaining challenges, and future opportunities. Curr Opin Chem Eng 2:191–199. https://doi.org/10.1016/j.coche.2013.03.005

Joo J, Kwon SG, Yu T, Cho M, Lee J, Yoon J, Hyeon T (2005) Large-scale synthesis of TiO2 nanorods via non hydrolytic sol-gel ester elimination reaction and their application to photocatalytic inactivation of E. coli. J Phys Chem B 109:15297–15302. https://doi.org/10.1021/jp052458z

Kandiel TA, Dillert R, Feldhoff A, Bahnemann DW (2009) Enhanced photocatalytic production of molecular hydrogen on TiO2 modified with Pt-polypyrrol nanocomposites. Photochem Photobiol Sci 8:683–690. https://doi.org/10.1039/c2pp25217a

Kirk-Davidoff DB, Keith DW (2007) On the climate impact of surface roughness anomalies. J Atmos Sci 65:2215–2234. https://doi.org/10.1175/2007JAS2509.1

Kocí K, Obalová L, Matejová L, Plachá D, Lacný Z, Jirkovský J, Šolcová O (2009) Effect of TiO2 particle size on the photocatalytic reduction of CO2. Appl Catal B Environ 89:494–502. https://doi.org/10.1016/j.apcatb.2009.01.010

Kočí K, Matějka V, Kovář P, Lacný Z, Obalová L (2011) Comparison of the pure TiO2 and kaolinite/TiO2 composite as catalyst for CO2 photocatalytic reduction. Catal Today 161 (1):105–109. https://doi.org/10.1016/j.cattod.2010.08.026

Kohno Y, Hayashi H, Takenaka S, Tanaka T, Funabiki T, Yoshida S (1999) Photo-enhanced reduction of carbon dioxide with hydrogen over Rh/TiO2. J Photochem Photobiol A Chem 126:117–124. https://doi.org/10.1016/S1010-6030(99)00113-6

Kong XY, Tan WL, Ng BJ et al (2017) Harnessing Vis–NIR broad spectrum for photocatalytic CO_2 reduction over carbon quantum dots-decorated ultrathin Bi_2WO_6 nanosheets. Nano Res 10:1720–1731. https://doi.org/10.1007/s12274-017-1435-4

Kudo A, Misekj Y (2009) Heterogenous photocatalyst material for water splitting. Chem Soc Rev 38:253–278. https://doi.org/10.1039/B800489G

Kumar B, Llorente M, Froehlich J, Dang T, Sathrum A, Kubiak CP (2012) Photochemical and photoelectrochemical reduction of CO2. Annu Rev Phys Chem 63:541–569. https://doi.org/10.1146/annurev-physchem-032511-143759

Kumar S, Isaacs MA, Trofimovaite R, Durndell L, Parlett CMA, Douthwaite RE, Coulson B, Cockett MCR, Wilson K, Lee AF (2017) P25@CoAl layered double hydroxide heterojunction nanocomposites for CO2 photocatalytic reduction. Appl Catal B Environ 209:394–404. https://doi.org/10.1016/j.apcatb.2017.03.006

Lee C, Wei X, Kysar JW, Hone J (2008) Measurement of the elastic properties and intrinsic strength of monolayer graphene. Science 321:385–388. https://doi.org/10.1126/science.1157996

Lee JC, Kim JH, Chang WS, Pak D (2012) Biological conversion of CO2 to CH4 using hydrogenotrophic methanogen in a fixed bed reactor. J Chem Technol Biotechnol 87 (6):844–847. https://doi.org/10.1002/jctb.3787

Lei J, Chen Y, Wang L, Liu Y, Zhang J (2015) Highly condensed g-C3N4-modified TiO2 catalysts with enhanced photodegradation performance toward acid orange 7. J Mater Sci 50 (9):3467–3476. https://doi.org/10.1007/s10853-015-8906-3

Li Y, Wang WN, Zhan Z, Woo H, Wu CY, Biswas P (2010) Photocatalytic reduction of CO2 with H2O on mesoporous silica supported cu/TiO2 catalysts. Appl Catal B Environ 100:386–392. https://doi.org/10.1016/j.apcatb.2010.08.015

Li Q, Guo B, Yu J, Ran J, Zhang B, Yan H, Gong JR (2011) Highly efficient visible-light-driven photocatalytic hydrogen production of CdS-cluster-decorated graphene Nanosheets. J Am Chem Soc 133:10878–10884. https://doi.org/10.1021/ja2025454

Li X, Zhuang Z, Li W, Pan H (2012a) Photocatalytic reduction of CO2 over noble metal-loaded and nitrogen-doped mesoporous TiO2. Appl Catal A Gen 429–430:31–38. https://doi.org/10.1016/j.apcata.2012.04.001

Li X, Liu H, Luo D, Li J, Huang Y, Li H, Fang Y, Xu Y, Zhu L (2012b) Adsorption of CO2 on heterostructure CdS(Bi2S3)/TiO2 nanotube photocatalysts and their photocatalytic activities in the reduction of CO2 to methanol under visible light irradiation. Chem Eng J 180:151–158. https://doi.org/10.1016/j.cej.2011.11.029

Li X, Wang Q, Zhao Y, Wu W, Chen J, Meng H (2013) Green synthesis and photo-catalytic performances for ZnO-reduced graphene oxide nanocomposites. J Colloid Interface Sci 411:69–75. https://doi.org/10.1016/j.jcis.2013.08.050

Li P, Xu J, Jing H, Wu C, Peng H, Lu J et al (2014) Wedged N-doped CuO with more negative conductive band and lower overpotential for high efficiency photoelectric converting CO2 to methanol. Appl Catal B Environ 156–157:134–140. https://doi.org/10.1016/j.apcatb.2014.03.011

Li H, Zhou L, Wang L, Liu Y, Lei J, Zhang J (2015) In situ growth of TiO2 nanocrystals on gC3N4 for enhanced photocatalytic performance. Phys Chem Chem Phys 17(26):17406–17412. https://doi.org/10.1039/c5cp02554k

Liang YT, Vijayan BK, Gray KA, Hersam MC (2011) Minimizing graphene defects enhances titania nanocomposite-based photocatalytic reduction of CO2 for improved solar fuel production. Nano Lett 11:2865–18670. https://doi.org/10.1021/nl2012906

Lin J, Pan Z, Wang X (2013) Photochemical reduction of CO2 by graphitic carbon nitride polymers. ACS Sustain Chem Eng 2(3):353–358. https://doi.org/10.1021/sc4004295

Liu L, Li Y (2014) Understanding the reaction mechanism of photocatalytic reduction of CO2 with H2O on TiO2-based photocatalysts: a review. Aerosol Air Qual Res 14:453–469. https://doi.org/10.4209/aaqr.2013.06.0186

Liu Z, Sun DD, Guo P, Leckie JO (2007) An efficient Bicomponent TiO2/SnO2 nanofiber Photocatalyst fabricated by electrospinning with a side-by-side dual spinneret method. Nano Lett 7:1081–1085. https://doi.org/10.1021/nl061898e

Liu L, Zhao H, Andino JM, Li Y (2012) Photocatalytic CO2 reduction with H2O on TiO2 nanocrystals: comparison of anatase, rutile, and brookite polymorphs and exploration of surface chemistry. ACS Catal 2(8):1817–1828. https://doi.org/10.1021/cs300273q

Liu L, Zhao C, Zhao H, Pitts D, Li Y (2013) Porous microspheres of MgO-patched TiO2 for CO2 photoreduction with H2O vapor: temperature-dependent activity and stability. Chem Commun 49:3664–3666. https://doi.org/10.1039/c3cc39054c

Liu L, Zhao C, Pitts D, Zhao H, Li Y (2014) CO2 photoreduction with H2O vapor by porous MgO–TiO2 microspheres: effects of surface MgO dispersion and CO2 adsorption–desorption dynamics. Cat Sci Technol 4:1539–1546. https://doi.org/10.1039/C3CY00807J

Liu LJ, Zhao CY, XuJY LY (2015) Integrated CO2 capture and photocatalytic conversion by a hybrid adsorbent/photocatalyst material. Appl Catal B Environ 179:489–499. https://doi.org/10.1016/j.apcatb.2015.06.006

Low JX, Yu JG, Ho WK (2015) Graphene-based photocatalysts for CO2 reduction to solar fuel. J Phys Chem Lett 6:4244–4251. https://doi.org/10.1021/acs.jpclett.5b01610

Lu X, Wang Q, Cui D (2010) Preparation and photocatalytic properties of g-C3N4/TiO2 hybrid composite. J Mater Sci Technol 26:925–930. https://doi.org/10.1016/S1005-0302(10)601-1

Luo D, Bi Y, Kan W, Zhang N, Hong S (2011) Copper and cerium co- doped titanium dioxide on catalytic photo reduction of carbon dioxide with water: experimental and theoretical studies. J Mol Struct 994:325. https://doi.org/10.1016/j.molstruc.,03.044

Maka KF, Sfeir MY, Misewich JA, Heinz TF (2010) The evolution of electronic structure in few-layer graphene revealed by optical spectroscopy. Proc Natl Acad Sci 107 (34):14999–15004. https://doi.org/10.1073/pnas.1004595107

Man QB, Hofmann JP, Litke A, Hensen EJM (2015) Cu2O photoelectrodes for solar water splitting: tuning photoelectrochemical performance by controlled faceting. Sol Energy Mater Sol Cells 141:178–186. https://doi.org/10.1016/j.solmat.2015.05.025

Mao Y, He J, Sun X, Li W, Lu X, Gan J, Liu Z, Gong L, Chen J, Liu P, Tong Y (2012) Electrochemical synthesis of hierarchical Cu2O stars with enhanced photo-electrochemical properties. Electrochim Acta 62:1–7. https://doi.org/10.1016/j.electacta.2011.10.106

Mao J, Peng T, Zhang X, Li K, Ye L, Zan L (2013) Effect of graphitic carbon nitride microstructures on the activity and selectivity of photocatalytic CO2 reduction under visible light. Cat Sci Technol 3:1253–1260. https://doi.org/10.1039/c3cy20822b

Martín A, Larrazábal J, Pérez-Ramírez GO (2015) Towards sustainable fuels and chemicals through the electrochemical reduction of CO2: lessons from water electrolysis. Green Chem 17 (12):5114–5130. https://doi.org/10.1039/c5gc01893e

Martin NM, Velin P, Skoglundh M, Bauer M, Carlsson P-A (2017) Catalytic hydrogenation of CO2 to methane over supported Pd, Rh and Ni catalysts. Catal Sci Technol 7:1086–1094. https://doi.org/10.1039/C6CY02536F

Matejová L, Kocí K, Reli M, Capek L, Hospodková A, Peikertová P, Matej Z, Lucie Obalová L, Wach A, Kuśtrowski P, Kotarba A (2014) Preparation, characterization and photocatalytic properties of cerium doped TiO2: on the effect of Ce loading on the photocatalytic reduction of carbon dioxide. Appl Catal B Environ 152–153:172–183. https://doi.org/10.1016/j.apcatb.2014.01.015

McMichael T, Costello A (2012) Health risks and future, from global climate change. Spotlight Clim Chang 344:1–5. https://doi.org/10.1136/bmj.e.1359

Mitchel JFB (1989) The greenhouse effect and climate change. American Geophysical Union. Rev Geophys 27:115–139

Mitoraj D, Beranek R, Kisch H (2010) Mechanism of aerobic visible light formicacid oxidation catalyzed by poly(tri-s-triazine) modified titania. Photochem Photobiol Sci 9:31–38. https://doi.org/10.1039/B9PP00052F

Mora C, Frazier AG, Longman RJ, Dacks RS, Walton MM, Tong EJ, Sanchez JJ, Kaiser LR, Stender YO, Anderson JM, Ambrosino CM, Fernandz-Silva I, Giambelluca TW (2013) The projected timing of climaye departure from recent variability. Nature 502:183–187. https://doi.org/10.1038/nature12540

Morales-Torres S, Pastrana-Martínez LM, Figueiredo JL, Faria JL, Silva AMT (2013) Graphene oxide-P25 photocatalysts for degradation of diphenhydramine pharmaceutical and methyl orange dye. Appl Surf Sci 275:361–368. https://doi.org/10.1016/j.apsusc.2012.11.157

Moya A, Cherevan A, Marchesan S, Gebhardt P, Prato M, Eder D, Vilatela JJ (2015) Oxygen vacancies and interfaces enhancing photocatalytic hydrogen production in mesoporous CNT/TiO2 hybrids. Appl Catal B Environ 179:574–582. https://doi.org/10.1016/j.apcatb.2015.05.052

Nakata K, Fujishima A (2012) TiO2 photocatalysis: design and applications. J Photochem Photobiol Photochem Rev 13:169–189. https://doi.org/10.1016/j.jphotochemrev.2012.06.001

Neto AHC, Guinea F, Peres NMR, Novoselov KS, Geim AK (2009) The electronic properties of graphene. The electronic properties of graphene. Rev Mod Phys 81:109–162. https://doi.org/10.1103/RevModPhys.81.109

Novoselov KS, Geim AK, Morozov SV, Jiang D, Zhang Y, Dubonos SV, Grigorieva IV, Firsov AA (2004) Electric field effect in atomically thin carbon films. Science 306:666–669. https://doi.org/10.1126/science.1102896

Ohno T, Murakami N, Koyanagi T, Yang Y (2014). Photocatalytic reduction of CO2 over a hybrid photocatalyst composed of WO3 and graphitic carbon nitride (g-C3N4) under visible light. J CO_2 Util 6:17–25. https://doi.org/10.1016/j.jcou.2014.02.002

Ola O, Maroto-Valer MM (2015) Transition metal oxide based TiO2 nanoparticles for visible light induced CO2 photoreduction. Appl Catal A Gen 502:114–121. https://doi.org/10.1016/j.apcata.2015.06.007

Paracchino A, Laporte V, Sivula K, Grätzel M, Thimsen E (2011) Highly active oxide photocathode for photoelectrochemical water reduction. Nat Mater 10:456–461. https://doi.org/10.1038/nmat3017

Paracchino A, Mathews N, Hisatomi T, Stefik M, Tilley S, Grätzel M (2012) Ultrathin films on copper(i) oxide water splitting photocathodes: a study on performance and stability. Energy Environ Sci 5(9):8673–8681. https://doi.org/10.1039/c2ee22063f

Pathak P, Meziani MJ, Li Y et al (2004) Improving photoreduction of CO2 dispersed nanoscale TiO2, into liquid products with homogeneously catalysts. Chem Commun 10(10):1234–1235. https://doi.org/10.1039/b400326h

Patz JA, Campbell-Lendrum D, Holloway T, Foley JA (2005) Impact of regional climate change on human health. Rev Nat 438:310–317. https://doi.org/10.1038/nature04188

Pelaez M, Nolan NT, Pillai SC, Seery MK, Falaras P, Kontos AG, Dunlop PSM, Hamilton JWJ, Byrne JA, O'Shea K, Entezari MH, Dionysiou DD (2012) A review on the visible light active titanium dioxide photocatalysts for environmental applications. Appl Catal B 125:331–349

Peng T, Zhao D, Dai K, Shi W, Hirao K (2005) Synthesis of titanium dioxide nanoparticles with mesoporous anatase wall and high photocatalytic activity. J Phys Chem B 109:4947–4952. https://doi.org/10.1021/jp044771r

Perkins KM, Munguin N, Moure-Eraso R, Delakowitz B, Giannetti BF, Gengyuan L, Nurunnabi M, Will M, Velazquez L (2018) International perspectives on the pedagogy of climate change. J Clean Prod 200:1043–1052. https://doi.org/10.1016/j.jclepro.2018.07.296

Putri LK, Ong WJ, Chang WS, Chai SP (2015) Heteroatom doped graphene in photocatalysis: a review. Appl Surf Sci 358:2–14. https://doi.org/10.1016/j.apsusc.2015.08.177

Qin S, Xin F, Liu Y, Yin X, Ma W (2011) Photocatalytic reduction of CO2 in methanol to methyl formate over CuO-TiO2 composite catalysts. J Colloid Interface Sci 356:257–261. https://doi.org/10.1016/j.jcis.2010.12.034

Rajeshwar K, de Tacconi NR, Ghadimkhani G, Chanmanee W, Janáky C (2013) Tailoring copper oxide semiconductor Nanorod arrays for Photoelectrochemical reduction of carbon dioxide to methanol. ChemPhysChem 14:2251–2259. https://doi.org/10.1002/cphc.201300080

Reñones P, Moya A, Fresno F, Collado L, Vilatela J J, de la Peña O'Shea VA (2016). Hierarchical TiO2 nanofibres as photocatalyst for CO2 reduction: influence of morphology and phase composition on catalytic activity. J CO_2 Util 15:24–31. https://doi.org/10.1016/j.jcou.2016.04.002

Royer DL, Berner RA, Park J (2007) Climate sensitivity constrained by CO2 concentration over the past 420 million years. Lett Nat 446:530–532. https://doi.org/10.1038/nature05699

Sasirekha N, Basha SJS, Shanthi K (2006) Photocatalytic performance of Ru doped anatase mounted on silica for reduction of carbon dioxide. Appl Catal Environ 62(1–2):169–180. https://doi.org/10.1016/j.apcatb.2005.07.009

Schulte KL, DeSario PA, Gray KA (2010) Effect of crystal phase composition on the reductive and oxidative abilities of TiO2 nanotubes under UV and visible light. Appl Catal B 97:354–360. https://doi.org/10.1016/j.apcatb.2010.04.017

Sun H, Dong B, Su G, Gao R, Liu W, Song L, Cao L (2015a) Modified of TiO2 nanotubes by WO3 species for improving their photocatalytic activity. Appl Surf Sci 343:181–187. https://doi.org/10.1016/j.apsusc.2015.148

Sun J, Song J, Gondal M, Shi S, Lu Z, Xu Q, Chang X, Xiang D, Shen K (2015b) Preparation of g-C3N4/BiOX (X = Cl, Br, I) composites, and their photocatalytic activity under visible light irradiation. Res Chem Intermed 41(10):6941–6955

Tahir M, Amin NS (2013) Advances in visible light responsive titanium oxide-based photocatalysts for CO2 conversion to hydrocarbon fuels. Energy Convers Manag 76:194–214. https://doi.org/10.1016/j.enconman.2013.07.046

Tahir M, Thair B, Amin NAS (2015) Gold-nanoparticle-modified TiO2 nanowires for plasmon-enhanced photocatalytic CO2 reduction with H2 under visible light irradiation. Appl Surf Sci 356:1289–1299. https://doi.org/10.1016/j.apsusc.2015.08.231

Tan LL, Ong WJ, Chai SP, Mohamed AR (2013) Reduced graphene oxide-TiO2 nanocomposite as a promising visible-light-active photocatalyst for the conversion of carbon dioxide. Nanoscale Res Lett 8:465. https://doi.org/10.1186/1556-276X-8-465

Tang B, Hu GX (2013) Three-dimensional graphene network assisted high performance dye sensitized solar cells. J Power Sources 234:60–68. https://doi.org/10.1016/j.jpowsour.2013.01.130

Tang B, Hu GX, Gao HY (2010) Raman spectroscopic characterization of graphene. Appl Spectrosc Rev 45:369–407. https://doi.org/10.1080/05704928.2010.483886

Tang B, Hu GX, Gao HY (2015) Application of graphene as filler to improve thermal transport property of epoxy resin for thermal interface materials. Int J Heat Mass Transf 85:420–429. https://doi.org/10.1016/j.ijheatmasstransfer.2015.01.141

Tian GH, Fu HG, Jing LQ et al (2009) Synthesis and photocatalytic activity of stable nanocrystalline TiO2 with high crystallinity and large surface area. J Hazard Mater 161(2–3):1122–1130. https://doi.org/10.1016/j.jhazmat.2008.04.065

Tian B, Tian B, Smith B, Scott MC, Hua R, Lei Q, Tian Y (2018) Supported black phosphorus nanosheets as hydrogen-evolving photocatalyst achieving 5.4% energy conversion efficiency at 353 K. Nat Commun 9. https://doi.org/10.1038/s41467-018-03737-4

Tran PD, Batabyal SK, Pramana SS, Barber J, Wong LH, Chye S, Loo S (2012a) A cuprous oxide–reduced graphene oxide (Cu2O–rGO) composite photocatalyst for hydrogen generation:

employing GO as an electron acceptor to enhance the photocatalytic activity and stability of Cu2O. Nanoscale 4:3875–3878. https://doi.org/10.1039/C2NR30881A

Tran PD, Wong LH, Barber J, Loo JSC (2012b) Recent advances in hybrid photocatalysts for solar fuel production. Energy Environ Sci 5:5902–5918. https://doi.org/10.1039/C2EE02849B

Tu WG, Zhou Y, Liu Q, Tian Z, Gao J, Chen X, Zhang H, Liu J, Zou Z (2012) Robust hollow spheres consisting of alternating Titania Nanosheets and graphene Nanosheets with high photocatalytic activity for CO2 conversion into renewable fuels. Adv Funct Mater 22:1215. https://doi.org/10.1002/adfm.201102566

Tuckett RP (2009) Chapter 1: The role of atmospheric gases in global warming. In: Climate change: observed impacts on planet earth. Elsevier, Amsterdam

Wang CC, Zhang Z, Ying JY (1997) Photocatalytic decomposition of halogenated organics over nanocrystalline titania. Nanostruct Mater 9:58–586. https://doi.org/10.1016/s0965-9773(97)00130-x

Wang H, Wu Y, Xu BQ (2005) Preparation and characterization of nanosized anatase TiO2 cuboids for photocatalysis. Appl Catal B Environ 59:139–146. https://doi.org/10.1016/j.apcatb.2005.02.001

Wang X, Maeda K, Thomas A, Takanabe K, Xin G, Carlsson JM, Domen K, Antonietti M (2009) A metal-free polymeric photocatalyst for hydrogen production from water under visible light. Nat Mater 8:76–80. https://doi.org/10.1038/nmat2317

Wang ZY, Chou HC, Wu JCS, Tsai DP, Mul G (2010) CO2 photoreduction using NiO/InTaO4 in optical-fiber reactor for renewable energy. Appl Catal A Gen 380:172–177. https://doi.org/10.1016/j.apcata.2010.03.059

Wang C, Ma X, Li J, Xu L, Zhang F (2012a) Chemical reduction of CO2 aqueous solution by using photosensitized-TiO2 nanotube catalysts modified by supramolecular metalloporphyrins-ruthenium (II) polypyridyl complexes. J Mol Catal A 363–364:108–114. https://doi.org/10.1016/j.molcata.2012.05.023

Wang S, Lian JS, Zheng WT, Jiang Q (2012b) Photocatalytic property of Fe doped anatase and rutile TiO2 nanocrystal particles prepared by sol – gel technique. Appl Surf Sci 263:260–265. https://doi.org/10.1016/j.apsusc.2012.09.040

Wang WN, An WJ, Ramalingam B et al (2012c) Size and structure matter: enhanced CO2 photoreduction efficiency by size-resolved ultrafine Pt nanoparticles on TiO2. J Am Chem Soc 134(27):11276–11281. https://doi.org/10.1021/ja304075b

Wang Y, Wang X, Antonietti M (2012d) Polymeric graphitic carbon nitride as a heterogeneous organocatalyst: from photochemistry to multipurpose catalysis to sustainable chemistry. Angew Chem Int Ed 51:68–89. https://doi.org/10.1002/anie.201101182

Wang PQ, Bai Y, Luo PY, Liu JY (2013a) Graphene–WO3 nanobelt composite: Elevated conduction band toward photocatalytic reduction of CO2into hydrocarbon fuels. Catal Commun 38:82–85. https://doi.org/10.1016/j.catcom.2013.04.020

Wang Y, Chen Y, Zuo Y, Wang F, Yao J, Li B, Kang S, Li X, Cui L (2013b) Hierarchically mesostructured TiO2/graphitic carbon composite as a new efficient photocatalyst for the reduction of CO2 under simulated solar irradiation. Cat Sci Technol 3:3286–3291. https://doi.org/10.1039/C3CY00524K

Wang Y, Li B, Zhang C, Cui L, Kang S, Li X, Zhou L (2013c) Ordered mesoporous CeO2-TiO2 composites: highly efficient photocatalysts for the reduction of CO2 with H2O under simulated solar irradiation. Appl Catal B Environ 130-131:277–284. https://doi.org/10.1016/j.apcatb.2012.11.019

Wang K, Li Q, Liu B, Cheng B, Ho W, Yu J (2015) Sulfur-doped g-C3N4 with enhanced photocatalyticCO2-reduction performance. Appl Catal B 176–177:44–52. https://doi.org/10.1016/j.apcatb.2015.03.045

Wang Y, Zhoa L, Otto A, Robinius M, Stolten D (2017) A review of post-combustion CO2 capture technologies from Coal-fired power plants. Energy Procedia 114:650–665. https://doi.org/10.1016/j.egypro.2017.03.1209

Wojciechowski BW (1988) Kinetics of the Fischer–Tropsch synthesis. Catal Rev 30(4):629–702. https://doi.org/10.1080/01614948808071755

Won DH, Choi CH, Chung J, Woo SI (2014) Photoelectrochemical production of formic acid and methanol from carbon dioxide on metal-decorated CuO/Cu2O-layered thin films under visible light irradiation. Appl Catal B Environ 158–159:217–223

Woolerton TW, Sheard S, Reisner E, Pierce E, Ragsdale SW, Armstrong FA (2010) Efficient and clean Photoreduction of CO2 to CO by enzyme-modified TiO2 nanoparticles using visible light. J Am Chem Soc 132:2132. https://doi.org/10.1021/ja910091z

Wu JM, Zhang TW, Zeng YW, Hayakawa S, Tsuru K, Osaka A (2005) Large-scale preparation of ordered titaniana norods with enhanced photocatalytic activity. Langmuir 21:6995–7002. https://doi.org/10.1021/la0500272

Xian T, Yang H, Di LJ, Dai JF (2013) Graphene-assisted enhancement of photocatalytic activity of bismuth ferrite nanoparticles. Res Chem Intermed 41(1):433–441. https://doi.org/10.1007/s11164-013-1204-2

Xiang Q, Cheng B, Yu J (2015) Graphene-based Photocatalysts for solar-fuel generation. Angew Chem Int Ed Engl 54:11350–11366. https://doi.org/10.1002/anie.201411096. Epub 2015 Jun 16

Xie Y, Wang Q, Zhang W, Fan W, Deng Y (2013a) Wang, photocatalytic reduction of CO2 with H2O: significant enhancement of the activity of Pt-TiO2 in CH4 formation by addition of MgO. Chem Commun 49:2451–2453. https://doi.org/10.1039/c3cc00107e

Xie S, Wang Y, Zhang Q, Fan W, Deng W, Wang Y (2013b) Photocatalytic reduction of CO2 with H2O: significant enhancement of the activity of Pt-TiO2 in CH4 formation by addition of MgO. Chem Commun 49:2451–2453. https://doi.org/10.1039/c3cc00107e

Xie S, Zhang Q, Liu G, Wang Y (2016) Photocatalytic and photoelectrocatalytic reduction of CO2 using heterogeneous catalysts with controlled nanostructures. Chem Commun 52:35–59. https://doi.org/10.1039/c5cc07613g

Xi-Liu Y, Qing-Xian G (2018) Contributions of natural systems and human activity to greenhouse gas emissions. Adv Clim Chang Res 9:243. https://doi.org/10.1016/j.accre.2018.12.003

Xiong Z, Zhao Y, Zhang J, Zheng C (2014) Efficient photocatalytic reduction of CO2 into liquid products over cerium doped titania nanoparticles synthesized by a sol-gel auto-ignited method. Fuel Process Technol 135:6–13. https://doi.org/10.1016/j.fuproc.2014.09.017

Xiong Z, Zhao Y, Zhang J, Zheng C (2015) Efficient photocatalytic reduction of CO2 into liquid products over cerium doped titania nanoparticles synthesized by a sol-gel auto-ignited method. Fuel Process Technol 135:6–13. https://doi.org/10.1016/j.fuproc.2014.09.017

Xiong W, Dai W, Hu X, Yang L, Wang T, Luo YX, Zou J (2018) Enhanced photocatalytic reduction of CO2 into alcohols on Z-scheme ag/Ag3PO4/CeO2 driven by visible light. Mater Lett 232:36. https://doi.org/10.1016/j.matlet.2018.08.072

Xu G, Liang F, Yang Y, Hu Y, Zhang K, Liu W (2014) An improved CO2 separation and purification system based on cryogenic separation and distillation theory. Energies 7:3484–3502. https://doi.org/10.3390/en7053484

Yan H, Yang H (2011) TiO2–g-C3N4 composite materials for photocatalytic H2 evolution under visible light irradiation. J Alloys Compd 509:L26–L29. https://doi.org/10.1016/j.jallcom.2010.09.201

Yang MQ, Xu YJ (2016) Photocatalytic conversion of CO2 over graphene-based composites: current status and future perspective. Nanoscale Horiz 1:185–200. https://doi.org/10.1039/C5NH00113G

Yang HG, Liu G, Qiao SZ, Sun CH, Jin YG, Smith SC, Zou J, Cheng HM, Lu GQ(Max) (2009) Solvothermal synthesis and Photoreactivity of Anatase TiO2 Nanosheets with dominant {001} facets. J Am Chem Soc 131(11):4078–4083. https://doi.org/10.1021/ja808790p

Yang CC, Vernimmen J, Meynen V, Cool P, Mul G (2011) Mechanistic study of hydrocarbon formation in photocatalytic CO2 reduction over Ti-SBA-15. J Catal 284:1–8. https://doi.org/10.1016/j.jcat.2011.08.005

Yong Z, Mata V, Rodriguez AE (2001) Adsorption of carbon dioxide onto hydrotalcite-like compounds (HTlcs) at high temperatures. Ind Eng Chem Res 40:204–209. https://doi.org/10.1021/ie000238w

Yu J, Yu H, Cheng B, Zhao X, Zhang Q (2006) Preparation and photocatalytic activity of mesoporous anatase TiO2 nanofibers by a hydrothermal method. J Photochem Photobiol A Chem 182:121–127. https://doi.org/10.1016/j.jphotochem.2006.01.022

Yu J, Jin J, Cheng B, Jaroniec M (2014a) A noble metal-free reduced graphene oxide–CdS nanorod composite for the enhanced visible-light photocatalytic reduction of CO2 to solar fuel. J Mater Chem A 2:3407–3416. https://doi.org/10.1039/C3TA14493C

Yu J, Wang K, Xiao W, Cheng B (2014b) Photocatalytic reduction of CO2 into hydrocarbon solar fuels over g-C3N4–Pt nanocomposite photocatalysts. Phys Chem Chem Phys 16:11492–11501. https://doi.org/10.1039/C4CP00133H

Yu W, Xu D, Peng T (2015) Enhanced photocatalytic activity of g-C_3N_4 for selective CO_2 reduction to CH_3OH via facile coupling of ZnO: a direct Z-scheme mechanism. J Mater Chem A. https://doi.org/10.1039/C5TA05503B

Yuan Y, Yu Z, Zhang J, Zou Z (2012) A copper (I) dye-sensitised TiO2-based system for efficient light harvesting and photoconversion of CO2 into hydrocarbon fuel. Dalton Trans:1–8. https://doi.org/10.1039/C2DT30865G

Yuan L, Han C, Yang MQ, Xu YJ (2016) Photoctalytic water splitting for solar hydrogen generation: fundamentals and recent advancement. Int Rev Phys Chem 35:1. https://doi.org/10.1080/014435.2015.1127027

Zhang Z, Wang CC, Zakaria R, Ying JY (1998) Role of particle size in nanocrystalline TiO2-based photocatalysts. J Phys Chem B 102:10871. https://doi.org/10.1021/jp982948

Zhang QH, Han WD, Hong YJ, Yu JG (2009) Photocatalytic reduction of CO_2 with H_2O on Pt-loaded TiO_2 catalyst. Catal Today 148(3):335–340. https://doi.org/10.1016/j.cattod.2009.07.081

Zhang Q, Lima DQ, Lee I, Zaera F, Chi M, Yin Y (2011) A highly active titanium dioxide based visible-light photocatalyst with nonmetal doping and plasmonic metal decoration. Angew Chem Int Ed 50(31):7088–7092. https://doi.org/10.1002/anie.201101969

Zhang Y, Shen B, Huang H, He Y, Fei B, Lv F (2014) BiPO4/reduced graphene oxide composites photocatalyst with high photocatalytic activity. Appl Surf Sci 319:272–277. https://doi.org/10.1016/j.apsusc.2014.07.052

Zhang G, Lan ZA, Wang X (2015) Merging surface organometallic chemistry with graphitic carbon nitride photocatalysis for CO2 Photofixation. ChemCatChem 7:1422–1423. https://doi.org/10.1002/cctc.201500133

Zhao X, Wang P, Li B (2010) CuO/ZnO core/shell heterostructure nanowire arrays: synthesis, optical property, and energy application. Chem Commun 46:6768. https://doi.org/10.1039/c0cc01610

Zhao C, Liu L, Zhang Q, Wang J, Li Y (2012) Photocatalytic conversion of CO2 and H2O to fuels by nanostructured Ce–TiO2/SBA-15 composites. Cat Sci Technol 2:2558. https://doi.org/10.1039/c2cy20346d

Zhao H, Xu J, Liu L, Rao G, Zhao C, Li, Y. (2016). CO2 photoreduction with water vapor by Ti-embedded MgAl layered double hydroxides. J CO_2 Util 15:15–23. https://doi.org/10.1016/j.jcou.2016.04.004

Zhou P, Yu J, Jaroniec M (2014) All-solid-state Z-scheme photocatalytic systems. Adv Mater 26(29):4920–4935. https://doi.org/10.1002/adma.201400288

Zhu J, Xiao P, Li H, Carabineiro SA (2014) Graphitic carbon nitride: synthesis, properties, and applications in catalysis. ACS Appl Mater Interfaces 6:16449–16465. https://doi.org/10.1021/am502925j

Chapter 5
Electrochemical Reduction of Carbon Dioxide to Methanol Using Metal-Organic Frameworks and Non-metal-Organic Frameworks Catalyst

Fayez Nasir Al-Rowaili and Aqil Jamal

Contents

5.1 Introduction 92
5.2 Challenges Involved in Methanol Production from Carbon Dioxide Electrocatalytic Reduction 96
5.3 Homogeneous and Heterogeneous Electrocatalysts for Electroreduction of Carbon Dioxide 99
5.3.1 Homogeneous Catalysts for Electroreduction of Carbon Dioxide 99
5.3.2 Heterogeneous Catalysis for Electroreduction of Carbon Dioxide 100
5.4 Kinetics of Electroreduction of Carbon Dioxide into Methanol 101
5.5 Formation of Carbon Dioxide Anion Radical 102
5.6 Formation of Methanoate from the Electroreduction of Carbon Dioxide 103
5.7 Formation of Carbon Monoxide from Electroreduction of Carbon Dioxide 104
5.8 Formation of Methanol from Electroreduction of Carbon Dioxide 105
5.9 Hydrogen Evolution Reaction 107
5.10 Benchmark Non-metal-Organic Framework-Based Catalysts for Carbon Dioxide Reduction 108
5.11 Metal-Organic Frameworks as Catalysts for the Carbon Dioxide Reduction Reaction 112
5.12 Conclusion and Recommendations 121
References 123

Abstract The world today is grappling with the need to come up with renewable, cheap, and sustainable energy sources with adequate energy density to replace the dwindling and sometimes harmful fossil fuel. Carbon dioxide, for instance, is a

F. N. Al-Rowaili (✉)
Research and Development Center, Saudi Aramco, Dhahran, Saudi Arabia

Department of Chemical Engineering, King Fahd University of Petroleum and Minerals, Dhahran, Saudi Arabia
e-mail: ruwailfn@aramco.com

A. Jamal
Research and Development Center, Saudi Aramco, Dhahran, Saudi Arabia

Inamuddin et al. (eds.), *Conversion of Carbon Dioxide into Hydrocarbons Vol. 1 Catalysis*, Environmental Chemistry for a Sustainable World 40,
https://doi.org/10.1007/978-3-030-28622-4_5

major contributor to the issue of global warming; therefore any process that will help convert carbon dioxide to any less harmful reusable chemical such as methanol is a priority area for research in our world today. A unique method of obtaining methanol through carbon dioxide conversion as an alternative source of energy is an electrocatalytic electrochemical reduction. Presently, significant efforts are channeled toward bettering the efficiency and understanding of electrocatalysts. For carbon dioxide (CO_2) electrochemical reduction to methanol, metal-organic frameworks, metal oxides, metals without impurities, and composite materials are employed. However, of all the above materials, metal-organic frameworks remain the preferred option for its outstanding structural features, very simple nature, and higher surface area. For the past few years, metal-organic frameworks and their products have often been employed to electrocatalize hydrogen, oxygen, carbon dioxide reduction, and hydrogen evolution.

The focus therefore of this review is to unite the advantages offered by composite materials, non-metal-organic frameworks, and metal-organic frameworks for use in electroreduction of carbon dioxide to methanol. In view of several methods being proposed by different studies on how to improve metal-organic framework efficiencies as electrocatalysts in the chemical transformation of carbon dioxide into methanol, this review emphasizes the basic need to research new and more efficient metal-organic frameworks that will offer better chemical stability and conductivity.

Keywords Heterogeneous catalysis · Green fuel · Carbon dioxide · Metal-organic frameworks · Artificial photosynthesis · Electroreduction · Electrocatalysts · Non-metal-organic frameworks · Methanol · and Electrochemical

5.1 Introduction

An often-important end-product of combustion of biomass and hydrocarbons is carbon dioxide (CO_2). Deforestation and subsequently industrial revolution have also contributed largely to the increase in the amount of carbon dioxide in our environment (von der Assen et al. 2016; Davis et al. 2010; Cherubini et al. 2011; Woodwell et al. 1983; Brovkin et al. 2004). Since the 1700s the amount of carbon dioxide in the atmosphere has been on a steady rise by at least 30% as a result of industrial revolution (Ganesh 2014).

Global warming occurs when greenhouse gases such as carbon dioxide prevent infrared radiation from outer space entering the earth surface from being reflected back away from the earth thereby raising the temperature of the earth immensely (Rodhe 1990; Schneider 1989). As the production of carbon dioxide from human and industrial activities keeps increasing, it has become paramount that scientific procedures for reducing, capturing, and consumption of carbon dioxide be researched. Many countries have adopted the United Nations Kyoto Protocol which came up with guidelines for reducing/ending climate change as a result of carbon dioxide emission (Bodansky 2010; Lau et al. 2012). The alternative to reducing carbon dioxide is to employ scientific methods that completely eliminate carbon dioxide from the main source, take, for example, converting carbon dioxide released from thermal power plants into useful chemicals and fuels, a process known as carbon dioxide capture and reuse (Ng et al. 2013; Markewitz et al. 2012).

Such captured carbon dioxide can be reused to produce intermediate and essential chemical compounds. Moreover, an industry that is self-sustaining would go a long way toward eliminating the carbon dioxide concentration in the environment. Carbon dioxide itself is nontoxic; hence, its conversion into other compounds is an exciting prospect as the gas is very readily available all around us, and it also possesses an interesting physiochemical characteristic (Albo et al. 2015a). As of today, plastics made from polycarbonates, urea, and salicylic acid have all been manufactured from the conversion of carbon dioxide (Kleij et al. 2017; Perathoner and Centi 2014). Besides all the above uses of carbon dioxide, it is also used as a supercritical fluid solvent in the separation, purification, and preparation process of many compounds (Darensbourg et al. 2010; Peters et al. 2011; Omae 2012). Carbon dioxide further promises to be a valuable ingredient in the process of oil and gas recovery as well (Mathisen and Skagestad 2017). Figure 5.1 shows a list of chemicals derived from carbon dioxide and their respective Gibbs free energies. Although many methods have been stated above for carbon dioxide reduction, more ingenious methods need to be discovered if the about 25 gigatons yearly production of carbon dioxide is to be greatly diminished to its barest minimum (Ganesh 2014).

Most of the world's energy is presently supplied by fossil fuels; the reserve levels of the fossil fuels, however, have been greatly depleted (Capellán-Pérez et al. 2014; McGlade and Ekins 2015). The growing world population and industrialization are even putting more pressure and driving home the urgencies and essence of alternative energy sources to meet the present energy demands. This alternative energy source should be carbon-based, easily scaled up, reusable, and sustainable. The

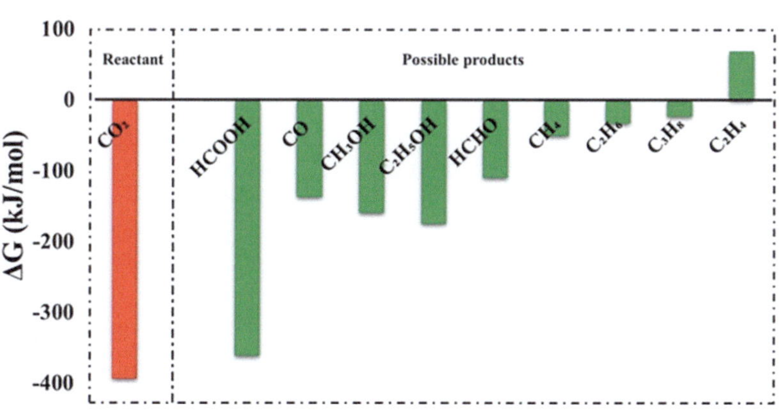

Fig. 5.1 The chemical compounds derived through CO_2 reduction along with their Gibbs free energies. CO_2 is thermodynamically stable with very low Gibbs energy such as $\Delta G = -394.4$ kJ/mol which indicates that methanol formation with much higher Gibbs free energy such as $\Delta G = -159.2$ kJ/mol is an endothermic process. (Reprinted from Fayez et al. (2018) with permission of ACS sustainable chemistry and engineering)

carbon dioxide in the atmosphere can be reduced by collecting most of it and converting it into useful methanol thereby reducing the world's dependence on fossil fuels (Ma et al. 2009). The needed energy to convert carbon dioxide to methanol is easily obtained from various renewable sources of energy like the geothermal and wind energy like the normal photosynthesis (Barton et al. 2008; Mikkelsen et al. 2010; Aresta et al. 1996). Methanol offers a comparative advantage over other fuel sources such as gasoline as gasoline has twice the energy density of methanol, as well as methanol, being more environmentally friendly compared to gasoline (Bromberg and Cheng 2010; Pourkhesalian et al. 2010; Olah 2013). In the preparation of essential compounds such as resins, aldehydes, and olefins, one of the important intermediate products is methanol (Omae 2012; Ott et al. 2012). Methanol alone or in addition with other chemicals when used in engines offer an advance fuel with improved octane rating, better combustion, and reduced, cleaner carbon dioxide emission. Another advantage of methanol is that its high-octane value makes it suitable for storage at atmospheric pressure and direct utilization in internal combustion engines (Olah 2005; Li and Faghri 2013) and fuel cells. The reasons, therefore, emphasize the fact that carbon dioxide conversion is a critical research area. Although much work has been done in this area, more still needs to be done to arrive at industrial level technologies that are not only economical but that are also durable.

Many researchers have shown instances where methanol synthesis was achieved by using C_1-carbon sourced from carbon dioxide, for instance, photochemical, electrochemical, hydrothermal, chemical, and biochemical processes (Olah et al. 2009; Sun et al. 2009; Huang et al. 2017; Yuan and Xu 2015; Kim et al. 2011; Albo and Irabien 2016). Electrocatalytic reduction of carbon dioxide still offers an edge above the other approach for carbon dioxide reduction; few among the numerous advantages offered by this method are the following: (a) minimal use of chemicals makes recycling of the accompanying electrolyte possible; (b) carbon dioxide reduction is possible using energy sources that are renewable; (c) the process can be regulated using the reaction temperature and a potential difference of the electrodes; and (d) implementing changes in electrochemical cells is an easy process (Albo et al. 2015a). The result obtained from the electrochemical reduction of carbon dioxide depends essentially on two factors, namely, nature of electrolyte and catalyst used. The type of electrode and state of the reaction medium, i.e., aqueous, gaseous, or non-aqueous, will determine the pathway the electrochemical reduction of carbon dioxide will follow; this can involve two, four, six, and eight electrons (Albo et al. 2015a; Qiao et al. 2014; Li et al. 2017a).

The steps for synthesizing methanol from the electrocatalytic reduction of carbon dioxide are outlined in Fig. 5.2.

Among the numerous products gotten from the reduction of carbon dioxide, methanol still by far surpasses the others in benefits despite the difficulties associated with the process.

With a free Gibbs energy value, $\Delta G = -394{,}400$ J/mol carbon dioxide is a chemically, highly stable and passive molecule and easily converted back endothermically into methanol, a compound with high Gibbs free energy of

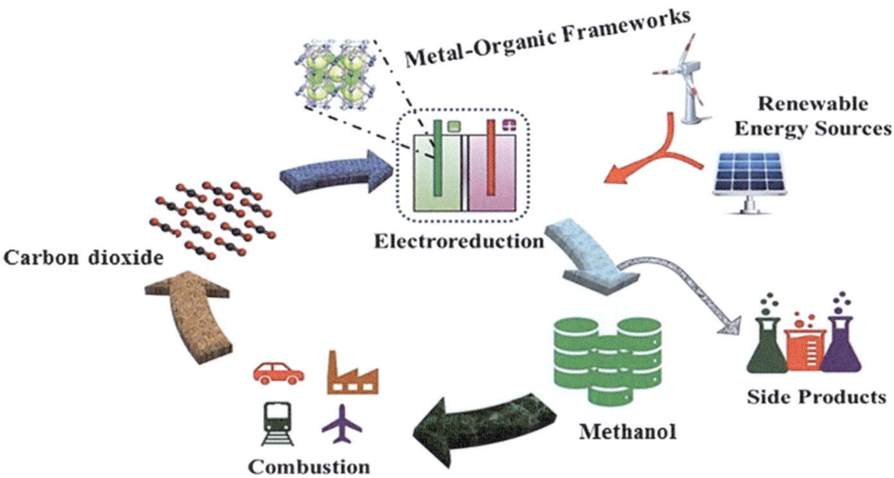

Fig. 5.2 Schematic diagram of electrochemical reduction of carbon dioxide and possible applications of fuel product. (Reprinted from Fayez et al. (2018) with permission of ACS sustainable chemistry and engineering)

$\Delta G = -159{,}200$ J/mol. The major challenge with the reduction of carbon dioxide is its high overpotential, catalyst instability, and reduced efficiency (Ganesh 2014). As a result of these challenges, redox electrocatalysts are added to the process to lower the potential barrier and improve the reaction efficiency (Back et al. 2015; Kuhl et al. 2014). However, much has been achieved in understanding the reaction pathway and enhancing the effectiveness of the catalyst. Most of the research done in the area of activating carbon dioxide has relied on metal centers of complex chemicals. Hence, metal-organic chemicals have been the focus of most of the recent development in the electroreduction of carbon dioxide (Peterson and Nørskov 2012). Various kinds of metals have been tested as electrocatalysts primarily because of their impressive electrocatalytic activity, but ligands remain an important enhancer to obtain the best performance from each of these metals. The result of improved comprehension of how ligands affect carbon dioxide activation has opened the door to many more interesting discoveries that have further improved the overall outcome of the catalysts addition to process (Finn et al. 2012). Transition metals, conducting polymers, enzymes, ionic liquids, metal-organic frameworks, and many other materials have been proposed by various literatures as electrocatalysts in carbon dioxide reduction (Qiao et al. 2014; Bourrez et al. 2011; Lin et al. 2015a; Zhang et al. 2014a; Sun et al. 2014; Schlager et al. 2016; Hod et al. 2015).

Over the last few years, metal-organic material groups that are porous, especially metal-organic frameworks, have found extensive applicability in various areas and as modern materials with various functions (Zhou and Kitagawa 2014; Zhou et al. 2012). Metal-organic frameworks are formed from the combination of metals, organic linkers, and pore spaces making them hybrids (Zhou and Kitagawa 2014; Zhou et al. 2012). Metal-organic frameworks (MOF) work as good catalytic

Table 5.1 Stepwise reactions for the electrochemical oxidation-reduction process of carbon dioxide. Methanol production from carbon dioxide electrochemical conversion

Cathode	$CO_2 + 6H^+ + 6e^-$	$\rightleftharpoons$	$CH_3OH + H_2O$	0.024 V (versus SHE)
Anode	$3H_2O$	$\rightleftharpoons$	$1.5O_2 + 6H^+ + 6e^-$	1.234 V (versus SHE)
Overall reaction	$CO_2 + 2H_2O$	$\rightleftharpoons$	$CH_3OH + 1.5O_2$	1.454 V (versus SHE)

where SHE means Standard Hydrogen Electrode

materials because organic linkers are present in their structure (Wu et al. 2005; Millward and Yaghi 2005). Moreover, the porous nature of metal-organic frameworks coupled with their large surface areas, impressive conductivity, and definite structural features makes them perfect choice for carbon dioxide electroreduction, adsorption, and separation (Lin et al. 2015a; Senthil Kumar et al. 2012; Liu et al. 2012; Albo et al. 2016; Lin et al. 2015b).

In the last decade, metal-organic frameworks and their composites have been used as catalysts as reported in some articles (Qiao et al. 2014; Kortlever et al. 2015; Sen et al. 2014). However, there is yet to be any report showing how metal-organic frameworks can be employed for producing methanol from carbon dioxide reduction. To fill this research void, a comprehensive study of the dynamics of carbon dioxide electroreduction, the essential electrocatalysts needed for the process, recent developments in the applications and utilization of metal-organic frameworks, and the reactions involving electrochemically electroreduced carbon dioxide to generate methanol has been presented. This study aimed toward acting as the basis for future development of improved electrocatalysts from metal-organic frameworks (Table 5.1).

5.2 Challenges Involved in Methanol Production from Carbon Dioxide Electrocatalytic Reduction

The electroreduction process of the carbon dioxide to methanol is feasible thermodynamically but nonspontaneous (Barton et al. 2008). Furthermore, carbon dioxide has an equilibrium reduction potential of 0.02 V (versus Standard Hydrogen Electrode, SHE) in aqueous electrolytic solution which is higher than the potential for water reduction which results in HER (hydrogen evolution reaction) (Kaneco et al. 1999). The equation shown below gives the energy needed for carbon dioxide conversion into methanol:

$$2H_2O_{(l)} + CO_{2(g)} \rightleftharpoons CH_3OH_{(l)} + 3/2O_{2(g)} \Delta H = -727 KJ/mol^{-1} \text{and } \Delta G^o = -703 \ KJmol^{-1} \quad (5.1)$$

Basically, at −1.90 V (versus Standard Hydrogen Electrode, SHE), carbon dioxide is converted to CO_2^{2-} as a result of tremendous amount of energy needed to reorganize the linearly structured carbon dioxide to bent CO_2^{2-}. The observed hindrance in transforming the linear carbon dioxide to methanol can be observed in the grouping of the nuclei to transform the simple linear carbon dioxide molecule to the more complexly active methanol compound (Schouten et al. 2011). Through reaction kinetics, the electroreduction of carbon dioxide can significantly be derived. The mechanistic approach reveals that carbon dioxide is firstly reduced into carbon monoxide and methanoate ion which requires the application of potential less than −1.12 V (versus SHE (Standard Hydrogen Electrode)) which may then proceed to methanol being formed. The challenge is at about −1.35 V (versus SHE (Standard Hydrogen Electrode)), hydrocarbons begin to form in a greater quantity compared to the quantity of carbon monoxide and formate ion being formed (Jitaru et al. 1997; Francke et al. 2018). The probable pathways for reduction of carbon dioxide into various products electrochemically are shown in Table 5.2.

Moreover, applying a greater negative potential to the electrolyte while in the aqueous state will cause the Hydrogen Electrode Reaction (HER) to begin to compete along with the carbon dioxide reduction process. This competition is the reason for a lowered Faradaic efficiency as both reactions occur at about equal potential difference (Zhang et al. 2014b; Ooka et al. 2017). From the foregoing, it goes without saying that the main challenge in carbon dioxide electroreduction is finding a way to compensate formation of methanol and at the same time discouraging the formation of Hydrogen Electrode Reaction (HER); this is both a thermodynamic and a kinetic problem which can be solved by introducing an electrocatalyst. The possible channels for electrochemically reducing carbon dioxide to obtain methanol are exemplified in Fig. 5.3 (Sen et al. 2014; Schouten et al. 2011; Chinchen et al. 1987; Joo et al. 1999).

Table 5.2 Possible electrochemical reduction pathways of carbon dioxide

$CO_{2(g)} + 2H^+ + 2e^-$	$\rightleftharpoons$	$HCOOH_{(l)}$	−0.25 V (versus SHE)
$CO_{2(g)} + H_2O_{(l)} + 2e^-$	$\rightleftharpoons$	$HCOO^-_{(aq)} + OH^-$	−1.08 V (versus SHE)
$CO_{2(g)} + 2H^+ + 2e^-$	$\rightleftharpoons$	$CO_{(g)} + H_2O_{(l)}$	−0.11 V (versus SHE)
$CO_{2(g)} + H_2O_{(l)} + 2e^-$	$\rightleftharpoons$	$CO_{(g)} + 2OH^-$	−0.93 V (versus SHE)
$CO_{2(g)} + 4H^+ + 4e^-$	$\rightleftharpoons$	$CH_2O_{(l)} + H_2O_{(l)}$	−0.07 V (versus SHE)
$CO_{2(g)} + 3H_2O_{(l)} + 4e^-$	$\rightleftharpoons$	$CH_2O_{(g)} + 4OH^-$	−0.90 V (versus SHE)
$CO_{2(g)} + 6H^+ + 6e^-$	$\rightleftharpoons$	$CH_3OH_{(l)} + H_2O_{(l)}$	+0.02 V (versus SHE)
$CO_{2(g)} + 5H_2O_{(l)} + 6e^-$	$\rightleftharpoons$	$CH_3OH_{(l)} + 6OH^-$	−0.81 V (versus SHE)
$CO_{2(g)} + 8H^+ + 8e^-$	$\rightleftharpoons$	$CH_{4(g)} + 2H_2O_{(l)}$	+0.17 V (versus SHE)
$CO_{2(g)} + 6H_2O_{(l)} + 8e^-$	$\rightleftharpoons$	$CH_{4(g)} + 8OH^-$	−0.66 V (versus SHE)
$2CO_{2(g)} + 2H^+ + 2e^-$	$\rightleftharpoons$	$H_2C_2O_{4(aq)}$	−0.50 V (versus SHE)
$2CO_{2(g)} + 2e^-$	$\rightleftharpoons$	$C_2O_4^{2-}{}_{(aq)}$	−0.59 V (versus SHE)
$2CO_{2(g)} + 12H^+ + 12e^-$	$\rightleftharpoons$	$CH_2CH_{2(g)} + 4H_2O_{(l)}$	+0.06 V (versus SHE)
$2CO_{2(g)} + 12H^+ + 12e^-$	$\rightleftharpoons$	$CH_3CH_2OH_{(g)} + 3H_2O_{(l)}$	+ 0.08 V (versus SHE)

where SHE means Standard Hydrogen Electrode

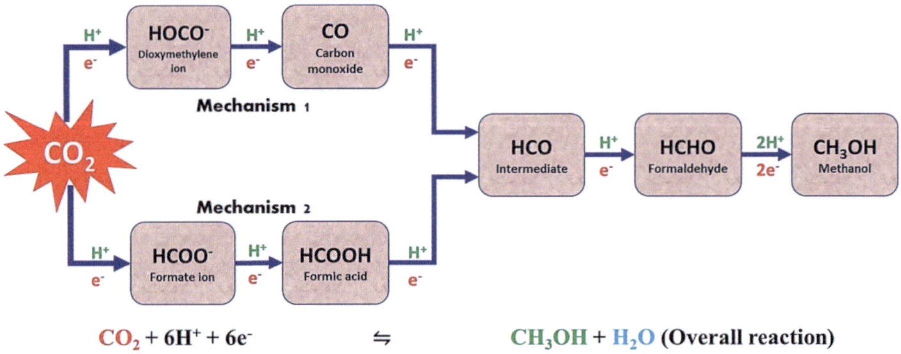

Fig. 5.3 Possible pathways for methanol formation as a product of carbon dioxide electrochemical reduction. The mechanism 1 yields formate ion and carbon monoxide. Mechanism 2 yields formate ion and formic acid. Both mechanisms involved the formation of intermediate as an intermediate product and formaldehyde then methanol. (Reprinted from Fayez et al. (2018) with permission of ACS sustainable chemistry and engineering)

The first pathway results in the formation of dioxymethylene ion formate ion and carbon monoxide. The second mechanism produces both formic acid (HCOOH) and formate ion ($HCOO^-$). In the two processes, HCO and HCHO (formaldehyde) are formed as intermediates. The preferred pathway either 1 or 2 that the carbon dioxide reduction will proceed with is based on the kind of metals that make up the metal-organic frameworks and non-metal-organic frameworks. For example, the first mechanism is favored when the metals Zn (zinc) and Cu (copper) are employed as electrodes as they favor carbon-coordinated joining to carbon dioxide for the formation of methanol, while the second mechanism is preferred when oxygen-coordinated joining of carbon dioxide is done with a metal involving high hydrogen potential. Following the electroreduction of carbon dioxide, not just methanol is formed but other compounds such as ethylene, glyoxal, propanol, glycol, carbon monoxide, methane, etc. are also formed (Kortlever et al. 2015; Min and Kanan 2015; Ren et al. 2015, 2016; Torelli et al. 2016; Lim et al. 2014). As earlier stated, carbon dioxide is a very stable molecule thermodynamically speaking, possessing extremely strong two double bonds between the carbon and the two oxygen atoms resulting in very low Gibbs free energy. Changing this stable carbon dioxide molecule into methanol a compound with enormous energy would mean that an external energy source is used. This could be in the form of electrons, heat, photons, or irradiation that would kinetically slow down the reaction. As depicted in the mechanisms in Fig. 5.3, methanol formation from carbon dioxide is a process that is kinetically slow and makes use of six electrons (Olah et al. 2009; Whipple and Kenis 2010). Hence, a catalyst with unique qualities of being able to reduce the overpotential for carbon dioxide reduction, improve selectivity, and increase efficiency is required to boost the overall formation of methanol.

5.3 Homogeneous and Heterogeneous Electrocatalysts for Electroreduction of Carbon Dioxide

The catalysts used for electrochemical carbon dioxide reduction can be grouped into two broad groups:

1. Homogeneous catalysts
2. Heterogeneous catalysts

5.3.1 Homogeneous Catalysts for Electroreduction of Carbon Dioxide

Homogeneous catalysts involved in the reduction of carbon dioxide are mostly molecular compounds (Benson et al. 2009; Saveant and Tard 2016; Froehlich and Kubiak 2015). The molecular compounds transport electrons between carbon dioxide and the electrode bringing about indirect electrolysis as explained in Fig. 5.4. The potential necessary for carbon dioxide reduction under the influence of homogeneous catalysts is different from that necessary in their absence; therefore

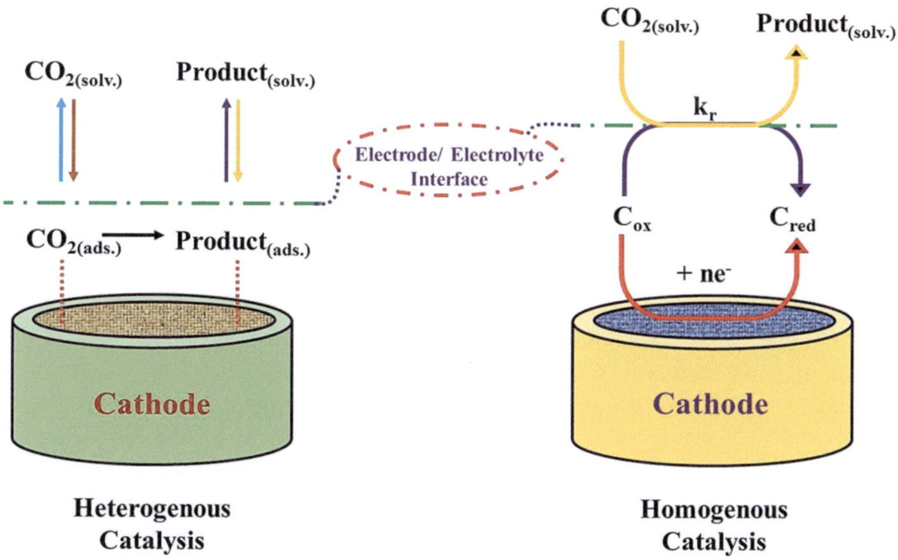

Fig. 5.4 The schematic diagram for homogeneous and heterogeneous catalysis. (Reprinted from Fayez et al. (2018) with permission of ACS sustainable chemistry and engineering)

homogeneous catalysts require a less negative potential when in use (Dridi et al. 2017). There are unique features a material must possess to qualify as a homogeneous catalyst for carbon dioxide reduction. Some of these qualities include high stability of its reduced form, increased rate of catalyst influenced reaction leading to improved yield within a reduced time interval, and the range of the lowered potential must fall within the needed range. This range should be near the thermodynamic potential needed to reduce carbon dioxide electrochemically. All these characteristics can be easily modified by modifying metal-organic framework structures with appropriate metals and suitable ligands.

5.3.2 Heterogeneous Catalysis for Electroreduction of Carbon Dioxide

In a bid to surmount the problem of overpotential for electrochemical carbon dioxide reduction, different electrode materials that can chemically adsorb the carbon dioxide on their surfaces such as copper, silver, and platinum are used as catalysts (Kumar et al. 2016). The selectivity of these electrode materials is greatly affected by pH, type or nature of electrode, nature of solvent, and temperature. The chemical composition of these catalysts ascertains the binding energy existing between the surface of these catalysts and the intermediates of the reactions (Hod et al. 2015). The quantity of carbon dioxide that will bind to the surface of the material relies on the catalytic material's surface area, and this also controls the catalytic reactivity. The nanostructured materials possess an extensive ratio of surface-to-volume which gives numerous sites of coordination for the catalyst to react upon or with (Ren et al. 2015; Wang et al. 2016). To upgrade the specificity and electroactivity, the composite materials are utilized in order to stir a cooperative effect for the efficiency of the catalyst and promote the selectivity process contrary to the view when a single material is used for the carbon dioxide electrochemical reduction to methanol (Studt et al. 2014; Gusain et al. 2016; Li et al. 2017b).

Both homogeneous and heterogeneous types of carbon dioxide electrochemical reduction catalysis come with advantages and disadvantages, hence making it difficult to make an excellent choice of catalyst for the reduction process. Degradation of the catalyst is an obvious problem that applies to the two classes of catalysis reactions, which hampers the longevity of the catalyst (Costentin et al. 2013). Noteworthy research developments are being channeled toward expanding the properties of homogeneous or heterogeneous catalysts to appropriately shape the two processes for industrial applications. To that effect, this review will focus on the heterogeneous catalysis.

5.4 Kinetics of Electroreduction of Carbon Dioxide into Methanol

As stated earlier, carbon dioxide is thermodynamically stable and has a very low Gibbs free energy as a result of the energetic double bonded carbon and the two atoms of oxygens. On the other hand, methanol has higher Gibbs free energy which indicates that the stability is lower than that of carbon dioxide (Albo et al. 2015a) as shown in Fig. 5.1. External sources of energy such as fast-moving electrons, irradiations, or photons and heat are required to convert highly nonreactive carbon dioxide molecule to such an energy-efficient methanol compound (Agarwal et al. 2016). Furthermore, to lower the energy barriers, catalysts are used (Kortlever et al. 2015). Figure 5.5 shows that the carbon dioxide-methanol conversion is an endothermic process with $\Delta G^o = -4.1$ KCal/Mol.

The huge geometrical differences between carbon dioxide and its radical anion, CO_2^-, ensure that the barrier emanating from the electron of the outer-sphere is quite high. The kinetic barriers involved in carbon dioxide reduction, as well as the large gap between the Lowest Unoccupied Molecular Orbital (LUMO) and the Highest Occupied Molecular Orbital (HOMO) energies, mean that a lot of overpotentials are needed for the process. This is as depicted in Fig. 5.6 (Kuhl et al. 2014).

The general pathway for electrochemical reduction of carbon dioxide follows some basic steps, the steps, however, entails charge transfer; a process which has been robustly studied using different metal surfaces as electrodes.

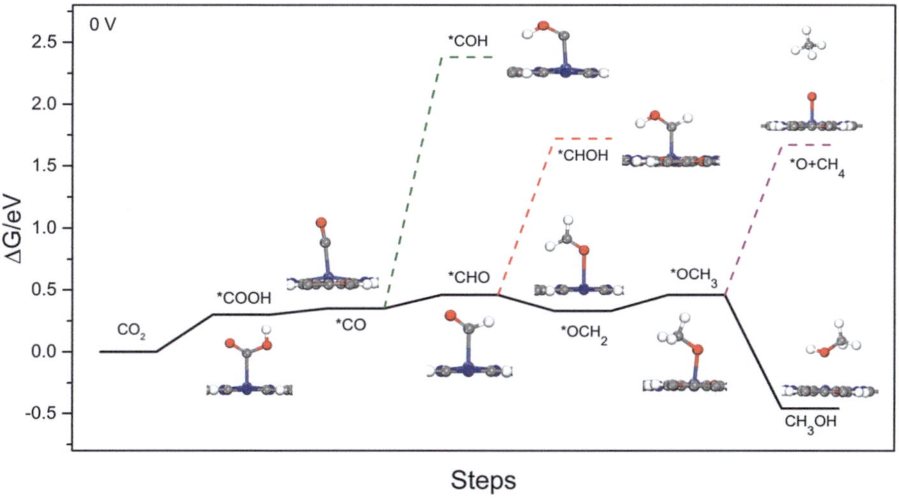

Fig. 5.5 Thermodynamically favorable reduction pathway shown in solid line from carbon dioxide to carbon monoxide and to methanol, other competing pathways shown in dash lines are also depicted. (Reprinted from Yao et al. (2018) with permission of chemistry: A European Journal)

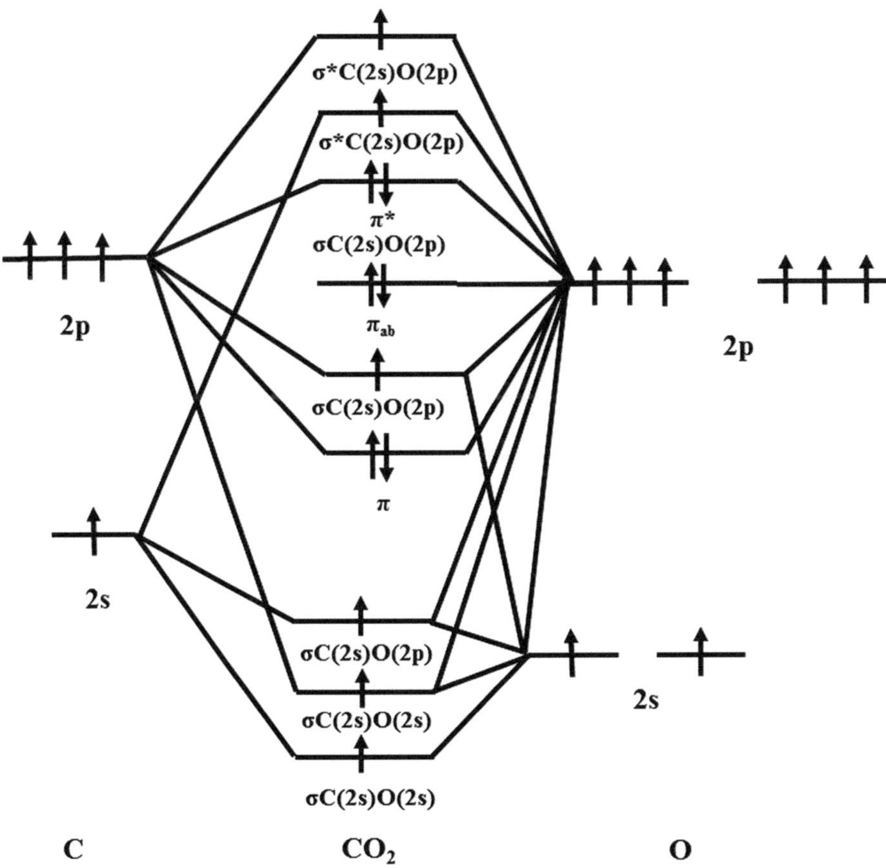

Fig. 5.6 Molecular orbital energy level diagram of carbon dioxide. It shows the combinations of the atomic orbitals of carbon and oxygen atom to form covalently bonded carbon dioxide. It also shows the different atomic and molecular orbital energy levels of the atoms and the molecule formed

5.5 Formation of Carbon Dioxide Anion Radical

Amphoteric carbon dioxide undergoes reduction firstly by chemisorption which occurs at the catalyst surface. The kind of electrode will determine the nature of the interaction (Wang et al. 2007). The orientation that carbon dioxide takes when binding with metal is a yet to be understood phenomena; the guess, however, is that the electrode binds to the carbon dioxide via a carbon atom and/or an oxygen atom (Burghaus 2014) as shown in Fig. 5.7.

Reducing carbon dioxide into carbon dioxide anion radical through one-electron pathway is a reaction that requires a huge amount of energy, without which the reaction will not proceed. In an aqueous media, −1.90 V (versus SHE (Standard Hydrogen Electrode)) is required for synthesis of carbon dioxide anion radical, while

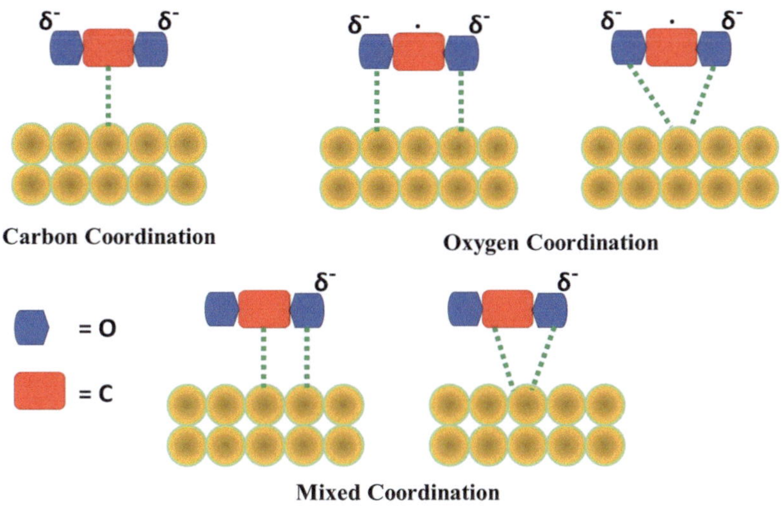

Fig. 5.7 Illustrative representation of carbon dioxide interaction with the metal surface. Though the actual binding positions of carbon dioxide on the metal are unknown; possible orientations are (1) carbon coordination, where carbon dioxide is attached to the metal through its carbon atom; (2) oxygen coordination, where carbon dioxide is attached to the metal through its two oxygen atoms of which either each of the oxygen atom is attached on different sites of the metal or the two oxygen atoms are attached on the same site of the metal; and (3) mixed coordination, where carbon dioxide is attached to the metal through both carbon and oxygen atoms of which either both carbon and oxygen atoms are attached on different or the same site of the metal as shown on the figure. (Reprinted from Fayez et al. (2018) with permission of ACS sustainable chemistry and engineering)

the lower overvoltage of 0.67 is the necessary transfer coefficient (Sheng et al. 2018; Schröder et al. 1999). During the nucleophilic interaction of carbon dioxide anion radical, carbon atom performs the task of the reactant. This has been shown both theoretically and experimentally by different studies. In free-state carbon dioxide anion radical can exist either as an aqueous or nonaqueous electrolytic solution (Kai et al. 2017; Chen et al. 2017). The type of catalyst used will determine if formate ion ($HCOO^-$) is formed when the carbon dioxide anion radical is reduced or if carbon monoxide (CO) is formed (Hori et al. 1994).

5.6 Formation of Methanoate from the Electroreduction of Carbon Dioxide

Methanoate ($HCOO^-$) is one of the basic products of carbon dioxide reduction on the surface of electrodes such as Sn and Hg (Zhang et al. 2014c). Carbon dioxide at a − 1.6 V potential (versus Standard Hydrogen Electrode, SHE) when placed to the surface of the metal, methanoate is formed from the atom of carbon in carbon dioxide takes a proton from water molecule (Hollingsworth et al. 2015). This reaction is as shown below:

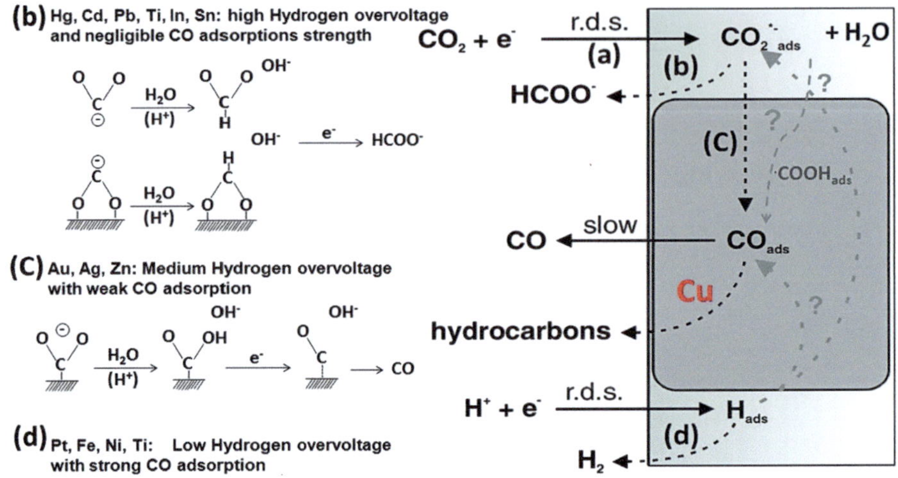

Fig. 5.8 The main reaction pathways at the electrode surface, with adsorbed blocking most of the surface and hydrocarbon products being formed by the further reduction of adsorbed CO. The formation of formate ion is favored by the oxygen-coordinated attachment of carbon dioxide anion radical with the electrocatalyst. The metal electrodes with high H_2 overvoltages such as Hg, Cd, Ti, In, and Sn demonstrate the formation of formate ion. (Reprinted from (Li 2010) with permission of ACS Symposium Series)

$$CO_{2ads}^{\blacksquare -} + H_2O \rightleftharpoons HCOO^{-}{}_{ads} + OH^{-} \tag{5.2}$$

$$HCOO^{-}{}_{ads} + e^{-} \rightleftharpoons HCOO^{-} \tag{5.3}$$

The atom of hydrogen adsorbed (intermediate of hydrogen evolution reaction, HER) on the surface of the catalyst can interact directly with the carbon dioxide anion radical to generate methanoate directly (Amatore and Saveant 1981).

$$CO_2{}^{-\cdot}{}_{ads} + H_{ads} \rightleftharpoons HCOO^{-} \tag{5.4}$$

With the help of the electrocatalyst, the right-hand side of the reaction above is favored as a result of the coordinative attachment of oxygen to the carbon dioxide anion radical. As can be seen in Fig. 5.8, metallic electrodes with high overvoltage of hydrogen such as cadmium, lead, titanium, and tin exhibit methanoate formation.

5.7 Formation of Carbon Monoxide from Electroreduction of Carbon Dioxide

The protonation and reduction steps resulted from the adsorbed carbon of the anionic radical on the metal surface. Due to back donation, the atom of carbon of the anionic radical which encourages the transfer of protons to the atoms of oxygen where water behaves as an electron-deficient center and COOH radical is generated in place of methanoate. Carbon monoxide, alongside the hydroxide ion, are the final products obtained (Lin et al. 2015a; Shaughnessy et al. 2017; Park et al. 2017) as illustrated in the equations below:

$$CO_2^{\cdot-}{}_{ads} + H_2O \rightleftharpoons {}^{\cdot}COOH_{ads} + OH^- \tag{5.5}$$

$$^{\cdot}COOH_{ads} + e^- \rightleftharpoons CO_{ads} + OH^- \tag{5.6}$$

The direct chemical interaction of the adsorbed hydrogen with carbon dioxide anion radical, on the other hand, results in the formation of carbon monoxide.

$$CO_2^{\cdot-}{}_{ads} + H_{ads} \rightleftharpoons CO_{ads} + OH^- \tag{5.7}$$

Carbon dioxide can be stabilized with metal electrodes such as Ga, Au, Zn, Pd, and Ag, but interactions between these metals are usually weak. The electrode surface is formed in the form of a gaseous molecule when carbon dioxide is desorbed. Furthermore, carbon monoxide is reduced into hydrocarbons and methanol fuel (Bagger et al. 2017) by Cu (copper)-based electrodes.

5.8 Formation of Methanol from Electroreduction of Carbon Dioxide

The yield of C_2H_4 and CH_4 increases drastically at 1.2 V which is a very high potential (versus SHE (Standard Hydrogen Electrode)) (Hirunsit et al. 2015; Shen et al. 2015), whereas the formate ion and carbon monoxide show visible results at low overpotentials (−0.9 V versus (SHE) Standard Hydrogen Electrode). The statistics above illustrates that formate ion and carbon monoxide serve as indicators in hydrocarbon and alcohol formation. Investigations such as Roman and Fourier transform infrared spectroscopy (FTIR) have shown that the copper electrode's surface at −0.6 V (versus Standard Hydrogen Electrode, SHE) is absorbed strongly by carbon monoxide, thus indicating that carbon monoxide is crucial to the formation of the useful and desired products of the reactions (Li and Kanan 2012; Gattrell et al. 2006).

Combining carbon monoxide to copper surface creates a strong bond which makes for the suppression of the competing HER (hydrogen evolution reaction) (Li et al. 2014; Ohya et al. 2009). Water molecules are removed due to the presence of carbon monoxide to produce formyl CHO intermediate. CH_2O^*, CH_3O^*, and CHO^* intermediates on the electrocatalyst surface became possible as a result of the reduction of absorbed carbon monoxide. Methanol is formed because the Cu (copper) metal electrodes accept or agree with the hydrogenation of the atom of the oxygen in CH_3O^* species. Figures 5.9 and 5.10 explains and shows the mechanistic carbon dioxide electroreduction on the surface of Copper (Nie et al. 2013).

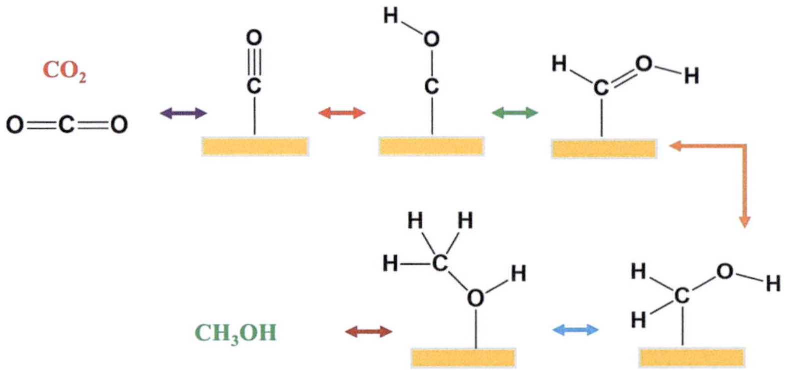

Fig. 5.9 A schematic representation of the reaction occurring on the surface of the Cu electrode. (Reprinted from Fayez et al. (2018) with permission of ACS sustainable chemistry and engineering)

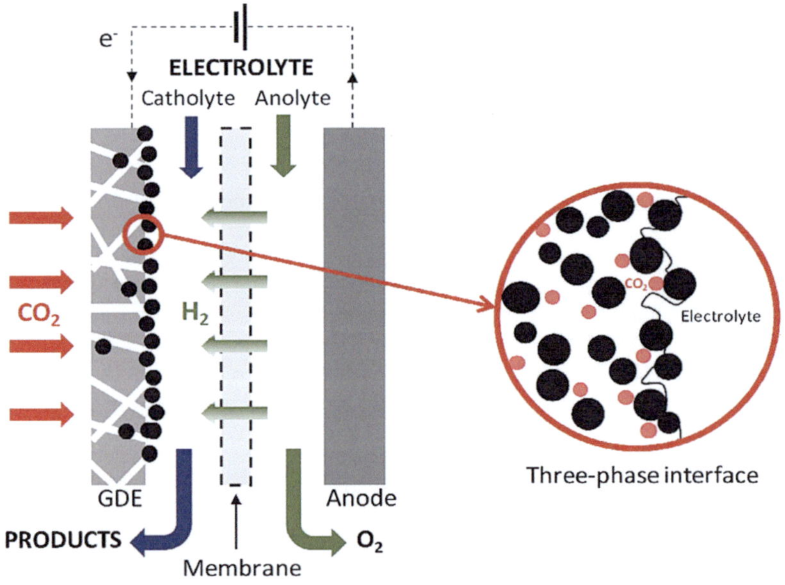

Fig. 5.10 Schematic diagram of the electrolytic cell configuration for the electroreduction of carbon dioxide to methanol supplied directly from the gas phase. The GDE (Gas Diffusion Electrode) allows the continuous striking of the carbon dioxide at a transfer rate for a constant reduction reaction as shown. Consequently, the flow rates of the electrolyte, e.g., catholyte and anolyte, can undergo optimization to obtain a better yield. (Reprinted from Albo and Irabien (2016) with permission of Journal of Catalysis)

5.9 Hydrogen Evolution Reaction

During the process of electrochemical carbon dioxide reduction to the formation of methanol, a significant reaction is observed or noticed in HER (hydrogen evolution reaction). The pH transformation of the aqueous electrolytic solution in the highly acidic region is because of the reaction of the kinetics of HER (Hydrogen Evolution Reaction), thereby making it a sustainable or successful process. The HER (hydrogen evolution reaction) is an independent solution pH in the region known to be neutral and alkaline and slightly acidic (Ooka et al. 2017; Strmcnik et al. 2013; Innocent et al. 2009; Dinh et al. 2018). During this process of the reaction of HER (hydrogen evolution reaction), atomic hydrogen is absorbed, or protons produce hydrogen. In some cases, metals such as electrodes, the H_{ads}, or H^+ is reduced to H_2 on certain applied potentials; the reaction of HER (hydrogen evolution reaction) can be written as:

$$H^+ + e^- \rightleftharpoons H_{ads} \tag{5.8}$$

$$2H_{ads} \rightleftharpoons H_2 \tag{5.9}$$

Or

$$H_{ads} + H^+ + e^- \rightleftharpoons H_2 \tag{5.10}$$

Electrochemical carbon dioxide reduction (Verma et al. 2018; Hall et al. 2015) selections are affected visibly by factors such as the pH. According to Hori et al., the differences in the pH surface have a major after effect on the choice of product, such as methane formation which relies on the pH of the electrolyte, but looking from a different perspective, the production of ethylene does not depend on the concentration of the proton (Hori et al. 1997; Gupta et al. 2006). As a result of the decoupled proton-electron transfer step, the basic electrolytic solution ethylene is an important product. In other words, methane is prominent in the acidic or neutral solution of the electrolyte. Efficiency of the selectivity of carbon dioxide reduction reaction is a direct consequence of the capacity of the buffer by regulating the pH. The lower that capacity of the buffer used as the electrolyte, the more likely ethylene is formed (Varela et al. 2016).

In addition, comparing with the bulk electrolyte as a result of the uninterrupted utilization of hydrogen ions and formation of hydroxide ions, OH^-, the pH closer to the electrodes surface is higher or more (Raciti et al. 2018a, b). Also, in comparison with the bulk electrolyte, the local pH is 6 units more or higher (Gupta et al. 2006).

Hence in the process of the reduction electrochemically of carbon dioxide, the electrolytes considered in the selection process are as a result of local pH. When the local pH which is high equals the carbon dioxide reduction reaction with the bicarbonates or carbonates, the transport of mass of carbon dioxide is hindered.

This makes the circulation or distribution of carbon dioxide to the surface of the electrode attain the high pH region (Raciti et al. 2018a). To comprehensively understand and come up with an electrocatalyst that is both effective and has high selectivity, an adequate understanding of the part played in the mass transfer of carbon dioxide by local pH needs to be well known in carbon dioxide electrochemical reduction.

The researchers opined that reduced local pH supports the carbon-1, C_1, products, and increasing local pH improves the process of selectivity of carbon dioxide reduction limiting HER (hydrogen evolution reaction) leading to more critical multicarbon compounds being derived (Dinh et al. 2018; Raciti et al. 2018a; Kas et al. 2015). As shown in Table 5.2, reduced carbon dioxide electrochemically consists of two, four, six, and eight electrons reduction process. As a result, formic acid, carbon monoxide, methanol, ethanol, etc. are derived.

In the process of reducing carbon dioxide, a combination of chemicals is gotten instead of one target chemical. However, the technique of reducing carbon dioxide electrochemically is succinctly explained in the above section.

More so, the concise knowledge of the likely reaction technique is the challenge in designing the best electrocatalyst. The inadequate reaction knowledge of the technique comes up with cogent challenges such as inadequate efficiency and lower selectivity which reduced the stability of the electrocatalyst. As a result of the issues explained above, a procedure which is dependable on a large scale for producing methanol from carbon dioxide is still a work in progress. The search for such electrocatalyst would be achieved if a thorough understanding of the reaction pathway, modeling, and extensive experimental studies are carried out. The new discoveries in the areas of the use of metal-organic frameworks and non-metal-organic frameworks as electrocatalysts for production of methanol from carbon dioxide reduction are discussed extensively in the proceeding sections for in-depth understanding and steering up interest in this field of research.

5.10 Benchmark Non-metal-Organic Framework-Based Catalysts for Carbon Dioxide Reduction

Recent studies into how important chemicals can be obtained from carbon dioxide reduction for use as energy stores or conversion are currently very inviting (Centi and Perathoner 2009).

The fact that selectivity is immensely critical for arriving at a target product is underscored by the fact that carbon dioxide reduction reaction produces compounds such as carbon monoxide, methane, $C_2O_4H_2$, etc. (Whipple and Kenis 2010; Costentin et al. 2013; Malik et al. 2017). The carbon dioxide reduction reaction mechanism involves water or protons thereby releasing hydrogen as its by-product.

The outcome of the reaction is premised on the metal used in the form of heterogeneous catalyst for the carbon dioxide reduction reaction process. Metals like Pb, Sn, and carbon monoxide show high Faradaic efficiency, FE, selectivity and the turnover frequency for the formation of carbon monoxide or methanoate from carbon dioxide RR (Kortlever et al. 2015; Whipple andKenis 2010; Kumar et al. 2016). Copper is an adequately researched material for use as a catalyst in the multi-electron reduction process leading up to the production of methane (CH_4), ethane (C_2H_4), and methanol (CH_3OH) (Schouten et al. 2011; Nie et al. 2013; Peterson et al. 2010). Notwithstanding arriving at the product in the presence of Cu (copper) is an uphill task during carbon dioxide reduction reaction. Zn, AU, and Ag have recently been tested as a catalyst during the formation of multi-electron products (Wu et al. 2017; Chen et al. 2016; Zhu et al. 2013). Comparable selectivity was also displayed by heterogeneous catalysts as juxtaposed with their homogeneous counterparts.

Complexes of iron have displayed very high turnover frequency and Faradaic efficiency for carbon dioxide reduction to carbon monoxide (Costentin et al. 2012). This process is however kinetically slow requiring that a catalyst is introduced to lower the overpotential necessary. Much work has been done toward arriving at homogeneous and heterogeneous catalysts employable in photochemical and electrochemical carbon dioxide reduction. Metal electrodes made from molybdenum (Mo) were used by Summers and associates for electrochemical reduction of carbon dioxide. At ambient temperature and relatively acidic media, methanol along with trace quantities of methane (CH_4) and carbon monoxide are produced. After many days of using molybdenum as an electrode, the significant reduction in Faradaic efficiency is observed for methanol formation.

This reduction in molybdenum efficiency as an electrode was suspected to result mainly due to variation in the physical configuration of the metal surface occasioned by the accumulation of corrosion and impurities including As and Hg from the electrolyte (Summers et al. 1986)·

Using Cu (copper) as a catalyst, Jianwei et al. employed a supercritical, and highly pressurized mixture such as 60 atm mixture of carbon dioxide and water to produce methanol. Current efficiency is improved by high pressure (Li 1997).

A new area of research is in the use of multimetallic or the right bimetallic catalyst for the reduction of carbon dioxide. Recent studies have also shown that by adding various kinds of atoms on the surface of metal electrodes, their selectivity can be greatly influenced. Improved designs and more efficient chemicals in terms of selectivity have been discovered from the findings of theoretical studies carried out on the reduction of carbon dioxide using a bimetallic catalyst. More effort is also being put into retaining or replicating the unique reducing feature of Cu (copper) and eliminating the HER (hydrogen evolution reaction) at the same time or at least reducing the onset potential in Cu-alloys (Karamad et al. 2014). The observable features of the Cu (copper) electrode were significantly altered when Sakata et al. studied the new characteristics of Cu alloy formed by combining it with Cd, Sn, Zn, Ni, and Ag. The rate of methanol formation increased as well as its selectivity when

Cu (copper) alloys were used as against using coper metal alone. Copper and Nickel metals used individually produced no methanol, but metals as an alloy reduced the overpotential required for methanol formation along with a 5% Faradaic efficiency (Watanabe et al. 1991).

In transforming carbon dioxide into methanol, Ru metal is discussed; in addition, atoms like Cd (cadmium) and Cu (copper) were further introduced. Carbon dioxide is reduced without the competition of HER (hydrogen evolution reaction) due to the properties of Ru such as its stability and that of its oxide to undergo vigorous hydrogen evolution as well as to adsorb hydrogen (Popic et al. 1997). Electrochemically deposing platinum electrode superficially modified with nanoporous copper films using gold alloy and copper enables us to get highly selective alcohols. Copper (Cu) (I) and gold (Au) (0) deposition can be confirmed by the exterior Cu-Au alloy characterization. At Cu-Au/NCF cathode, electrocatalytic reaction leads to the production of methanol and ethanol which was revealed by gas chromatographic analysis of products. Comparing the established surfaces of Cu-Au/NCF electrode with a copper (Cu) plate and Cu nanoparticle-based electrode, it could be observed that the formal shows a better Faradaic efficiency (FE) and greater selectivity to methanol production (Jia et al. 2014).

The decomposition of the comparable halides of metals on titanium (Ti) sheet via heat results in a composite of ruthenium (Ru) and titanium (Ti) metal oxides. The fate of reaction is determined by RuO_2 ratio titanium dioxide; such that hydrogen gas evolution emerges as the fundamental reaction when there is a rise in the quantity of titanium dioxide. Modifying the surface of the composite electrode with thin layers of copper further enhances the efficacy of electrocatalyst. However, the reduction potential for carbon dioxide (CO_2) remains unaffected while that of hydrogen (H_2) becomes more negative. The evolution of hydrogen becomes limited as a result of continuous purging of nitrogen gas in order to optimize the different factors affecting the electrode of the rotating disk; meanwhile electrolyte pH does not affect electroreduction of carbon dioxide (Bandi and Kiihne 1992). Other researchers such as Jianping et al. used ruthenium dioxide/titanium dioxide (RuO_2/TiO_2) nanotubes and nanoparticles to modify the platinum electrode, although the composite modified with RuO_2/TiO_2 nanotubes shows a better current efficacy when compared with that modified with nanoparticles of RuO_2/TiO_2. The titanium dioxide nanotubes have low binding energy, and it helps in the reacting species diffusion as such is more popular. However, the number of active sites for carbon dioxide reduction reaction is enhanced, and the electroactive area of the electrode is increased by the dispersed particles of ruthenium dioxide (Qu et al. 2005). The light harvesting features and the photocathode conduction capacity are enhanced by inducing functional group such as (3-aminopropyl)triethoxysilane ligands on the titanium dioxide. Consequently, more electrons are accessible for carbon dioxide electroreduction. In this study, the suggested mechanism shows that ground-state electrons (HOMO—Highest Occupied Molecular Orbital) of the dye molecules absorb proton and thereafter migrate to an excited state (LUMO—Lowest

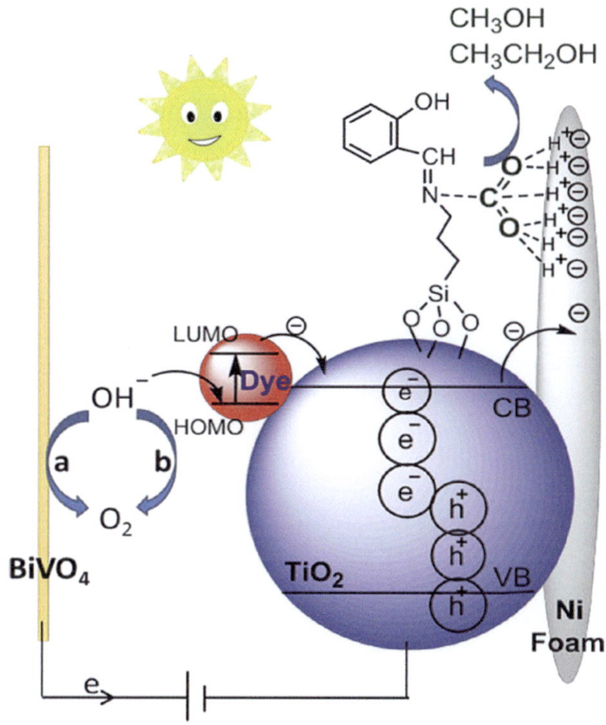

Fig. 5.11 Mechanism showing the photocatalytic process of carbon (IV) oxide reduction. The mechanism here shows that the electrons in the ground state (HOMO—Highest Occupied Molecular Orbital) of the dye molecules travel to an excited state (LUMO—Lowest Unoccupied Molecular Orbital) after the absorption of photos. Thereafter, the excited electron moves to the CB (Conduction Band) of TiO_2 and gets reduced after being conveyed to the nickel foam at VB (Valence Band). (Reprinted from Wang et al. (2017) with permission of Journal of Catalysis)

Unoccupied Molecular Orbital). Subsequently, the electron that is excited moves to the band of conduction of titanium dioxide (TiO_2) which is then moved to the nickel foam which becomes accessible to the reduction reaction. Figure 5.11 describes the above mechanism (Wang et al. 2017).

The prospect of p-type gallium phosphide solar powered electrode in the conversion of carbon dioxide electrochemically into methanol, has been previously studied and reported by several researchers. Some of these studies and reports revealed the nonstop of carbon dioxide reduction after formic acid production as an advantage of p-type gallium phosphide electrodes over other photo-driven electrodes. Meanwhile, methanol and formaldehyde are continually produced from the reduction reaction. It was observed that the optimum solar power wavelength highest quantum efficiency for carbon dioxide conversion into methanol was obtained at 365 nm (Halmann 1978). Emily and his co-workers in another study used photoelectrochemical reduction of an aqueous solution of carbon (IV) oxide. In the presence of pyridinium, p-type gallium phosphide electrode reveals a hundred percent Faradaic efficiency (FE). Pyridinium's catalytic activity clearly improves methanol production through the selective photoelectrochemical conversion of carbon dioxide reduction reaction (). The role of $[C_5H_5NH]^+$ as a homogeneous catalyst in the carbon dioxide reduction was previously evaluated by Gayatri et al.; even though on the surface of the cathode heterogeneous reaction occurs, yet the pyridinium presence enhances the reaction of homogeneous catalysis. The pyridinium and palladium cathode often serve as electrocatalyst as well as the hydrogen gas reduction reaction by poisoning the surface of the electrode (Seshadri et al. 1994). Interestingly, Weiran et al. investigated the effect of the size of the particle of palladium (Pd) nanoparticles on electrochemical reduction of carbon dioxide to methanol in acidified media. The introduction of the polyaniline to the surface of the nanoparticle stabilizes the palladium nanoparticles and mitigates the HER (hydrogen evolution reaction) (Zheng et al. 2017). A brief description of non-metal-organic framework electrocatalysts for carbon dioxide reduction reaction as reported in the other literature is shown in Table 5.3.

5.11 Metal-Organic Frameworks as Catalysts for the Carbon Dioxide Reduction Reaction

Interaction of metal ions such as iron, copper, cobalt, nickel, etc. and organic ligands comprising of carbon, hydrogen, oxygen, and nitrogen results in the production of metal-organic frameworks. Properties of the resultant compounds are a high surface area of $500m^2/g$–$6240\ m^2/g$; evenly distributed pore size of 3.0 A°–3.5 A°; distinct crystal structure; extensively low density; and tangible volume of pores of $0.4cm^3/g$–$3.6cm^3/g$. In the 1990s, *Yaghi* (Yaghi et al. 1995) initially explains metal-organic frameworks; thereafter more than 20,000 kinds of metal-organic frameworks have so far been prepared and characterizations conducted for various applications (Mahmood et al. 2016). Among some of the distinct characteristics that allow them to fit into large varieties of applications include the ability to capture the gas (Yazaydın et al. 2009), the ability to purify and separate the gas (Rodenas et al. 2015), the tendency to exhibit luminescence (Xiao et al. 2018), the ability to exhibit magnetic property (Zhang et al. 2015), conduction of ion and proton (Wong et al. 2017), and the ability to recognize molecule (Bai et al. 2015). Catalytic properties and uses of metal-organic frameworks in chemical convention reactions have been greatly studied and reviewed by many (Lee et al. 2009; Choi et al. 2015). It is important to note the vital roles of metals or "open metal sites" contributing to the catalytic applications of metal-organic frameworks.

Even though metal-organic frameworks are a very useful electrocatalytic carbon dioxide reduction to methanol, it is noteworthy to mention that there is an important hurdle such as low electrical conductance and low thermal and chemical stability in an aqueous environment that is highly basic or acidified which must be overcome. Doping with different heteroatoms or substituting metal centers with other metallic

Table 5.3 Summary of non-metal organic framework electrocatalysts for carbon dioxide reduction reaction

Cathode	Electrocatalyst	Electrode potential	Electrolyte	Catalysis type	Catalytic performance	Reaction time	Remarks	References
Molybdenum metal	Molybdenum metal	−0.8 V (versus SCE)	0.2 M Na_2SO_4 (pH 4.2)	Heterogeneous	$FE_{CH_3OH} = 84\%$ $j = 120\ \mu Acm^{-2}$	23.3 h	Passivation of electrode during extended electrolysis reaction	Summers et al. (1986)
Cu	Cu	−1.1 V (versus SCE)	C_2H_5OH-H_2O-LiCl	Heterogeneous	$FE_{CH_3OH} = 40\%$ $j = 9\ mAcm^{-2}$	8 h	High pressure needed to maintain supercritical CO2 solution	Li (1997)
Co-protoporphyrin/ pyrolytic graphite electrode	Co-protoporphyrin	−0.5 V (versus RHE)	$HClO^{4-}$$NaClO^{4-}$NaOH-borate	Heterogeneous	$FE_{CO} = 60\%$ $FE_{CH_4} = 3\%$	1 h	A mixture of products is obtained	Shen et al. (2015)
Electrodeposited Cu_2O film	Electrodeposited Cu_2O film	−1.1 V (versus SCE)	0.5 M $KHCO_3$	Heterogeneous	$FE_{CH_3OH} = 38\%$	10 min	Reduction of Cu (I) into Cu (0)	Le et al. (2011)
Carbon paper	Cu_2O Cu_2O/ZnO	−1.3 V (versus Ag/AgCl)	0.5 M $KHCO_3$	Heterogeneous	$FE_{CH_3OH} = 45.7$ $j_{total} = 6.93\ mAcm^{-2}$	5 h	Limited mass transfer	Albo et al. (2015b)
Carbon paper	Cu_2O Cu_2O/ZnO	−1.39 V (versus Ag/AgCl)	0.5 M $KHCO_3$	Heterogeneous	$FE_{CH_3OH} = 54.8$ $j_{total} = 10\ mAcm^{-2}$	20 h	–	Albo and Irabien (2016)
Cu_2O/ZnO	Cu_2O/ZnO and pyridine based co-catalyst	−0.8 V (versus Ag/AgCl)	10 mM 2-mPy	Homogeneous and heterogeneous	$FE_{CH_3OH} = 25.6\%$	5 h	Change in selectivity after longer reaction time	Albo et al. (2017)

(continued)

Table 5.3 (continued)

Cathode	Electrocatalyst	Electrode potential	Electrolyte	Catalysis type	Catalytic performance	Reaction time	Remarks	References
Ru/Cu	Ru/Cu	−0.8 V (versus SCE)	0.5 M Na_2HCO_3	Heterogeneous	FE_{CH_3OH} = 41.3%	4 h	Stable Ru catalyst/catalytic poisoning	Popic et al. (1997)
Cu-Au	Cu-Au	−1.0 V (versus SCE)	0.5 M $KHCO_3$	Heterogeneous	FE_{CH_3OH} = 15.9%		Formation of side products in a higher concentration other than methanol	Jia et al. (2014)
Ni/TiO_2	Ni/TiO_2	−0.6 V (versus SCE)	0.1 M $KHCO_3$	Heterogeneous	FE_{CH_3OH} = 100%	28 h		(Wang et al. 2017)
RuO_2-TiO_2	RuO_2-TiO_2	−0.95 V (versus SCE)	Buffer	Heterogeneous	FE_{CH_3OH} = 29.8%	–	–	Bandi and Kiihne (1992)
RuO_2-TiO_2 NTs	RuO_2-TiO_2 NTs	−0.8 V (versus SCE)	0.5 M $NaHCO_3$	Heterogeneous	FE_{CH_3OH} = 60.5%	10 min	–	Qu et al. (2005)
p-GaP	p-GaP	−1.4 V (versus SCE)	0.05 M K_2HPO_4/ KH_2PO_4 buffer solution	Heterogeneous	$FE_{CH_3OH} = 60\%$ $\Phi_{MeOH} = 0.61\%$ $j = 6.0$ mAcm^{-2}	24 h	Lower selectivity of electrocatalyst for methanol production	Halmann (1978)
p-GaP	Pyridinium	−0.52 V (versus SCE)	0.1 M acetate buffer +10 mM pyridine	Homogeneous	FE_{CH_3OH} = 100 (quantum efficiency) $\Phi_{MeOH} = 44\%$ $j_{total} = 0.2$ mAcm^{-2}	30 h	–	Barton et al. (2008)

Hydrogenated Pd	Pyridinium	−0.55 V (versus SCE)	0.5 M $NaClO_4$ + 10 mM pyridine	Homogeneous and heterogeneous	$FE_{CH_3OH} = 30$	19 h	Limited energy conversion efficiency	Seshadri et al. (1994)
PdNP-PANI	PdNP-PANI	−0.9 V (versus Ag/AgCl)	0.5 M H_2SO_4	Heterogeneous	$FE_{CH_3OH} = 5.4\%$	–	Very low FE for methanol production	Zheng et al. (2017)

where *SCE* Saturated calomel electrode, *RHE* Reversible hydrogen electrode, *FE* Faradaic efficiency, *MeOH* Methanol, *PdNP-PANI* Palladium nanoparticle-polyaniline

atoms could be performed to improve electrocatalytic activities of metal-organic frameworks. Additionally, in order to achieve better compositional and regioselective control over the metal-organic frameworks, heterometallic metal-organic frameworks or sophisticated nanomorphologies could also be accomplished. Producing the composites by functionalization of metal-organic frameworks with various materials helps to establish the significant features to the metal-organic frameworks. At isolated form, metal-organic frameworks are not possible to attain; hence the composites of metal-organic frameworks and modification materials show improved characteristics because of their synergetic effect. Several materials like metal NP (nanoparticles) (Ma et al. 2016), liquified ions (Cao et al. 2015), graphene (Wen et al. 2015), and polymeric materials (Zhao et al. 2017), to mention but a few, are used to modify the metal-organic frameworks. The specific size pores and shape carried by metal-organic frameworks encourage their high selectivity for the molecules used in this study. More so, metal-organic frameworks can easily be reused; it is highly stable and active. Metal-organic frameworks electrocatalytic activity can be further enhanced by modifying their structure; introducing other metals or nanoparticles is one of the ways of modifying metal-organic frameworks. Literature has reported numerous mechanisms for developing metal-organic frameworks on surfaces as highly conductive as those of graphene and carbon nanotubes (Perfecto-Irigaray et al. 2018; Fayez et al. 2018). The properties revealed earlier favor the catalytic application of metal-organic frameworks and their composites.

To get oxide-derived copper/carbon electrocatalyst, Zhao et al. carbonized the Cu-based metal-organic framework (HKUST-1). The reduction of carbon dioxide electrochemically to methanol and C_2H_5OH at −0.1 V to −0.7 V (versus Reversible Hydrogen Electrode, RHE) is selectively catalyzed by oxide-derived Cu/C, and its electrocatalyst produces ethanol at lowest overpotential of −0.1 V (versus Reversible Hydrogen Electrode, RHE) as compared to other catalysts reported by other researchers. Due to the confinement effect, there was a reduction in overpotential which was accredited to the strong carbon (IV) oxide binding with the electrocatalyst surface. The comparative outcome of soaring scattered copper in a porous matrix of carbon matrix was the reason for the high conductivity of oxide-derived copper/carbon in comparison with the porous metallic materials such as copper and carbon. More so, the porous structure further assists in the effective alcohol's circulation to the solutions and transfer of carbon dioxide to the electrocatalyst. Comparing porous copper with oxide-derived (OD) Cu/C, the latter gives a sizeable electrochemically active area and structurally stable material for delayed carbon dioxide electroreduction to alcohols. The oxide-derived copper/carbon material stability was credited in favor of the presence of a matrix of carbon matrix preventing the copper from deactivating in the act of electrochemically reducing carbon (IV) oxide. To compare the rates of production and stability of oxide-derived (OD) Cu/C with porous copper, five consecutive batches were made, each batch lasting for at least

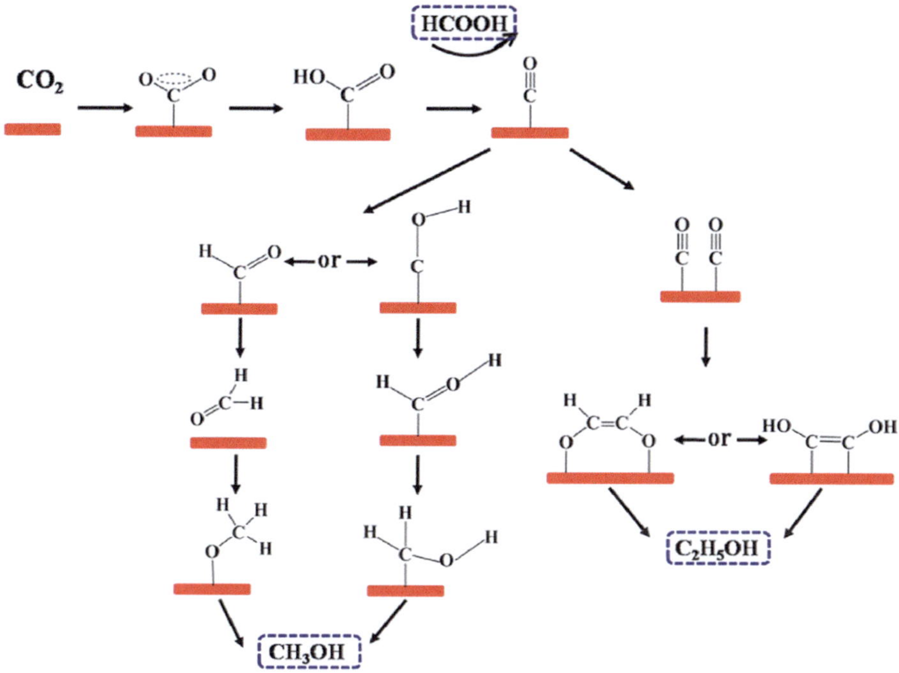

Fig. 5.12 The proposed reaction mechanism for carbon dioxide electroreduction on oxidized-derived (OD) Cu/C. (Reprinted from Zhao et al. (2017) with permission of ACS Applied Materials and Interfaces)

3 h. The earlier investigation on the capacity of the reaction shows how the production rate of methanol and ethanol can range from 5.1 to 12.4 and 3.7 to 13.4 $mg^{-1}\ h^{-1}$ likewise because of the dissimilarity in the activity of the oxide-derived copper/carbon composite which depends on different temperatures of carbonization. Shown in Fig. 5.12 is the electrochemical reduction reaction pathway of carbon dioxide (Zhao et al. 2017).

In similar research, gas diffusion electrode (GDE) contains various copper-based metal-organic frameworks like:

1. HKUST-1 MOF
2. CuAdeAce MOF
3. CuDTA MOA
4. CuZnDTA MOA

All of which have been employed in the electrochemical carbon dioxide reduction. Of all the materials used, HKUST-1 study revealed porosity that is very high, surface area that is very large, and greatest conductivity, as confirmed through voltammetry study and characterization of the surface. The HKUST-1 begins with the electroreduction of carbon dioxide at −1.0 V (versus Ag/AgCl) after which methanol is produced by 6 electrons pathway, followed by 12 electrons mechanism to produce ethanol. Figure 5.13 shows the significant role the applied current density plays in composition and electrocatalytic activity of the obtained product.

The optimal flow rate value of the electrolyte was selected at 3 mLmin^{-1} cm^{-2}, while that of gas was selected at 20 mLmin^{-1} cm^{-2} to attain maximum alcohol production, even though there is good electrocatalytic activity displayed by the metal-organic framework-based electrodes toward carbon (IV) oxide reduction. Nevertheless, uninterrupted activity loss and Faradaic efficiency (FE) drop of 65% for CuAdeAce, 98% for CuDTA, and 51% for CuZnDTA were revealed by the study of Fourier transform infrared spectroscopy (FTIR) and powder X-ray diffractometers (PXRD) in a 5-h reaction in gas diffusion electrode (GDE). The decrease observed in activity can be attributed to the porosity leaching of the MOF material coming from the gas diffusion electrode support of the carbon. Primarily, the cause of the drop-in

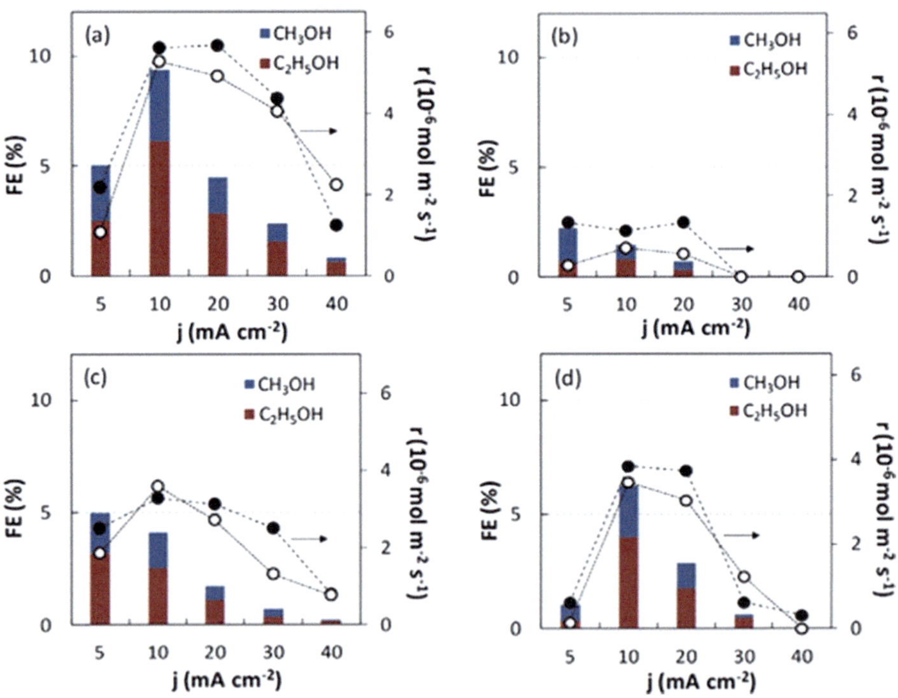

Fig. 5.13 The formation rate and faradaic efficiencies for methanol and ethanol during electrocatalytic carbon (IV) oxide reduction as a function of the current densities applied with (**a**) HKUST-1, (**b**) CuAdeAce, (**c**) CuDTA, and d) CuZnDTA. As shown, HKUST-1 outperforms the other metal-organic framework materials investigated in terms of Faradaic efficiency and formation rate of methanol and ethanol at the same range of current density. (Reprinted from Albo et al. (2016) with permission of ChemSusChem)

product yield can also be accredited to catalytic layer defects which help the formation of hydrogen during the preparation of metal organic porous material-gas diffusion electrodes. Alternatively, upon 17 h that was apportioned to the conservation of the metal-organic framework structure, HKUST-1 shows a rate of formation that is comparably stable structure despite remarkable loss in crystallinity during the reaction. Meanwhile, prior to the application of the electrocatalysis on a large scale, it is important to address major issues such as conversion effectiveness, level of selectivity, long-term stability, and the understanding of the mechanism of the reaction over catalytic system (Albo et al. 2016).

To increase the percentage yield of reduction products, researches designed a covalent organic framework to encourage the high concentration carbon dioxide around the center of reduction. Studying carbon (IV) oxide storage capacity, it showed that pores of the covalent organic framework can retain as high as 24 molecules of the carbon (IV) oxide with the greatest concentration at the center of the pores. Electrochemical reduction of carbon (IV) oxide is also made suitable by the greater gravimetric capacity of the covalent organic framework at normal pressure. The below equation shows the relationship between Gibbs free energy and the carbon dioxide concentration:

$$\Delta G = \Delta G^{\theta} - 2.303 \cdot (CO_2) \tag{5.11}$$

where ΔG^{θ} gives the free energy gotten by the density functional theory calculations. When compared to the aqueous electrolyte, the action rate rises by 97.5 times as well as the reaction overpotential lowered from 0.39 V to 0.27 V (versus RHE) due to the high concentration of carbon dioxide (Yao et al. 2018). Figure 5.14 shows the integrated electrode system design.

Cardoso and co-workers recently engrafted 50 nm ZIF-8 nanoparticles on the titanium/titanium dioxide nanotubes making use of the layer-by-layer process.

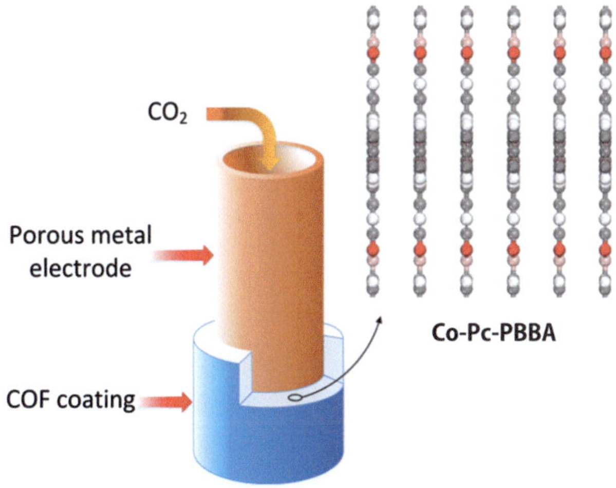

Fig. 5.14 Suggested electrode design by deposition of the covalent organic framework (COF) layer onto the hollow porous metal tube. (Reprinted from (Yao et al. 2018) with permission of Chemistry A European Journal)

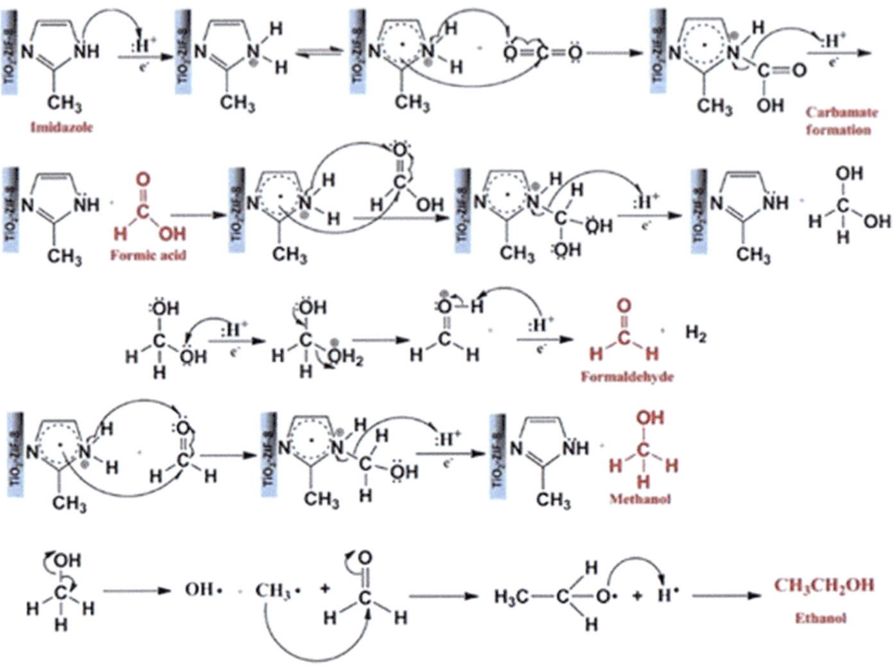

Fig. 5.15 Proposed mechanism for alcohol formation using Ti/TiO_2-NT-ZIF-8 electrodes operated under UV–vis irradiation and with the application of the external potential

As illustrated in Fig. 5.15, the resulting Ti/TiO_2-ZIF-8 electrodes demonstrate the ability to absorb carbon dioxide in the carbamate form. Subsequently, there is a production of methanol and ethanol as a result of the photoelectrocatalytic reduction under ultraviolet-visible irradiation at ambient temperature. It is worthy of note that in the conversion of carbon dioxide to products with greater values on the surface of Ti/TiO_2-ZIF-8 electrode, the bias potential plays significant roles. However, in comparison to photocatalysis, electrolysis, and photolysis, the outcome showed that applying a 0.1 V potential (bias) significantly improves the yield of the alcohol. Figure 5.15 shows the proposed mechanism for the reaction.

Aside methanol other traces of side products were produced as the reaction description proposed this. However, the conversion of methanol to ethanol as a product is greatly reduced due to the availability of the photoelectrons. In electroreduction of carbon dioxide, for about 50 h, Ti/TiO_2-ZIF-8 showed extraordinary stability of the photocurrent. At the end of 50-h reaction, the scanning electron microscope (SEM) image revealed that the material structure was changed. The IR spectra of the Ti/TiO_2-ZIF-8 photoelectrode affirms the decrease in the intensity of carbon dioxide absorption band at 2338 and 2359 cm^{-1}, while a strong band at 1727 and 3330 cm^{-1} reveals the presence of carbamate species (Cardoso et al. 2018). The solvent-less partial doping of the ruthenium (III), zinc (II), and palladium (II) to an electrocatalyst HKUST-1 (Cu) was carried out to produce heterometallic metal-organic framework. In comparison to pristine HKUST-1 metal-organic frameworks material, raise in the concentration of the dopant significantly improves the ethanol selectivity. However, the intense interaction between the reactants and dopant metals resulted in the ethanol selectivity and invariably supports C-C coupling to C_2 in ethanol as a product. The best yield of methanol and ethanol was displayed by Ru (III)-based electrode with 47.2% Faradaic efficiency (FE) in comparison with other dopants. Due to the repeated loss of metal-organic framework particles from the electrode surface, with time, there was a continuous decay of the gas diffusion electrode (GDE) activity (Perfecto-Irigaray et al. 2018). The summary of the metal-organic framework-based electrolysis for carbon dioxide reduction reaction was reported in Table 5.4.

5.12 Conclusion and Recommendations

It is important to note here that electroreduction of carbon dioxide to methanol is an interesting and upcoming research area as such; the evolution of remarkably efficient and low-priced catalysts is promising. Hence, by using renewable energy sources like wind and solar power, electrochemical carbon dioxide reduction reaction electricity requirement bias can be provided (Costentin et al. 2013). Although, the main interest of this study was to provide electrocatalyst that is proficient and effective in converting energies from the renewable energy sources, ideal features such as; minimal cost, availability, high activity of the catalyst, structural porosity, extensive surface area, massive active sites & mass transport of the interesting compounds, are of paramount requirement in considering materials that can act as catalyst (Whipple and Kenis 2010).

Finally, environmental hazard as a result of carbon dioxide release can be extremely reduced by low carbon fuels from carbon dioxide electroreduction. We strongly recommend that a well-articulated effort is essential in order to enhance basic and resourceful researches on the electrocatalytic carbon dioxide reduction to methanol on the metal-organic frameworks surface.

Table 5.4 Summary of metal organic framework (MOF)-based electrocatalysts for CO_2 reduction reaction

Cathode	Electrocatalyst	Electrode potential	Electrolyte	Catalysis type	Catalytic performance	Reaction Time	Remarks	References
Cu/C-derived from metal organic framework (MOF)	Cu/C	−0.1 V (versus RHE)	0.1 M $KHCO_3$	Heterogeneous	$FE_{CH_3OH} =$ 43.2%	5 h	Relatively low yield of methanol	Zhao et al. (2017)
					$FE_{C_2H_5OH} = 34.8\%$			
HKUST-1	HKUST-1	−1.0 V (versus Ag/AgCl)	0.5 M $KHCO_3$	Heterogeneous	$FE_{CH_3OH} =$ 5.6%	17 h	Very low FE	Albo et al. (2016)
					$FE_{C_2H_5OH} = 10.3\%$			
					$j = 10$ $mAcm^{-2}$			
Co-Pc-PBBA	Co-Pc-PBBA	−0.27 V (versus RHE)	–	Heterogeneous	–	24 h	Not specific for the methanol production	Yao et al. (2018)
Ti/TiO_2-ZIF-8	Ti/TiO_2-ZIF-8	+0.1 V (versus Ag/AgCl)	0.1 M Na_2SO_4	Heterogeneous	–	50 h	Relatively low yield of methanol	Cardoso et al. (2018)
Ru doped HKUST-1	Ru doped HKUST-1	−2.0 V (versus Ag/AgCl)	0.5 M $KHCO_3$	Heterogeneous	$FE_{CH_3OH} + FE_{C_2H_5OH} =$ 47.2%	60 min	Unstable activity for longer reaction	Perfecto-Irigaray et al. (2018)

References

Agarwal AS, Rode E, Sridhar N, Hill D (2016) Conversion of CO_2 to value-added chemicals: opportunities and challenges. In: Handbook of climate change mitigation and adaptation. Springer, New York, pp 1–29. https://doi.org/10.1007/978-3-319-14409-2_86

Albo J, Irabien A (2016) Cu_2O-loaded gas diffusion electrodes for the continuous electrochemical reduction of CO_2 to methanol. J Catal 343:232–239. https://doi.org/10.1016/j.jcat.2015.11.014

Albo J, Alvarez-Guerra M, Castaño P, Irabien A (2015a) Towards the electrochemical conversion of carbon dioxide into methanol. Green Chem 17(4):2304–2324. https://doi.org/10.1039/C4GC02453B

Albo J, Sáez A, Solla-Gullón J, Montiel V, Irabien A (2015b) Production of methanol from CO_2 electroreduction at Cu_2O and Cu_2O/ZnO-based electrodes in aqueous solution. Appl Catal B Environ 176–177:709–717. https://doi.org/10.1016/j.apcatb.2015.04.055

Albo J, Vallejo D, Beobide G, Castillo O, Castano P, Irabien A (2016) Copper-based metal-organic porous materials for CO_2 electrocatalytic reduction to alcohols. ChemSusChem 9:1–11. https://doi.org/10.1002/cssc.201600693

Albo J, Beobide G, Castaño P, Irabien A (2017) Methanol electrosynthesis from CO_2 at Cu_2O/ZnO prompted by pyridine-based aqueous solutions. J CO2 Util *18*:164–172. https://doi.org/10.1016/j.jcou.2017.02.003

Amatore C, Saveant JM (1981) Mechanism and kinetic characteristics of the electrochemical reduction of carbon dioxide in media of low proton availability. J Am Chem Soc 103 (17):5021–5023. https://doi.org/10.1021/ja00407a008

Aresta M, Dibenedetto A, Angelini A (1996) The use of solar energy can enhance the conversion of carbon dioxide into energy-rich products: stepping towards artificial photosynthesis. Philos Trans R Soc A Math Phys Eng Sci 2013(371):20120111–20120111. https://doi.org/10.1098/rsta.2012.0111

Back S, Kim H, Jung Y (2015) Selective heterogeneous CO_2 electroreduction to methanol. ACS Catal 5(2):965–971. https://doi.org/10.1021/cs501600x

Bagger A, Ju W, Varela AS, Strasser P, Rossmeisl J (2017) Electrochemical CO_2 reduction: a classification problem. ChemPhysChem 18(22):3266–3273. https://doi.org/10.1002/cphc.201700736

Bai L, Wang P, Bose P, Li P, Zou R, Zhao Y (2015) Macroscopic architecture of charge transfer-induced molecular recognition from electron-rich polymer interpenetrated porous frameworks. ACS Appl Mater Interfaces 7(9):5056–5060. https://doi.org/10.1021/am5089549

Bandi A, Kiihne H (1992) Electrochemical reduction of carbon dioxide in water: analysis of reaction mechanism on ruthenium-titanium-oxide. J Electrochem Soc 139(6):1605. https://doi.org/10.1149/1.2069464

Barton EE, Rampulla DM, Bocarsly AB (2008) Selective solar-driven reduction of co_2 to methanol using a catalyzed p-gap based photoelectrochemical cell. J Am Chem Soc 130(20):6342–6344. https://doi.org/10.1021/ja0776327

Benson EE, Kubiak CP, Sathrum AJ, Smieja JM (2009) Electrocatalytic and homogeneous approaches to conversion of CO_2 to liquid fuels. Chem Soc Rev 38(1):89–99. https://doi.org/10.1039/B804323J

Bodansky D (2010) The Copenhagen climate change conference: a postmortem. Am J Int Law 104 (2):230. https://doi.org/10.5305/amerjintelaw.104.2.0230

Bourrez M, Molton F, Chardon-Noblat S, Deronzier A (2011) [Mn(Bipyridyl)$(CO)_3$Br]: an abundant metal carbonyl complex as efficient electrocatalyst for CO_2 reduction. Angew Chemie 123 (42):10077–10080. https://doi.org/10.1002/anie.201103616

Bromberg L, Cheng WK (2010) Methanol as an alternative transportation fuel in the US: options for sustainable and/or energy-secure transportation. Final Report, Sloan Automotive Laboratory, Massachusetts Institute of Technolohy, Cambridge, MA

Brovkin V, Sitch S, von Bloh W, Claussen M, Bauer E, Cramer W (2004) Role of land cover changes for atmospheric CO_2 increase and climate change during the last 150 years. Glob Chang Biol 10(8):1253–1266. https://doi.org/10.1111/j.1365-2486.2004.00812.x

Burghaus U (2014) Surface chemistry of CO_2 – adsorption of carbon dioxide on clean surfaces at ultrahigh vacuum. Prog Surf Sci 89(2):161–217. https://doi.org/10.1016/j.progsurf.2014.03.002

Cao X, Zheng B, Shi W, Yang J, Fan Z, Luo Z, Rui X, Chen B, Yan Q, Zhang H (2015) Reduced graphene oxide-wrapped moo_3 composites prepared by using metal-organic frameworks as precursor for all-solid-state flexible supercapacitors. Adv Mater 27(32):4695–4701. https://doi.org/10.1002/adma.201501310

Capellán-Pérez I, Mediavilla M, de Castro C, Carpintero Ó, Miguel LJ (2014) Fossil fuel depletion and socio-economic scenarios: an integrated approach. Energy 77:641–666. https://doi.org/10.1016/j.energy.2014.09.063

Cardoso JC, Stulp S, de Brito JF, Flor JBS, Frem RCG, Zanoni MVB (2018) Metal-organic frameworks based on ZIF-8 deposited on TiO nanotubes increase the surface adsorption of CO and its photoelectrocatalytic reduction to alcohols in aqueous media. Appl Catal B Environ 225:563–573. https://doi.org/10.1016/j.apcatb.2017.12.013

Centi G, Perathoner S (2009) Opportunities and prospects in the chemical recycling of carbon dioxide to fuels. Catal Today 148(3–4):191–205. https://doi.org/10.1016/j.cattod.2009.07.075

Chen LD, Urushihara M, Chan K, Nørskov JK (2016) Electric field effects in electrochemical CO_2 reduction. ACS Catal 6(10):7133–7139. https://doi.org/10.1021/acscatal.6b02299

Chen L, Li F, Zhang Y, Bentley CL, Horne M, Bond AM, Zhang J (2017) Electrochemical reduction of carbon dioxide in a monoethanolamine capture medium. ChemSusChem 10 (20):4109–4118. https://doi.org/10.1002/cssc.201701075

Cherubini F, Peters GP, Berntsen T, Stronnan AH, Hertwich E (2011) CO_2 emissions from biomass combustion for bioenergy: atmospheric decay and contribution to global warming. GCB Bioenergy 3(5):413–426. https://doi.org/10.1111/j.1757-1707.2011.01102.x

Chinchen GC, Spencer MS, Waugh KC, Whan DA (1987) Promotion of methanol synthesis and the water-gas shift reactions by adsorbed oxygen on supported copper catalysts. J Chem Soc Faraday Trans 1 *83*(7):2193. https://doi.org/10.1039/F19878302193

Choi KM, Na K, Somorjai GA, Yaghi OM (2015) Chemical environment control and enhanced catalytic performance of platinum nanoparticles embedded in nanocrystalline metal–organic frameworks. J Am Chem Soc 137(24):7810–7816. https://doi.org/10.1021/jacs.5b03540

Costentin C, Drouet S, Robert M, Saveant J-M (2012) A local proton source enhances co_2 electroreduction to CO by a molecular Fe catalyst. Science 338(6103):90–94. https://doi.org/10.1126/science.1224581

Costentin C, Robert M, Savéant J-M (2013) Catalysis of the electrochemical reduction of carbon dioxide. Chem Soc Rev 42(6):2423–2436. https://doi.org/10.1039/C2CS35360A

Darensbourg DJ, Andreatta JR, Moncada AI (2010) Polymers from carbon dioxide: polycarbonates, polythiocarbonates, and polyurethanes. In: Carbon dioxide as chemical feedstock. Wiley-VCH Verlag GmbH & Co. KGaA, Weinheim, pp 213–248. https://doi.org/10.1002/9783527629916.ch8

Davis SJ, Caldeira K, Matthews HD (2010) Future CO_2 emissions and climate change from existing energy infrastructure. Science 329(5997):1330–1333. https://doi.org/10.1126/science.1188566

Dinh C-T, Burdyny T, Kibria MG, Seifitokaldani A, Gabardo CM, García de Arquer FP, Kiani A, Edwards JP, De Luna P, Bushuyev OS et al (2018) CO_2 electroreduction to ethylene via hydroxide-mediated copper catalysis at an abrupt interface. Science 360(6390):783–787. https://doi.org/10.1126/science.aas9100

Dridi H, Comminges C, Morais C, Meledje J-C, Kokoh KB, Costentin C, Savéant J-M (2017) Catalysis and inhibition in the electrochemical reduction of CO_2 on platinum in the presence of protonated pyridine. New insights into mechanisms and products. J Am Chem Soc 139 (39):13922–13928. https://doi.org/10.1021/jacs.7b08028

Fayez NA, Aqil J, Mohammed SB, Azeem R (2018) A review on recent advances for electrochemical reduction of carbon dioxide to methanol using metal organic framework (MOF) and non-MOF catalysts: challenges and future prospects. ACS Sustain Chem Eng. https://doi.org/10.1021/acssuschemeng.8b03843

Finn C, Schnittger S, Yellowlees LJ, Love JB (2012) Molecular approaches to the electrochemical reduction of carbon dioxide. Chem Commun 48(10):1392–1399. https://doi.org/10.1039/C1CC15393E

Francke R, Schille B, Roemelt M (2018) Homogeneously catalyzed electroreduction of carbon dioxide—methods, mechanisms, and catalysts. Chem Rev 118(9):4631–4701. https://doi.org/10.1021/acs.chemrev.7b00459

Froehlich JD, Kubiak CP (2015) The homogeneous reduction of CO_2 by $[Ni(Cyclam)]^+$: increased catalytic rates with the addition of a CO scavenger. J Am Chem Soc 137(10):3565–3573. https://doi.org/10.1021/ja512575v

Ganesh I (2014) Conversion of carbon dioxide into methanol – a potential liquid fuel: fundamental challenges and opportunities (a review). Renew Sust Energ Rev 31:221–257. https://doi.org/10.1016/j.rser.2013.11.045

Gattrell M, Gupta N, Co A (2006) A review of the aqueous electrochemical reduction of co_2 to hydrocarbons at copper. J Electroanal Chem 594(1):1–19. https://doi.org/10.1016/j.jelechem.2006.05.013

Gupta N, Gattrell M, MacDougall B (2006) Calculation for the cathode surface concentrations in the electrochemical reduction of CO_2 in $KHCO_3$ solutions. J Appl Electrochem 36(2):161–172. https://doi.org/10.1007/s10800-005-9058-y

Gusain R, Kumar P, Sharma OP, Jain SL, Khatri OP (2016) Reduced graphene oxide–cuo nanocomposites for photocatalytic conversion of CO_2 into methanol under visible light irradiation. Appl Catal B Environ 181:352–362. https://doi.org/10.1016/j.apcatb.2015.08.012

Hall AS, Yoon Y, Wuttig A, Surendranath Y (2015) Mesostructure-induced selectivity in CO_2 reduction catalysis. J Am Chem Soc 137(47):14834–14837. https://doi.org/10.1021/jacs.5b08259

Halmann M (1978) Photoelectrochemical reduction of aqueous carbon dioxide on p-type Gallium phosphide in liquid junction solar cells. Nature 275(5676):115–116. https://doi.org/10.1038/275115a0

Hirunsit P, Soodsawang W, Limtrakul J (2015) CO_2 electrochemical reduction to methane and methanol on copper-based alloys: theoretical insight. J Phys Chem C 119(15):8238–8249. https://doi.org/10.1021/acs.jpcc.5b01574

Hod I, Sampson M, Deria P, Kubiak C, Farha O, Hupp J (2015) Fe-porphyrin-based metal-organic framework films as high-surface concentration, heterogeneous catalysts for electrochemical reduction of CO_2. ACS Catal 5(11):6302–6309. https://doi.org/10.1021/acscatal.5b01767

Hollingsworth N, Taylor SFR, Galante MT, Jacquemin J, Longo C, Holt KB, de Leeuw NH, Hardacre C (2015) Reduction of carbon dioxide to formate at low overpotential using a superbase ionic liquid. Angew Chemie Int Ed 54(47):14164–14168. https://doi.org/10.1002/anie.201507629

Hori Y, Wakebe H, Tsukamoto T, Koga O (1994) Electrocatalytic process of co selectivity in electrochemical reduction of CO_2 at metal electrodes in aqueous media. Electrochim Acta 39 (11–12):1833–1839. https://doi.org/10.1016/0013-4686(94)85172-7

Hori Y, Takahashi R, Yoshinami Y, Murata A (1997) Electrochemical reduction of CO at a copper electrode. J Phys Chem B 101(36):7075–7081. https://doi.org/10.1021/jp970284i

Huang C, Mao D, Guo X, Yu J (2017) Microwave-assisted hydrothermal synthesis of CuO-ZnO-ZrO_2 as catalyst for direct synthesis of methanol by carbon dioxide hydrogenation. Energy Technol 5(11):2100–2107. https://doi.org/10.1002/ente.201700190

Innocent B, Liaigre D, Pasquier D, Ropital F, Léger J-M, Kokoh KB (2009) Electro-reduction of carbon dioxide to formate on lead electrode in aqueous medium. J Appl Electrochem 39 (2):227–232. https://doi.org/10.1007/s10800-008-9658-4

Jia F, Yu X, Zhang L (2014) Enhanced selectivity for the electrochemical reduction of CO_2 to alcohols in aqueous solution with nanostructured Cu–Au alloy as catalyst. J Power Sources 252:85–89. https://doi.org/10.1016/j.jpowsour.2013.12.002

Jitaru M, Lowy DA, Toma M, Toma BC, Oniciu L (1997) Electrochemical reduction of carbon dioxide on flat metallic cathodes. J Appl Electrochem 27(8):875–889. https://doi.org/10.1023/A:1018441316386

Joo O-S, Jung K-D, Moon I, Rozovskii AY, Lin GI, Han S-H, Uhm S-J (1999) Carbon dioxide hydrogenation to form methanol via a reverse-water-gas-shift reaction (the CAMERE process). Ind Eng Chem Res 38(5):1808–1812. https://doi.org/10.1021/ie9806848

Kai T, Zhou M, Duan Z, Henkelman GA, Bard AJ (2017) Detection of $CO_2^{\bullet -}$ in the electrochemical reduction of carbon dioxide in N,N -dimethylformamide by scanning electrochemical microscopy. J Am Chem Soc 139(51):18552–18557. https://doi.org/10.1021/jacs.7b08702

Kaneco S, Iiba K, Hiei N, Ohta K, Mizuno T, Suzuki T (1999) Electrochemical reduction of carbon dioxide to ethylene with high faradaic efficiency at a Cu electrode in CsOH/methanol. Electrochim Acta 44(26):4701–4706. https://doi.org/10.1016/S0013-4686(99)00262-5

Karamad M, Tripkovic V, Rossmeisl J (2014) Intermetallic alloys as CO electroreduction catalysts – the role of isolated active sites. ACS Catal 4:2268–2273. https://doi.org/10.1021/cs500328c

Kas R, Kortlever R, Yılmaz H, Koper MTM, Mul G (2015) Manipulating the hydrocarbon selectivity of copper nanoparticles in CO_2 electroreduction by process conditions. ChemElectroChem 2(3):354–358. https://doi.org/10.1002/celc.201402373

Kim J, Henao CA, Johnson TA, Dedrick DE, Miller JE, Stechel EB, Maravelias CT (2011) Methanol production from CO_2 using solar-thermal energy: process development and techno-economic analysis. Energy Environ Sci 4(9):3122. https://doi.org/10.1039/C1EE01311D

Kleij AW, North M, Urakawa A (2017) CO_2 catalysis. ChemSusChem 10(6):1036–1038. https://doi.org/10.1002/cssc.201700218

Kortlever R, Shen J, Schouten KJP, Calle-Vallejo F, Koper MTM (2015) Catalysts and reaction pathways for the electrochemical reduction of carbon dioxide. J Phys Chem Lett 6 (20):4073–4082. https://doi.org/10.1021/acs.jpclett.5b01559

Kuhl KP, Hatsukade T, Cave ER, Abram DN, Kibsgaard J, Jaramillo TF (2014) Electrocatalytic conversion of carbon dioxide to methane and methanol on transition metal surfaces. J Am Chem Soc 136(40):14107–14113. https://doi.org/10.1021/ja505791r

Kumar B, Brian JP, Atla V, Kumari S, Bertram KA, White RT, Spurgeon JM (2016) New trends in the development of heterogeneous catalysts for electrochemical CO_2 reduction. Catal Today 270:19–30. https://doi.org/10.1016/j.cattod.2016.02.006

Lau LC, Lee KT, Mohamed AR (2012) Global warming mitigation and renewable energy policy development from the kyoto protocol to the copenhagen accord—a comment. Renew Sust Energ Rev 16(7):5280–5284. https://doi.org/10.1016/j.rser.2012.04.006

Le M, Ren M, Zhang Z, Sprunger PT, Kurtz RZ, Flake JC (2011) Electrochemical reduction of CO_2 to CH_3OH at copper oxide surfaces. J Electrochem Soc *158*(5):E45–E49. https://doi.org/10.1149/1.3561636

Lee J, Farha OK, Roberts J, Scheidt KA, Nguyen ST, Hupp JT (2009) Metal–organic framework materials as catalysts. Chem Soc Rev 38(5):1450. https://doi.org/10.1039/b807080f

Li J (1997) Electrochemical synthesis of methanol from CO_2 in high-pressure electrolyte. J Electrochem Soc 144(12):4284. https://doi.org/10.1149/1.1838179

Li W (2010) Electrocatalytic reduction of CO_2 to small organic molecule fuels on metal catalysts:55–76. https://doi.org/10.1021/bk-2010-1056.ch005

Li X, Faghri A (2013) Review and advances of direct methanol fuel cells (DMFCs) part i: design, fabrication, and testing with high concentration methanol solutions. J Power Sources 226:223–240. https://doi.org/10.1016/j.jpowsour.2012.10.061

Li CW, Kanan MW (2012) CO_2 reduction at low overpotential on Cu electrodes resulting from the reduction of thick Cu_2O films. J Am Chem Soc 134(17):7231–7234. https://doi.org/10.1021/ja3010978

Li CW, Ciston J, Kanan MW (2014) Electroreduction of carbon monoxide to liquid fuel on oxide-derived nanocrystalline copper. Nature 508(7497):504–507. https://doi.org/10.1038/nature13249

Li Y, Cui X, Dong K, Junge K, Beller M (2017a) Utilization of CO_2 as a C_1 building block for catalytic methylation reactions. ACS Catal 7(2):1077–1086. https://doi.org/10.1021/acscatal.6b02715

Li Q, Fu J, Zhu W, Chen Z, Shen B, Wu L, Xi Z, Wang T, Lu G, Zhu J et al (2017b) Tuning sn-catalysis for electrochemical reduction of CO_2 to CO via the core/shell cu/sno_2 structure. J Am Chem Soc 139(12):4290–4293. https://doi.org/10.1021/jacs.7b00261

Lim RJ, Xie M, Sk MA, Lee J-M, Fisher A, Wang X, Lim KH (2014) A review on the electrochemical reduction of CO_2 in fuel cells, metal electrodes and molecular catalysts. Catal Today 233:169–180. https://doi.org/10.1016/j.cattod.2013.11.037

Lin S, Diercks C, Zhang Y-B, Yaghi O, Chang C, Kornienko N, Nichols E, Zhao Y, Paris A, Kim D et al (2015a) Covalent organic frameworks comprising cobalt porphyrins for catalytic CO_2 reduction in water. Science 349(6253):1208–1213. https://doi.org/10.1126/science.aac8343
Lin S, Diercks CS, Zhang Y-B, Kornienko N, Nichols EM, Zhao Y, Paris AR, Kim D, Yang P, Yaghi OM et al (2015b) Covalent organic frameworks comprising cobalt porphyrins for catalytic CO_2 reduction in water. Science 349(6253):1208–1213. https://doi.org/10.1126/science.aac8343
Liu Y, Wang ZU, Zhou H-C (2012) Recent advances in carbon dioxide capture with metal-organic frameworks. Greenhouse Gases Sci Technol 2(4):239–259. https://doi.org/10.1002/ghg.1296
Ma J, Sun N, Zhang X, Zhao N, Xiao F, Wei W, Sun Y (2009) A short review of catalysis for CO_2 conversion. Catal Today 148(3–4):221–231. https://doi.org/10.1016/j.cattod.2009.08.015
Ma J, Ying Y, Guo X, Huang H, Liu D, Zhong C (2016) Fabrication of mixed-matrix membrane containing metal–organic framework composite with task-specific ionic liquid for efficient CO separation. J Mater Chem A 4(19):7281–7288. https://doi.org/10.1039/C6TA02611G
Mahmood A, Guo W, Tabassum H, Zou R (2016) Metal-organic framework-based nanomaterials for electrocatalysis. Adv Energy Mater 6(17):1600423. https://doi.org/10.1002/aenm.201600423
Malik K, Singh S, Basu S, Verma A (2017) Electrochemical reduction of CO_2 for synthesis of green fuel. Wiley Interdiscip Rev Energy Environ 6(4):e244. https://doi.org/10.1002/wene.244
Markewitz P, Kuckshinrichs W, Leitner W, Linssen J, Zapp P, Bongartz R, Schreiber A, Müller TE (2012) Worldwide innovations in the development of carbon capture technologies and the utilization of CO_2. Energy Environ Sci 5(6):7281. https://doi.org/10.1039/C2EE03403D
Mathisen A, Skagestad R (2017) Utilization of CO_2 from emitters in Poland for CO_2-EOR. Energy Procedia 114:6721–6729. https://doi.org/10.1016/j.egypro.2017.03.1802
McGlade C, Ekins P (2015) The geographical distribution of fossil fuels unused when limiting global warming to 2°C. Nature 517(7533):187–190. https://doi.org/10.1038/nature14016
Mikkelsen M, Jørgensen M, Krebs FC (2010) The Teraton challenge. A review of fixation and transformation of carbon dioxide. Energy Environ Sci 3(1):43–81. https://doi.org/10.1039/B912904A
Millward AR, Yaghi OM (2005) Metal−organic frameworks with exceptionally high capacity for storage of carbon dioxide at room temperature. J Am Chem Soc 127(51):17998–17999. https://doi.org/10.1021/ja0570032
Min X, Kanan MW (2015) Pd-catalyzed electrohydrogenation of carbon dioxide to formate: high mass activity at low overpotential and identification of the deactivation pathway. J Am Chem Soc 137(14):4701–4708. https://doi.org/10.1021/ja511890h
Ng KS, Zhang N, Sadhukhan J (2013) Techno-economic analysis of polygeneration systems with carbon capture and storage and CO_2 reuse. Chem Eng J 219:96–108. https://doi.org/10.1016/j.cej.2012.12.082
Nie X, Esopi MR, Janik MJ, Asthagiri A (2013) Selectivity of CO_2 reduction on copper electrodes: the role of the kinetics of elementary steps. Angew Chemie Int Ed 52(9):2459–2462. https://doi.org/10.1002/anie.201208320
Ohya S, Kaneco S, Katsumata H, Suzuki T, Ohta K (2009) Electrochemical reduction of CO_2 in methanol with aid of Cuo and Cu_2O. Catal Today 148(3–4):329–334. https://doi.org/10.1016/j.cattod.2009.07.077
Olah GA (2005) Beyond oil and gas: the methanol economy. Angew Chemie Int Ed 44(18):2636–2639. https://doi.org/10.1002/9783527627806
Olah GA (2013) Towards oil independence through renewable methanol chemistry. Angew Chemie Int Ed 52(1):104–107. https://doi.org/10.1002/anie.201204995
Olah GA, Goeppert A, Prakash GKS (2009) Chemical recycling of carbon dioxide to methanol and dimethyl ether: from greenhouse gas to renewable, environmentally carbon neutral fuels and synthetic hydrocarbons. J Org Chem 74(2):487–498. https://doi.org/10.1021/jo801260f
Omae I (2012) Recent developments in carbon dioxide utilization for the production of organic chemicals. Coord Chem Rev 256(13–14):1384–1405. https://doi.org/10.1016/j.ccr.2012.03.017

Ooka H, Figueiredo MC, Koper MTM (2017) Competition between hydrogen evolution and carbon dioxide reduction on copper electrodes in mildly acidic media. Langmuir 33(37):9307–9313. https://doi.org/10.1021/acs.langmuir.7b00696

Ott J, Gronemann V, Pontzen F, Fiedler E, Grossmann G, Kersebohm DB, Weiss G, Witte C (2012) Methanol. In: Ullmann's encyclopedia of industrial chemistry. Wiley-VCH Verlag GmbH & Co. KGaA, Weinheim. https://doi.org/10.1002/14356007.a16_465

Park H, Choi J, Kim H, Hwang E, Ha D-H, Ahn SH, Kim S-K (2017) AgIn dendrite catalysts for electrochemical reduction of CO_2 to CO. Appl Catal B Environ 219:123–131. https://doi.org/10.1016/j.apcatb.2017.07.038

Perathoner S, Centi G (2014) CO_2 recycling: a key strategy to introduce green energy in the chemical production chain. ChemSusChem 7(5):1274–1282. https://doi.org/10.1002/cssc.201300926

Perfecto-Irigaray M, Albo J, Beobide G, Castillo O, Irabien A, Pérez-Yáñez S (2018) Synthesis of heterometallic metal–organic frameworks and their performance as electrocatalyst for CO_2 reduction. RSC Adv 8(38):21092–21099. https://doi.org/10.1039/C8RA02676A

Peters M, Köhler B, Kuckshinrichs W, Leitner W, Markewitz P, Müller TE (2011) Chemical technologies for exploiting and recycling carbon dioxide into the value chain. ChemSusChem 4 (9):1216–1240. https://doi.org/10.1002/cssc.201000447

Peterson AA, Nørskov JK (2012) Activity descriptors for CO_2 electroreduction to methane on transition-metal catalysts. J Phys Chem Lett 3(2):251–258. https://doi.org/10.1021/jz201461p

Peterson AA, Abild-Pedersen F, Studt F, Rossmeisl J, Nørskov JK (2010) How copper catalyzes the electroreduction of carbon dioxide into hydrocarbon fuels. Energy Environ Sci 3(9):1311. https://doi.org/10.1039/C0EE00071J

Popic JP, Avramov-Ivic ML, Vukovic NB (1997) Reduction of carbon dioxide on ruthenium oxide and modified ruthenium oxide electrodes in 0.5 M $NHCO_3$. J Electroanal Chem *421*:105–110. https://doi.org/10.1016/S0022-0728(96)04823-1

Pourkhesalian AM, Shamekhi AH, Salimi F (2010) Alternative fuel and gasoline in an si engine: a comparative study of performance and emissions characteristics. Fuel 89(5):1056–1063. https://doi.org/10.1016/j.fuel.2009.11.025

Qiao J, Liu Y, Hong F, Zhang J (2014) A review of catalysts for the electroreduction of carbon dioxide to produce low-carbon fuels. Chem Soc Rev 43(2):631–675. https://doi.org/10.1039/C3CS60323G

Qu J, Zhang X, Wang Y, Xie C (2005) Electrochemical reduction of CO_2 on RuO_2/TiO_2 nanotubes composite modified Pt electrode. Electrochim Acta 50(16–17):3576–3580. https://doi.org/10.1016/j.electacta.2004.11.061

Raciti D, Mao M, Park JH, Wang C (2018a) Local pH effect in the CO_2 reduction reaction on high-surface-area copper electrocatalysts. J Electrochem Soc 165(10):F799–F804. https://doi.org/10.1149/2.0521810jes

Raciti D, Mao M, Wang C (2018b) Mass transport modelling for the electroreduction of CO_2 on cu nanowires. Nanotechnology 29(4):044001. https://doi.org/10.1088/1361-6528/aa9bd7

Ren D, Deng Y, Handoko AD, Chen CS, Malkhandi S, Yeo BS (2015) Selective electrochemical reduction of carbon dioxide to ethylene and ethanol on copper(i) oxide catalysts. ACS Catal 5 (5):2814–2821. https://doi.org/10.1021/cs502128q

Ren D, Wong NT, Handoko AD, Huang Y, Yeo BS (2016) Mechanistic insights into the enhanced activity and stability of agglomerated Cu nanocrystals for the electrochemical reduction of carbon dioxide to n -propanol. J Phys Chem Lett 7(1):20–24. https://doi.org/10.1021/acs.jpcc.6b07128

Rodenas T, Luz I, Prieto G, Seoane B, Miro H, Corma A, Kapteijn F, Llabrés i Xamena FX, Gascon J (2015) Metal–organic framework nanosheets in polymer composite materials for gas separation. Nat Mater 14(1):48–55. https://doi.org/10.1038/nmat4113

Rodhe H (1990) A comparison of the contribution of various gases to the greenhouse effect. Science 248(4960):1217–1219. https://doi.org/10.1126/science.248.4960.1217

Saveant J-M, Tard C (2016) Attempts to catalyze the electrochemical CO_2 -to-methanol conversion by biomimetic $2e^- + 2H^+$ transferring molecules. J Am Chem Soc 138(3):1017–1021. https://doi.org/10.1021/jacs.5b05992

Schlager S, Dumitru LM, Haberbauer M, Fuchsbauer A, Neugebauer H, Hiemetsberger D, Wagner A, Portenkirchner E, Sariciftci NS (2016) Electrochemical reduction of carbon dioxide to methanol by direct injection of electrons into immobilized enzymes on a modified electrode. ChemSusChem 9(6):631–635. https://doi.org/10.1002/cssc.201501496

Schneider SH (1989) The greenhouse effect: science and policy. Science 243(4892):771–781. https://doi.org/10.1126/science.243.4892.771

Schouten KJP, Kwon Y, van der Ham CJM, Qin Z, Koper MTM (2011) A new mechanism for the selectivity to C_1 and C_2 species in the electrochemical reduction of carbon dioxide on copper electrodes. Chem Sci 2(10):1902. https://doi.org/10.1039/C1SC00277E

Schröder D, Schalley CA, Harvey JN, Schwarz H (1999) On the formation of the carbon dioxide anion radical $CO_2^- \cdot$ in the gas phase. Int J Mass Spectrom 185–187:25–35. https://doi.org/10.1016/S1387-3806(98)14042-3

Sen S, Liu D, Palmore GTR (2014) Electrochemical reduction of CO_2 at copper nanofoams. ACS Catal 4(9):3091–3095. https://doi.org/10.1021/cs500522g

Senthil Kumar R, Senthil Kumar S, Anbu Kulandainathan M (2012) Highly selective electrochemical reduction of carbon dioxide using Cu based metal organic framework as an electrocatalyst. Electrochem Commun 25:70–73. https://doi.org/10.1016/j.elecom.2012.09.018

Seshadri G, Lin C, Bocarsly A (1994) A new homogeneous electrocatalyst for the reduction of carbon dioxide to methanol at low overpotential. J Electroanal Chem 372(1–2):145–150. https://doi.org/10.1016/0022-0728(94)03300-5

Shaughnessy CI, Jantz DT, Leonard KC (2017) Selective electrochemical CO_2 reduction to CO using in situ reduced in_2o_3 nanocatalysts. J Mater Chem A 5(43):22743–22749. https://doi.org/10.1039/C7TA06570A

Shen J, Kortlever R, Kas R, Birdja YY, Diaz-Morales O, Kwon Y, Ledezma-Yanez I, Schouten KJP, Mul G, Koper MTM (2015) Electrocatalytic reduction of carbon dioxide to carbon monoxide and methane at an immobilized cobalt protoporphyrin. Nat Commun 6(1):8177. https://doi.org/10.1038/ncomms9177

Sheng H, Oh MH, Osowiecki WT, Kim W, Alivisatos AP, Frei H (2018) Carbon dioxide dimer radical anion as surface intermediate of photoinduced CO_2 reduction at aqueous Cu and CdSe nanoparticle catalysts by rapid-scan FTIR spectroscopy. J Am Chem Soc 140(12):4363–4371. https://doi.org/10.1021/jacs.8b00271

Strmcnik D, Uchimura M, Wang C, Subbaraman R, Danilovic N, van der Vliet D, Paulikas AP, Stamenkovic VR, Markovic NM (2013) Improving the hydrogen oxidation reaction rate by promotion of hydroxyl adsorption. Nat Chem 5(4):300–306. https://doi.org/10.1038/nchem.1574

Studt F, Sharafutdinov I, Abild-Pedersen F, Elkjær CF, Hummelshøj JS, Dahl S, Chorkendorff I, Nørskov JK (2014) Discovery of a Ni-Ga catalyst for carbon dioxide reduction to methanol. Nat Chem 6(4):320–324. https://doi.org/10.1038/nchem.1873

Summers DP, Leach S, Frese KW (1986) The electrochemical reduction of aqueous carbon dioxide to methanol at molybdenum electrodes with low overpotentials. J Electroanal Chem Interfacial Electrochem 205(1–2):219–232. https://doi.org/10.1016/0022-0728(86)90233-0

Sun Q, Jiang Y, Jiang Z, Zhang L, Sun X, Li J (2009) Green and efficient conversion of CO_2 to methanol by biomimetic coimmobilization of three dehydrogenases in protamine-templated titania. Ind Eng Chem Res 48(9):4210–4215. https://doi.org/10.1021/ie801931j

Sun L, Ramesha GK, Kamat PV, Brennecke JF (2014) Switching the reaction course of electrochemical CO_2 reduction with ionic liquids. Langmuir 30(21):6302–6308. https://doi.org/10.1021/la5009076

Torelli DA, Francis SA, Crompton JC, Javier A, Thompson JR, Brunschwig BS, Soriaga MP, Lewis NS (2016) Nickel–gallium-catalyzed electrochemical reduction of CO_2 to highly reduced

products at low overpotentials. ACS Catal 6(3):2100–2104. https://doi.org/10.1021/acscatal.5b02888
Varela AS, Ju W, Reier T, Strasser P (2016) Tuning the catalytic activity and selectivity of Cu for CO_2 electroreduction in the presence of halides. ACS Catal 6(4):2136–2144. https://doi.org/10.1021/acscatal.5b02550
Verma S, Hamasaki Y, Kim C, Huang W, Lu S, Jhong H-RM, Gewirth AA, Fujigaya T, Nakashima N, Kenis PJA (2018) Insights into the low overpotential electroreduction of CO_2 to CO on a supported gold catalyst in an alkaline flow electrolyzer. ACS Energy Lett 3 (1):193–198. https://doi.org/10.1021/acsenergylett.7b01096
von der Assen N, Müller LJ, Steingrube A, Voll P, Bardow A (2016) Selecting CO_2 sources for CO_2 utilization by environmental-merit-order curves. Environ Sci Technol 50(3):1093–1101. https://doi.org/10.1021/acs.est.5b03474
Wang S-G, Liao X-Y, Cao D-B, Huo C-F, Li Y-W, Wang J, Jiao H (2007) Factors controlling the interaction of CO_2 with transition metal surfaces. J Phys Chem C 111(45):16934–16940. https://doi.org/10.1021/jp074570y
Wang Z-L, Li C, Yamauchi Y (2016) Nanostructured nonprecious metal catalysts for electrochemical reduction of carbon dioxide. Nano Today 11(3):373–391. https://doi.org/10.1016/j.nantod.2016.05.007
Wang L, Jia Y, Nie R, Zhang Y, Chen F, Zhu Z, Wang J, Jing H (2017) Ni-foam-supported and amine-functionalized TiO_2 photocathode improved photoelectrocatalytic reduction of co_2 to methanol. J Catal 349:1–7. https://doi.org/10.1016/j.jcat.2017.01.013
Watanabe M, Shibata M, Kato A, Azuma M, Sakata T (1991) Design of alloy electrocatalysts for CO_2 reduction. III . The selective and reversible reduction of CO_2 on Cu alloy electrodes. J Electrochem Soc 138(11):3382. https://doi.org/10.1149/1.2085417
Wen P, Gong P, Sun J, Wang J, Yang S (2015) Design and synthesis of ni-mof/cnt composites and RGO/carbon nitride composites for an asymmetric supercapacitor with high energy and power density. J Mater Chem A 3(26):13874–13883. https://doi.org/10.1039/C5TA02461G
Whipple DT, Kenis PJA (2010) Prospects of CO_2 utilization via direct heterogeneous electrochemical reduction. J Phys Chem Lett 1(24):3451–3458. https://doi.org/10.1021/jz1012627
Wong NE, Ramaswamy P, Lee AS, Gelfand BS, Bladek KJ, Taylor JM, Spasyuk DM, Shimizu GKH (2017) Tuning intrinsic and extrinsic proton conduction in metal–organic frameworks by the lanthanide contraction. J Am Chem Soc 139(41):14676–14683. https://doi.org/10.1021/jacs.7b07987
Woodwell GM, Hobbie JE, Houghton RA, Melillo JM, Moore B, Peterson BJ, Shaver GR (1983) Global deforestation: contribution to atmospheric carbon dioxide. Science 222 (4628):1081–1086. https://doi.org/10.1126/science.222.4628.1081
Wu C-D, Hu A, Zhang L, Lin W (2005) A homochiral porous metal−organic framework for highly enantioselective heterogeneous asymmetric catalysis. J Am Chem Soc 127(25):8940–8941. https://doi.org/10.1021/ja052431t
Wu Y, Jiang J, Weng Z, Wang M, Broere DLJ, Zhong Y, Brudvig GW, Feng Z, Wang H (2017) Electroreduction of co_2 catalyzed by a heterogenized Zn–porphyrin complex with a redox-innocent metal center. ACS Cent Sci 3(8):847–852. https://doi.org/10.1021/acscentsci.7b00160
Xiao Z-Z, Han L-J, Wang Z-J, Zheng H-G (2018) Three zn(ii)-based metal-organic frameworks for luminescence sensing of Fe^{3+} and $Cr_2O_7^{2-}$ ions. Dalton Trans 47(10):3298–3302. https://doi.org/10.1039/C7DT04659F
Yaghi OM, Li G, Li H (1995) Selective binding and removal of guests in a microporous metal–organic framework. Nature 378(6558):703–706. https://doi.org/10.1038/378703a0
Yao C, Li J, Gao W, Jiang Q (2018) An integrated design with new metal functionalized covalent organic frameworks for effective electro-reduction of CO_2. Chem – A Eur J. https://doi.org/10.1002/chem.201800363
Yazaydın AO, Snurr RQ, Park T-H, Koh K, Liu J, LeVan MD, Benin AI, Jakubczak P, Lanuza M, Galloway DB et al (2009) Screening of metal−organic frameworks for carbon dioxide capture

from flue gas using a combined experimental and modeling approach. J Am Chem Soc 131 (51):18198–18199. https://doi.org/10.1021/ja9057234
Yuan L, Xu Y-J (2015) Photocatalytic conversion of co_2 into value-added and renewable fuels. Appl Surf Sci 342:154–167. https://doi.org/10.1016/j.apsusc.2015.03.050
Zhang S, Kang P, Ubnoske S, Brennaman MK, Song N, House RL, Glass JT, Meyer TJ (2014a) Polyethylenimine-enhanced electrocatalytic reduction of CO_2 to formate at nitrogen-doped carbon nanomaterials. J Am Chem Soc 136(22):7845–7848. https://doi.org/10.1021/ja5031529
Zhang Y-J, Sethuraman V, Michalsky R, Peterson AA (2014b) Competition between CO_2 reduction and h_2 evolution on transition-metal electrocatalysts. ACS Catal 4(10):3742–3748. https://doi.org/10.1021/cs5012298
Zhang S, Kang P, Meyer TJ (2014c) Nanostructured tin catalysts for selective electrochemical reduction of carbon dioxide to formate. J Am Chem Soc 136(5):1734–1737. https://doi.org/10.1021/ja4113885
Zhang N, Zhang J-Y, Jia Q-X, Deng W, Gao E-Q (2015) Solvent-controlled structural diversity observed in three Cu(ii) metal-organic frameworks with a 2,2′-dinitro-biphenyl-4,4′-dicarboxylate ligand: synthesis, structures and magnetism. RSC Adv 5(87):70772–70780. https://doi.org/10.1039/C5RA10459A
Zhao K, Liu Y, Quan X, Chen S, Yu H (2017) CO_2 electroreduction at low overpotential on oxide-derived cu/carbons fabricated from metal organic framework. ACS Appl Mater Interfaces 9 (6):5302–5311. https://doi.org/10.1021/acsami.6b15402
Zheng W, Man HW, Ye L, Tsang SCE (2017) Electroreduction of carbon dioxide to formic acid and methanol over a palladium/polyaniline catalyst in acidic solution: a study of the palladium size effect. Energy Technol 5(6):937–944. https://doi.org/10.1002/ente.201600659
Zhou H-C"J", Kitagawa S (2014) Metal–organic frameworks (metal-organic frameworks). Chem Soc Rev 43(16):5415–5418. https://doi.org/10.1039/C4CS90059F
Zhou H-C, Long JR, Yaghi OM (2012) Introduction to metal–organic frameworks. Chem Rev 112 (2):673–674. https://doi.org/10.1021/cr300014x
Zhu W, Michalsky R, Metin Ö, Lv H, Guo S, Wright CJ, Sun X, Peterson AA, Sun S (2013) Monodisperse Au nanoparticles for selective electrocatalytic reduction of CO_2 to CO. J Am Chem Soc 135(45):16833–16836. https://doi.org/10.1021/ja409445p

Chapter 6
Photocatalytic Conversion of Carbon Dioxide into Hydrocarbons

Pramila Murugesan, Sheeba Narayanan, and Matheswaran Manickam

Contents

6.1 Introduction 135
6.2 General Principles of Artificial Photocatalysis 139
  6.2.1 Thermodynamic Theory of Carbon Dioxide Photoreduction 142
  6.2.2 General Criterion of Carbon Dioxide Photoconversion Systems 144
6.3 Photocatalytic Material for Carbon Dioxide Photoreduction 145
  6.3.1 Metal Oxide Photocatalyst for Carbon Dioxide Reduction 146
  6.3.2 Layered Double Hydroxide 146
  6.3.3 Metal Chalcogenides 147
  6.3.4 Carbon-Based Two-Dimensional Layered Material 148
6.4 Surface Modification of Photocatalyst for Carbon Dioxide Reduction 148
  6.4.1 Metal and Non-metal Doping Semiconductor 149
  6.4.2 Surface Sensitization of Semiconductor 150
  6.4.3 Hybridization with Another Semiconductor Material 151
6.5 Effect Operating Parameters on Carbon Dioxide Reduction 152
  6.5.1 Reaction Medium 152
  6.5.2 pH 153
  6.5.3 Wavelength and Light Intensity 153
  6.5.4 Amount of Catalyst 154
  6.5.5 Particle Size 154
  6.5.6 Pressure 154
  6.5.7 Temperature 155
6.6 Photoreactors for Carbon Dioxide Photoconversion 155
  6.6.1 Fluidized Bed Reactor 155
  6.6.2 Fixed Bed Reactor 156
6.7 Conclusions 157
References 158

P. Murugesan · S. Narayanan (✉) · M. Manickam
Department of Chemical Engineering, National Institute of Technology, Tiruchirappalli, Tamil Nadu, India
e-mail: sheeba@nitt.edu

Inamuddin et al. (eds.), *Conversion of Carbon Dioxide into Hydrocarbons Vol. 1 Catalysis*, Environmental Chemistry for a Sustainable World 40,
https://doi.org/10.1007/978-3-030-28622-4_6

Abstract The significant upward rise in atmospheric CO_2 level through combustion of fossil fuel, deforestation, and human activates could contribute to severe environmental issues like greenhouse effect and climate change. Conversion of CO_2 into chemical fuels or hydrocarbons is an effective way for mitigating the level of CO_2 in the atmosphere, producing value-added products and improving energy security in the light of depletion of carbon-based energy sources. Besides, carbon dioxide is an attractive starting precursor material for producing chemical fuel, owing to its abundance, low cost, and low toxicity. Due to its stability, extra energy is favored to transform CO_2 into hydrocarbon or chemical fuel. Various techniques have been employed for CO_2 conversion which includes chemical, thermal, biological, electrocatalytic, and photocatalytic conversion. Among such methods, the application of photocatalysis in carbon dioxide conversion plays a significant role to resolve energy crisis and global warming. The process employs light-driven photocatalytic transformation of carbon dioxide to value-added chemical fuels including methane, carbon monoxide, formic acid, formaldehyde, methanol, and ethanol.

The overall conversion efficacy can be enhanced through the fabrication of more efficient visible light active photocatalyst and suitable configuration of photoreactor. Numerous reports have been devoted to the fabrication of semiconductor-based photocatalysts for carbon dioxide conversion process. However, inadequate absorption of visible light and rapid photogenerated charge carrier recombination rate of single semiconductor-based photocatalyst will influence the CO_2 photoconversion efficiency. In order to improve the performance of photocatalytic material, different augmentation strategies are available which include doping with metal/non-metal sensitization and hybridization of photocatalysts. In this chapter, the fundamental aspect of the heterogeneous photocatalytic carbon dioxide conversion using various semiconductor-based photocatalysts is summarized. Moreover, different surface modification routes adapted in photocatalytic materials have been presented in detail. Additionally, the influence of various experimental parameters and different types of photoreactors for carbon dioxide photoconversion is described with applications.

Keywords Global warming · Carbon dioxide · Photocatalysis · Semiconductor · Reduction · Charge carrier · Doping · Sensitization · Hybridization · Photoreactor

Abbreviations

IEA	International Energy Agency
LDHs	Layered double hydroxides
RGO	Reduced graphene oxide

6.1 Introduction

The reckless utilization of nonrenewable fossil fuels has accelerated the atmospheric carbon dioxide concentration level in the recent years. Based on the combustion stoichiometry, 3.5 tons of carbon dioxide is released from the combustion of 1 ton of carbon in fossil fuels (Alper and Yuksel Orhan 2017). Burning fossil fuels and excessive carbon dioxide emission into the atmosphere cause energy crisis and global warming issues, respectively (Yu et al. 2014; Wang et al. 2016a, b; He et al. 2015). Figure 6.1 illustrates the emission level of carbon dioxide in different sectors. Currently, more than 80% of the energy consumed is from fossil fuels including coal, oil, and natural gas and their combustion processes (Ioelovich 2015). It is estimated that the atmospheric carbon dioxide concentration increase could result in the rise of global temperature by 1.9 °C (Li et al. 2014a, b). At present, the level of carbon dioxide in atmosphere has reached about 406 ppm (Herzog 2011). According to the IEA report, global energy associated with carbon dioxide emissions grew by 1.4% in 2017. The significant upward thrust in carbon dioxide level may create serious environmental issues like climate change, increment in earth temperature, sea level rise, environmental pollution, and destructive effects on crop cultivation and human health. So, regulating the atmospheric carbon dioxide concentration requires utmost attention in the twenty-first century. It is a vital task to

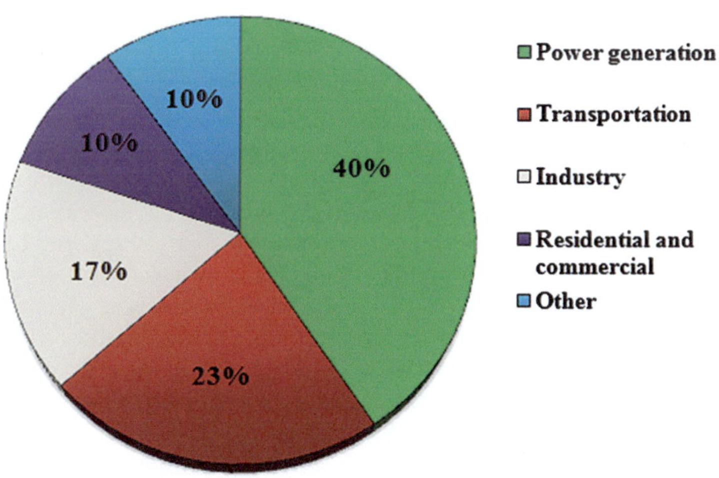

Fig. 6.1 Emission level of carbon dioxide representing various energy sectors. (Reprinted from Kiesgen et al. 2013)

reduce the usage of fossil fuels, thereby reducing the environmental burden. Carbon dioxide emission level is controlled through three different approaches such as:

- Direct controlling of carbon dioxide emission through the efficient use of fossil fuels
- Carbon dioxide capture and storage technology
- Utilization of carbon dioxide to chemical fuels

Although the increase in utilization efficiency of carbon-based energy source or carbon-free energy sources would dramatically lower the carbon dioxide emission, there is a steep rise in the emissions owing to the growing population. Carbon dioxide capture and storage technology are significant for controlling the carbon dioxide emission quickly, but the danger of potential leakage of carbon dioxide persists, and the energy required for compression of gas, transportation, and separation is huge. To date, promising attention has focused research activity for carbon dioxide photoconversion into value-added chemicals because carbon dioxide utilization technique can reduce the carbon dioxide level in atmosphere and generate value-added chemical fuels, thereby resolving the carbon dioxide storage issues and controlling the risk of global warming (Khoo et al. 2011; Dindi et al. 2019). Strategy to overcome the issues associated with environmental pollution and energy crisis, carbon dioxide transformation into value-added chemical fuels by sunlight seems to be the most substantial solutions among all the conversion technologies (Salehi-Khojin et al. 2013).

Many conversion methods such as catalytic hydrogenation, mineralization, electrochemical, thermochemical, plasma, and enzymatic conversion have been adapted for converting carbon dioxide into chemicals (Benson et al. 2009; Huang et al. 2015; Rezaei et al. 2015; Tian et al. 2015; Kothandaraman et al. 2016; Noji et al. 2017; Zhou et al. 2017) and are as listed in Table 6.1. Besides, a huge quantity of input energy is needed for photoconversion of carbon dioxide, because carbon dioxide is basically more stable with a centrosymmetric structure, which needs about 750 $kJmol^{-1}$ to cleave the C=O bond. Also, demand on the development of sustainable renewable energy resources (wind energy, solar energy, and hydroelectric power) to produce energy and mitigate the carbon dioxide concentration (Owusu and Asumadu-Sarkodie 2016) is high. Solar energy is the most plentiful energy source existing in the globe. Approximately, 1.3×10^5 terawatt solar energy is incident on the earth surface (Li et al. 2016). The solar energy for carbon dioxide conversion process can be utilized in three ways. The first one is the direct conversion by photovoltaic cell into electrical energy subsequently used for electrochemical carbon dioxide conversion on the surface of metal electrodes. The second way is the photoelectrochemical cell in which electrical energy generation and carbon dioxide conversion are combined into a single device (not separated), although external electrical potential bias facilitates the catalytic reduction process. The third way is the utilization of solar radiation to convert into chemical fuels (generation of hydrogen from water splitting reaction and hydrocarbon fuels from carbon dioxide photoconversion process) by photocatalysis process (Kalamaras et al. 2018; Liu et al. 2016; Gurudayal et al. 2017). Among these, carbon dioxide

Table 6.1 Process description, products, and comparative analysis of various methods for reduction of carbon dioxide

Routes	Process description	Major products	Advantages	Disadvantages
Hydrogenation process	The hydrogenation process reduces carbon dioxide by hydrogen or hydrocarbons. Methane is the main reactant for this process	Carbon monoxide and methanol	High efficiency	Cost of hydrogen
Mineralization process	Mineralization of carbon dioxide to chemicals $Mg + CO_2 \rightarrow MgO + CO$	Magnesium oxide, calcium oxide, magnesium carbonate, sodium bicarbonate, and calcium carbonate	Environmental friendly and thermal stability of products (carbonates) are high	Low carbonization rate and cost of energy
Electrochemical	Reduction of carbon dioxide is achieved via externally applying an electrical bias $CO_2 + Xe^- + XH^+ \xrightarrow{eV} CO, HCOOH, (COOH)_2$	Carbon monoxide, methane, methanol, formic acid, formaldehyde, and ethanol	Direct conversion of carbon dioxide into liquid long-chain liquid fuels (long-chain molecule)	Cost of electricity, solubility, and deactivation of electrodes
Thermochemical	Dissociation of carbon dioxide on a catalyst at high temperature $CO_2 \xrightarrow{\text{Metal oxide}, T>900^\circ C} CO + \frac{1}{2}O_2$	Carbon monoxide	High efficiency	Cost of energy (high-temperature process)
Enzymatic conversion	Carbon dioxide conversion occurs using microbes and algae via photosynthetically (Calvin cycle) or non-photosynthetically $CO_2 + 4H_2 \xrightarrow{\text{Bacteria}} CH_4 + 2H_2O$	Carbon monoxide, methane, methanol, formic acid, formaldehyde acetic acid, and sugar	Environmental friendly and high efficiency	Complicated and cumbersome biological processes
Plasma-chemical	Dissociation of carbon dioxide using plasma (ionized gas)	Carbon monoxide, methane, methanol, ethanol, and syngas	High efficiency	Initial investment cost and system dependency (difficult to maintain the

(continued)

Table 6.1 (continued)

Routes	Process description	Major products	Advantages	Disadvantages
				operation process parameters)
Photochemical conversion	Conversion of carbon dioxide using photon energy with the presence of photocatalyst $CO_2 \overset{h\nu \geq E_g}{\rightarrow} CH_3OH$	Carbon monoxide, methane, methanol, formic acid, formaldehyde, and ethanol	Environmental friendly and high efficiency	Stability and limited utilization of visible light

Note: Mg magnesium, CO_2 carbon dioxide, MgO magnesium oxide, CO carbon monoxide, HCOOH formic acid, $(COOH)_2$ oxalic acid, O_2 oxygen, H_2 hydrogen, CH_4 methane, H_2O water, CH_3OH methanol, T temperature, Xe^- number of electrons, XH^+ number of protons

photoconversion is an economically feasible and environmentally safe process, and it could be an effective supplement of other conventional technologies. Carbon dioxide photoconversion is a blend of both photophysical and chemical processes. At first, Inoue et al. (1979) conducted experiments on photoelectrocatalytic transformation of carbon dioxide into methane, carbon monoxide, formic acid, and methanol using various photocatalysts under the exposure of xenon and mercury light irradiation.

Generally, the total emitted solar light consists of 4% (<400 nm) of ultraviolet light and 43% (400–700 nm) and 52% (>700 nm) of visible light (Nikokavoura and Trapalis 2017). So the major goal for carbon dioxide photoconversion is the fabrication of visible light active photocatalyst with high enough conduction band position. In order to improve carbon dioxide photoconversion performance, approaches concerning light harvesting and separation of photogenerated charge carrier have been reported. Instead, photocatalytic conversion of carbon dioxide is a relatively challenging task since the process only requires reduction of carbon, proton transfer, and hydrogenation. Besides, most of the research work is focused on understanding the mechanism connecting carbon dioxide adsorption, carbon dioxide activation, product desorption, and possible reaction mechanism. This chapter will give a brief outline of the basic principle of photocatalysis of carbon dioxide and explore potential photocatalysts for the process. Finally, the influence of different operating parameters and type of photoreactors adapted in carbon dioxide photoconversion process are concisely outlined.

6.2 General Principles of Artificial Photocatalysis

Photocatalysis is the initiation of a photoreaction in the existence of an irradiated catalyst. The schematic representation of the basic photocatalytic reaction mechanism is presented in Fig. 6.2. When exposed to light, the photocatalytic material captures the photon and produces electron and hole pairs (charge carriers), whereas the light energy should be more or equal to the energy band gap of semiconductor material ($h\nu$ more than or equal to E_g).

$$\text{Photocatalyst} \xrightarrow{h\nu \text{ more than or equal to } E_g} e^- + h^+$$

where Eg, h, v, h^+, and e^- are the energy band gap, Planck constant ($h = 6.63.10^{-34}$ Js^{-1}), light frequency, valence band holes, and conduction band electron, respectively. Then, the excited electrons move from valence band to conduction band energy level. These excited species may undergo the following reactions including:

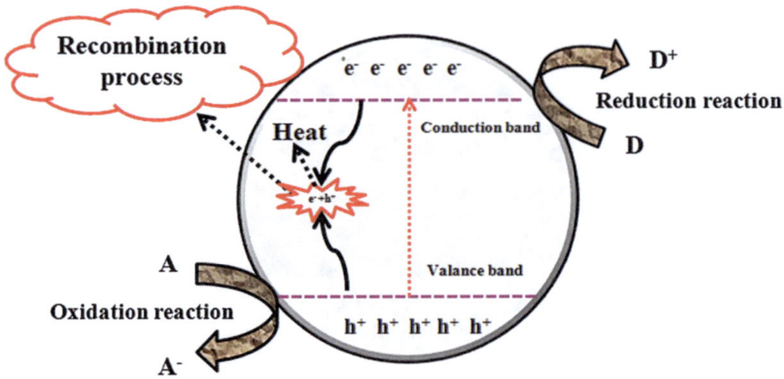

Fig. 6.2 Oxidation, reduction, and recombination process on photocatalytic semiconductor surface after absorption of solar light. Note: A,D reactant, A^- oxidized product, D^+ reduced product, e^- electrons, and h^+ holes

Recombination Process Recombination process generates only heat or light with no photocatalytic activity as represented below:

$$e^- + h^+ \rightarrow \text{heat}$$

Redox Reactions

(i) Oxidation reaction: photogenerated holes react with electron donor molecules adsorbed on the surface of photocatalytic material and generate oxidative products. For example, water oxidation:

$$2H_2O \xrightarrow{h\vartheta} 2H_2 + O_2$$

(ii) Reduction reaction: photogenerated electrons react with electron acceptor molecules present at the surface and produce reduced products. For example, carbon dioxide photoconversion:

$$H_2O + CO_2 \xrightarrow{h\vartheta} \text{Fuel} + O_2$$

Photocatalytic transformation of carbon dioxide to renewable chemical fuels mimics the natural photosynthesis reaction and is named as artificial photosynthesis. In the natural photosynthesis process, chlorophyll absorbs the photon energy and converts carbon dioxide to carbohydrates and oxygen with the assistance of a series of enzyme-mediated redox process steps. In artificial photosynthesis process, the

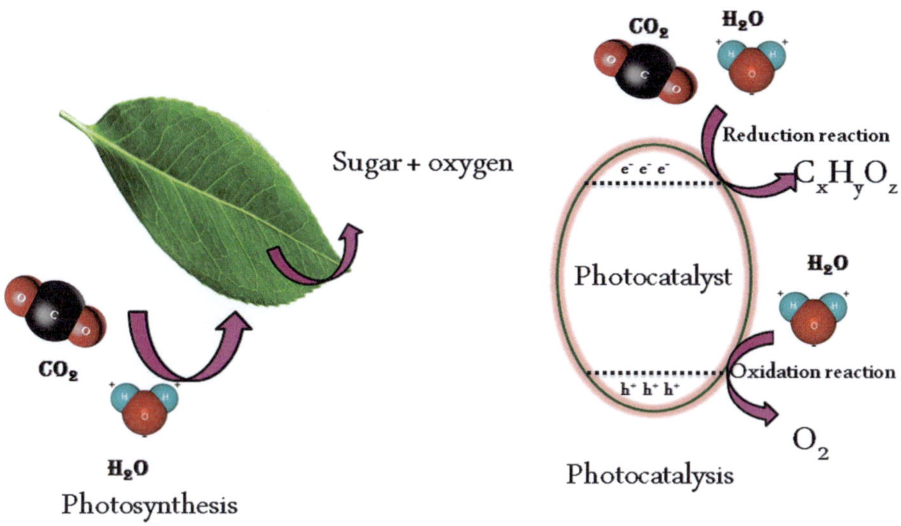

Fig. 6.3 Comparison of the natural photosynthesis of CO_2 to sugar in a plant and artificial photosynthesis of CO_2 to hydrocarbons over a photocatalyst

photocatalyst is used to convert carbon dioxide into other chemicals including carbon monoxide, methane, formaldehyde, formic acid, and ethanol. However, both photocatalysis and photosynthesis processes can utilize solar energy (Fig. 6.3). Generally, the photocatalytic efficiency depends on the intrinsic physico-chemical traits of photocatalyst (Dong et al. 2013). Two important species that participate in the carbon dioxide photoconversion are hydrogen radical and carbon dioxide anion radical. The photocatalytic reaction of carbon dioxide to carbonaceous fuel is as shown below:

$$CO_2 + H_2O \text{ or } H_2 \overset{h\vartheta,\ \text{Photocatalyst}}{\longrightarrow} \text{Carbonaceous fuel} + O_2$$

In particular, the carbon dioxide reaction products are influenced by physico-chemical properties of the photocatalysts and reaction conditions (Chang et al. 2016; Zhang et al. 2018). Some of the crucial factors that limit the overall effectiveness are as shown in Table 6.2.

The major advantages of carbon dioxide photoconversion process are as follows:

1. This process can be induced by solar energy.
2. Carbon dioxide is easily obtainable as a key carbon source.
3. The process can be conducted at room temperature and pressure.
4. The production of chemical fuels from carbon dioxide can efficiently compete with energy crisis issue.

Table 6.2 Factors influencing photoconversion of carbon dioxide and various strategies for improvement of photoconversion process

Factors	Augmentation strategies
Charge excitation	Increased visible light excitation such as doping with metal and non-metal, sensitization, and semiconductor heterojunctions
Charge transport	Improved charge transfer and separation such as nanostructured photocatalysts, semiconductor heterojunctions, and semiconductor/nano-carbon heterojunctions
Adsorption and activation of CO_2	Enhanced adsorption and activation of CO_2 such as increased surface area and basic sites for CO_2 adsorption
CO_2 reduction kinetics	Accelerated CO_2 reduction kinetics such as developing mesoporous photocatalysts, loading CO_2 reduction CO-catalysts, and improving water oxidation kinetics
Undesirable reaction	Suppressed undesirable reaction like inhibited H_2 and product oxidation

Reprinted from Li et al. (2014b)

6.2.1 Thermodynamic Theory of Carbon Dioxide Photoreduction

Carbon dioxide molecules are nonreactive and optically inert with a linear molecule and closed-shell electronic configuration (Wang et al. 2016a, b; Zhou et al. 2017). Also, they have a relatively high thermodynamic stability, because of its greater bond energy. The bond energies of C=O, C-H, C-O, and C-C are 750 kJ mol^{-1}, 411 kJ mol^{-1}, 327 kJ mol^{-1}, and 336 kJ mol^{-1}, respectively, and thus sufficient input energy from the photon is needed to cleave C=O bond into C-H bond. Additionally, more negative electrochemical potential of -1.9 eV is required for the generation of carbon dioxide anion from carbon dioxide (Li et al. 2015). The multi-electron reduction steps assisted by proton are much more promising while dealing with relatively lower redox (conduction band) potential, which is already practiced in various carbon dioxide photoconversion systems. The formation of different products takes place at a precise chemical potential against the normal hydrogen electrode as shown in Table 6.3. The product formation is governed by the number of electrons shifted between the conduction band, reaction pathway, and the adsorbed species in the aqueous and nonaqueous phase systems.

Still, the reaction pathway for carbon dioxide photoconversion is not completely understood. However, three possible reaction pathways for carbon dioxide photoconversion on semiconductor material have been proposed, namely, (a) carbene pathway, (b) formaldehyde pathway, and (c) glyoxal pathway (Dimitrijevic et al. 2011). In the case of carbine pathway, the adsorbed carbon dioxide on the surface of the catalyst surface is combined with anion radical to form chemical fuels.

Table 6.3 Redox potentials of various species against normal hydrogen electrode for carbon dioxide conversion into chemical fuels (Hong et al. 2013)

Chemical equation	Chemical potential (V vs. NHE)	Electrons consumed
$2H_2O + 4h^+ \rightarrow O_2 + 4H^+$	1.23	(Water oxidation)
$2H^+ + 2e^- \rightarrow H_2$	−0.82	$2e^-$
$CO_2 + e^- \rightarrow CO_2^-$	−1.49	e^-
$CO_2 + 2h^+ + 2e^- \rightarrow HCOOH$	−0.58	$2e^-$
$CO_2 + 2h^+ + 2e^- \rightarrow CO + H_2O$	−0.52	$2e^-$
$CO_2 + 4h^+ + 4e^- \rightarrow HCHO + H_2O$	−0.48	$4e^-$
$CO_2 + 6h^+ + 6e^- \rightarrow CH_3OH + H_2O$	−0.38	$6e^-$
$CO_2 + 8h^+ + 8e^- \rightarrow CH_4 + H_2O$	−0.24	$8e^-$
$2CO_2 + 8h^+ + 8e^- \rightarrow CH_3COOH + 2H_2O$	−0.31	$8e^-$
$2CO_2 + 12h^+ + 12e^- \rightarrow C_2H_5OH + 3H_2O$	−0.39	$12e^-$
$2CO_2 + 14h^+ + 14e^- \rightarrow C_2H_6 + 4H_2O$	−0.27	$14e^-$
$3CO_2 + 16h^+ + 16e^- \rightarrow CH_3COCH_3 + 5H_2O$	−0.31	$16e^-$
$3CO_2 + 16h^+ + 16e^- \rightarrow CH_3CH_2CHO + 5H_2O$	−0.32	$16e^-$
$3CO_2 + 18h^+ + 18e^- \rightarrow CH_3CH_2CH_2OH + 5H_2O$	−0.31	$18e^-$

Note: H_2O water, h^+ holes, O_2 oxygen, H_2 hydrogen, e- electron, CO_2 carbon dioxide, CO_2-carbon dioxide anion, HCOOH formic acid, HCHO formaldehyde, CO carbon monoxide, CH_4 methane, CH_3COOH acetic acid, C_2H_5OH ethanol, CH_3COCH_3 acetone, CH_3CH_2CHO propionaldehyde, V voltage, and NHE normal hydrogen electrode

$$CO_2 \xrightarrow{e^-} CO_2^- \rightarrow CO + OH^- \rightarrow .CO^- \rightarrow C^{\cdot} \rightarrow CH^{\cdot} \rightarrow CH_2 \rightarrow CH_3 \rightarrow CH_4 + CH_3OH$$

For formaldehyde pathway, the adsorbed carbon dioxide is reduced into carboxylic acid. Carboxylic acid will further react with hydrogen radical to produce formic acid and formaldehyde. The process continues with the formation of other products such as methanol.

$$CO_2 \xrightarrow{e^-} CO_2^- \rightarrow COOH \rightarrow HCOOH \rightarrow H_2OOC^{\cdot} \rightarrow HCOH \rightarrow H_2OHC^{\cdot} \rightarrow CH_3OH \rightarrow .CH_3 \rightarrow H_2OCH_4$$

In the glyoxal pathway, acetaldehyde is formed by reaction between the anion radical and carbon dioxide followed by dimerization process. Finally, methane is produced. The products from carbon dioxide photoconversion are strongly dependent on the number of formation of electrons and holes during the photocatalysis (Lee et al. 2017).

$$CO_2 \xrightarrow{e^-} CO_2^- \rightarrow CHO \xrightarrow{\text{Dimerization}} (CHO)_2 \rightarrow CH_3CHO \rightarrow .CH_3 \rightarrow CH_4$$

6.2.2 General Criterion of Carbon Dioxide Photoconversion Systems

Product Formation Rate

The efficiency of carbon dioxide photoconversion can be determined by the product rate formation (*R*) at a certain time period with exposure of light source per gram of catalyst. The unit of R can be defined as mol h^{-1} g^{-1}.

$$R = \frac{n\ (\text{product})}{\text{Time} \times m(\text{catalyst})} \tag{6.1}$$

Selectivity Percentage

The selectivity percentage of the catalyst is the molar ratio of the number of individual products from carbon dioxide reduction to the sum of overall products (Pang et al. 2017)

$$\text{Selectivity}\ (\%) = \frac{(n \times \text{Number of individual product}\ (\mu\text{mole}))}{(n \times \text{sum of all products}\ (\mu\text{mole}))} \times 100 \tag{6.2}$$

The selectivity value delivers quantitative data on the carbon dioxide reduction efficiency at precise experimental conditions.

Amount of Carbon Dioxide Converted

The amount of carbon dioxide photoconversion was determined by summing up the C amount in the significant C-containing products (Wang et al. 2016a, b).

$$\text{Amount } of \text{ CO}_2 \text{ conversion} = n \times \text{number of carbon in the products} \tag{6.3}$$

where *n* is the corresponding product yield in the unit of μmole/g of catalyst.

Apparent Quantum Efficiency

The apparent quantum efficiency of carbon dioxide photoconversion was calculated using the following equation (Chen et al. 2017; Tan et al. 2017):

$$\text{Apparent quantum efficiency} = \frac{\text{Number of reacted electrons}}{\text{Number of incident photons}} \times 100 \ (4)$$

$$\textit{Num}\text{ber of reacted electrons} = n \times \text{Sum of number of products } (\mu\text{mole}) \quad (6.5)$$

$$\text{Number of incident photons} = \frac{I \times S \times \lambda \times t}{h \times c \times N_A} \ (6)$$

where I, S, λ, t, h, c, and N_A are the light intensity, irradiation area, wavelength of light source, reaction time, Planck constant, light speed (3×10^8 m s^{-1}), and Avogadro number (6.022×10^{23} mol^{-1}), respectively.

Turnover Number

Turnover number can be determined as the molar ratio between the reduction products from carbon dioxide to the amount of catalyst.

$$TON = \frac{CO_2 \text{ reduction products}}{\text{amount of catalyst}} \quad (6.7)$$

6.3 Photocatalytic Material for Carbon Dioxide Photoreduction

Carbon dioxide photoconversion reaction mainly depends upon the types of the photocatalyst. As compared with conductor and insulator, semiconductor-based photocatalytic materials have been widely examined for photocatalysis process, because of appropriate electronic structure (medium band gap), light absorption property, and charge transport characteristics (Gaya 2014). The conductor (metal) materials display no band gap, because the valence band and conduction band overlap with each other. In the case of an insulator, high energy is required for the excitation process because the valence band is away from the conduction band (wide band gap). So electrical conductivity is not possible in the insulator. Several of single semiconductor-based photocatalytic materials including metal oxides, layered double hydroxides, metal chalcogenides, and carbon-based (non-metal) semiconductors have been employed as photocatalysts for carbon dioxide conversion reaction. An ideal semiconductor-based photocatalyst material should have the following properties (Anandan et al. 2010; Serpone and Emeline 2012; Lee et al. 2017):

(i) Appropriate band gap (1.7–3.1 eV) that can efficiently utilize the visible light. Usually, a low-energy band gap is favored over a larger one.

(ii) Appropriate location of the conduction band and valence band that deliver adequate redox potentials for the specific target oxidation-reduction reaction and an efficient transfer and separation of photogenerated electron-hole pairs.
(iii) The photocatalyst must be long term, photostable, chemically and biologically inert, easy to prepare, and less expensive.

6.3.1 Metal Oxide Photocatalyst for Carbon Dioxide Reduction

Metal oxide photocatalysts have gained attention toward carbon dioxide photoconversion owing to its benefits such as low cost, stable, and easy to handle (Martha et al. 2015; Yaipimai et al. 2015). Titanium oxide is the widely studied photocatalyst for photocatalysis. Pure titanium oxide exhibits in three different phases, such as brookite, anatase, and rutile. Different phases and morphological properties of titanium oxide photocatalysts influence the photoconversion efficiency (Habisreutinger et al. 2013). Liu et al. compared three different phases of titanium oxide for carbon dioxide photoconversion to carbon monoxide and methane (Liu et al. 2012). They found that the highest carbon dioxide conversion efficiency is observed in brookite than others. Additionally, a lot of different metal oxide and mixed metal oxide semiconductors including copper oxide, zinc oxide, tin oxide, cerium oxide, zirconium oxide, gallium oxide, tantalum pentoxide, strontium titanate, copper ferrite, sodium niobate, zinc gallate, and zinc germinate (Shi et al. 2011; Li et al. 2016; Lee et al. 2017) have been considered for carbon dioxide photoconversion. Most of them are active only in the ultraviolet region due to their large band gap.

6.3.2 Layered Double Hydroxide

Layered double hydroxides are natural clays and consist of positively charged divalent and trivalent metals with replaceable intercalated anions within the layers (Bukhtiyarova 2019). Owing to the fascinating properties in biocompatibility, compositional flexibility, and anion exchangeability, layered double hydroxides have attracted attention in several technologically substantial fields including drug storage-delivery agents, catalysis, and environmental applications (Gwak et al. 2016; Jenisha Barnabas et al. 2017; Zubair et al. 2017). At first, demonstrated the structure and properties of layered double hydroxide. Three main routes have been used for synthesizing layered double hydroxide material including coprecipitation route, reconstruction or rehydration route, and ion-exchange route. Recently, layered double hydroxides have been used as photocatalyst for carbon dioxide conversion process (Yang et al. 2019). Layered double hydroxide material absorbs higher

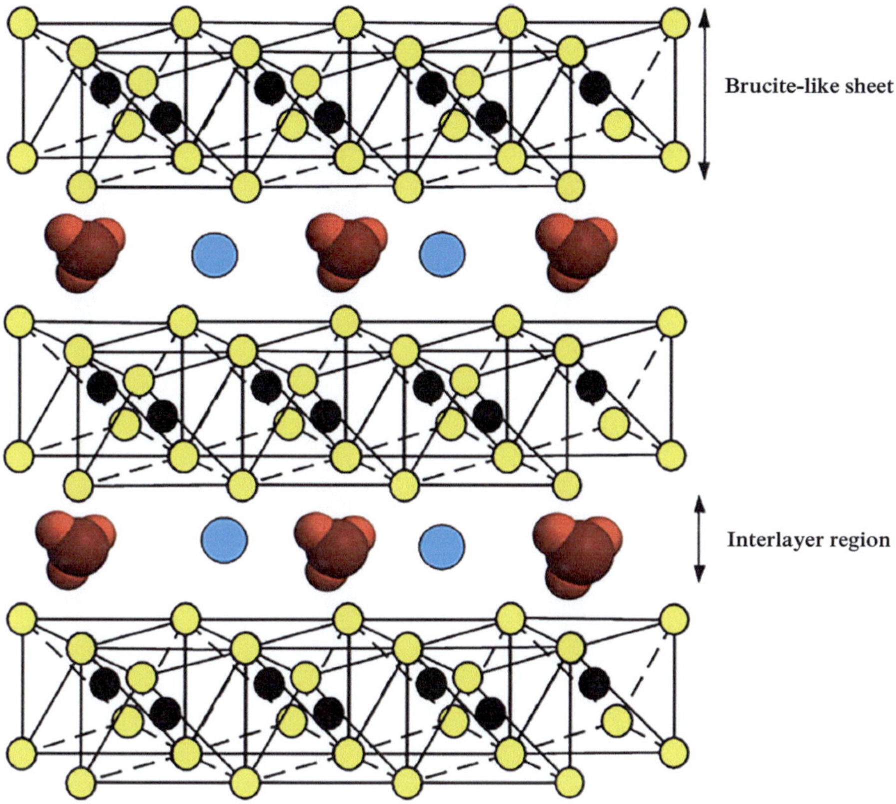

Fig. 6.4 Layered double hydroxide structure depicting brucite-like sheet structure with interlayer adsorption of carbonate ion. (Reprinted from Yang et al. 2019)

amount of carbon dioxide in the interlayer space and can act as a semiconductor by selecting suitable metal ion (Miyano et al. 2017). The adaptability of the layered double hydroxide for altering the composition of the hydroxyl layer, oxo-layer, and interlayer gap to optimize the activity of the solid as the photocatalyst is high (Xie et al. 2018). Additionally, the basicity of layered double hydroxide contributes majorly to the interaction with carbon dioxide (Fig. 6.4).

6.3.3 Metal Chalcogenides

Metal chalcogenides have lesser energy band gaps as compared with metal oxides and possess visible light absorption. A number of metal sulfide compounds have been explored for photocatalytic application, which includes ZnS, In_2S_3, $AgIn_5S_8$, and $ZnIn_2S_4$ (Koca and Musa 2002). For photocatalytic reaction the metal sulfide semiconductors are inappropriate due to their low stability as they readily undergo photocorrosion. Crystalline metal chalcogenides are of remarkable interest due to its ideal properties such as more stability in air and aqueous solution, tunable band gap, less expensive, good conductivity of electron holes, and improved resistance to the oxidation reaction during photocatalysis (Xiao et al. 2017). Due to their distinctive properties, metal chalcogenides have improved carbon dioxide adsorption, and its intermediates show favorable carbon dioxide conversion performance. High-temperature solid-state synthesis, surfactant-thermal method, ionothermal method, and hydro(solvo)thermal method have been commonly used for synthesizing metal chalcogenides. The carbon dioxide conversion efficiency depends on the composition of the chalcogenides (Zhou et al. 2018a).

6.3.4 Carbon-Based Two-Dimensional Layered Material

Carbon is the widely available element on the earth surface and has various applications including supercapacitors, batteries, fuel cells, and photocatalyst (Ping et al. 2015; Wang et al. 2015). Two-dimensional layered carbon-based materials have delocalized π-conjugated electrons that can enhance carbon dioxide adsorption. Afterward, the adsorbed carbon dioxide molecules can be reduced to chemical fuels by conduction band electrons (Low et al. 2016). Ascribing to their two-dimensional structure, graphene and graphitic carbon nitride can be incorporated as supporting material for another photocatalytic material since they can broaden the surface area and provide the surface active sites (Hsu et al. 2013; Sun et al. 2018). It was described that the incorporation of two-dimensional structured carbon-based materials can also expand the photocatalyst distribution, thus enhancing its surface area. The addition of carbon-based two-dimensional layered material can significantly inhibit and obstruct the photocorrosion behavior of low photostable photocatalysts (Nie and Zhang 2017). During the carbon dioxide photoconversion reaction, the two-dimensional structure carbon-based material facilitates the multi-electron reaction.

6.4 Surface Modification of Photocatalyst for Carbon Dioxide Reduction

The electrons from the conduction band of single photocatalytic material can easily interact with valence band holes and release heat, which may reduce the carbon dioxide conversion efficiency. Three different modification methods available to overcome the problems of low conversion efficiency of carbon dioxide for effective photocatalysts with visible light active and efficient photogenerated charge carrier separation are as follows (Table 6.4):

Table 6.4 Modification strategies for performance improvement of the photocatalyst and the commonly used materials for various strategies

Type of strategy	Advantages	Examples
Metal and non-metal doping	Enhance the surface electron excitation	Metals: Ag, Au, Pt, Pd, Cu, and Ni
	Enhance photoactivity	Non-metals: C, P, S, O, F, and B
	Reduce band gap energy	
	Reduce recombination rate of charge carriers	
	Act as sink for charge particles	
Sensitization	Enhance photo activity	Rhodamine blue, cobalt-phthalocyanine (CoP) dye, erythrosin B dye
	Extend the visible light utilization efficiency	
	Higher mobility of charge carriers	
	Hybrid catalyst for high yield	
Hybridization	Enhance photo activity	TiO_2, ZnO, In_2O_3, AgX (X = Br, Cl, and I), CeO_2
	Higher charge separation efficiency	
	Improves stability	

Note: Ag silver, Au gold, Pt platinum, Pd palladium, Cu copper, Ni nickel, C carbon, P phosphorous, S sulfur, O oxygen, F fluorine, B boron, TiO_2 titanium oxide, ZnO zinc oxide, AgX silver halide, Br bromine, Cl chlorine, I iodine, In_2O_3 indium oxide, CeO_2 cerium oxide

- Metal and non-metal doping of semiconductor material
- Surface sensitization of semiconductor material
- Hybridization with another semiconductor material

6.4.1 Metal and Non-metal Doping Semiconductor

Metal (cation) and non-metal (anion) doping technique can provide visible light active wide band gap semiconductor. The schematic view for the metal and non-metal doping is as shown in Fig. 6.5. Metal doping can alter the energy band gap of semiconductor material and broaden the spectral absorption toward the visible region and reduce the recombination rate as well as improve the charge separation efficiency. Generally, transition metal ions including copper, nickel, and strontium and noble metals such as platinum, silver, and gold are commonly employed as dopants to enhance photocatalytic efficiency (Pang et al. 2017; Zhou et al. 2018b). However, the process suffers from drawbacks such as photocorrosion of doped metals, and the doped metal itself would act as electron-hole recombination center.

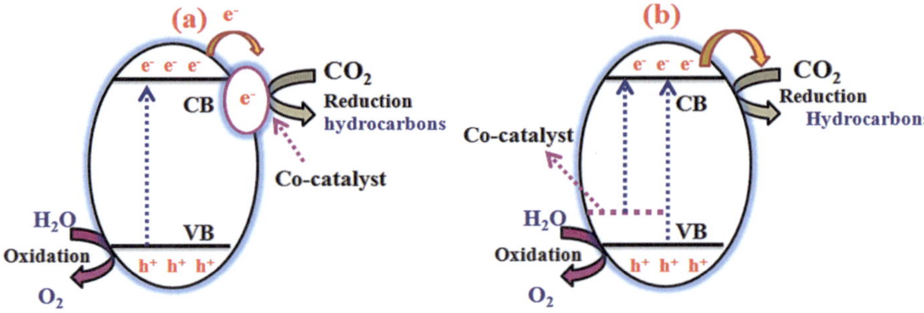

Fig. 6.5 Deposition of dopants on the semiconductor surface and corresponding oxidation and reduction reactions (**a**) metal doping and (**b**) non-metal doping for photocatalysis process. Note: CB conduction band, VB valence band, e^-, electrons, h^+ holes, CO_2 carbon dioxide, H_2O water, and O_2 oxygen

Non-metal dopant can create new valence band and narrow down the valence band level (Yu et al. 2018). Sulfur, carbon, fluorine, boron, phosphorous, and nitrogen are the commonly employed dopants (Li et al. 2014a, b). Due to its fewer recombination centers, non-metal doping efficiently harvests visible light as compared with the metal doping.

6.4.2 *Surface Sensitization of Semiconductor*

The wide band gap semiconductor-based photocatalysts are ultraviolet light-responsive. Sensitization of light-absorbing compounds (narrow band gap semiconductor materials or dye compounds) into a wide band gap semiconductor is a significant method to improve the visible light carbon dioxide conversion efficiency as illustrated in Fig. 6.6. Besides, the process can effectively diminish the photogenerated charge carrier's recombination. Typically, the light-absorbing particle can absorb the visible light and generates electron and hole pairs. Then the generated electrons are moved to conduction band of wide band gap semiconductors. Light-absorbing compounds including $((Ru(Bpy)_3]^{2+}$ dye, cobalt-phthalocyanine (CoPc) dye, erythrosin B dye, magnesium phthalocyanine (MgPc) dye, eosin Y dye, porphyrin, merocyanine, CdS, CdSe, and GaAs are commonly used as sensitizing material for improving the visible light utilization efficiency of semiconductors (Wang et al. 2010; Maeda et al. 2013; Cheng et al. 2014; Lin et al. 2014; Go and IshiTani 2015). The major disadvantages of the sensitization process are solubility of the dye in water or any other solvents, and coupled semiconductor may undergo photocorrosion, which affects the photocatalytic performance.

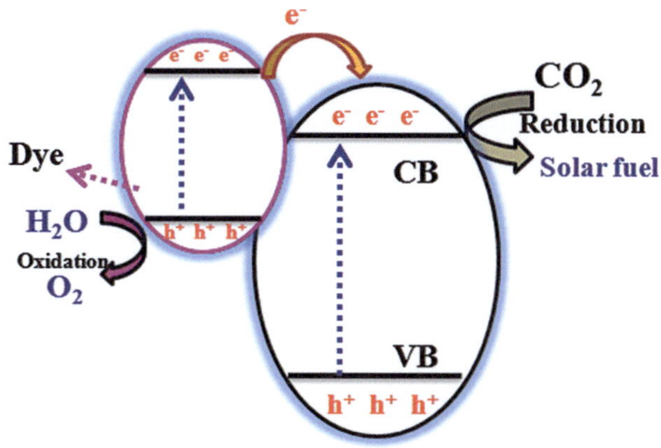

Fig. 6.6 Sensitization of semiconductor on the photocatalyst surface and corresponding oxidation and reduction reactions on the surface. Note: CB conduction band, VB valence band, e^- electrons, h^+ holes, CO_2 carbon dioxide, H_2O water, and O_2 oxygen

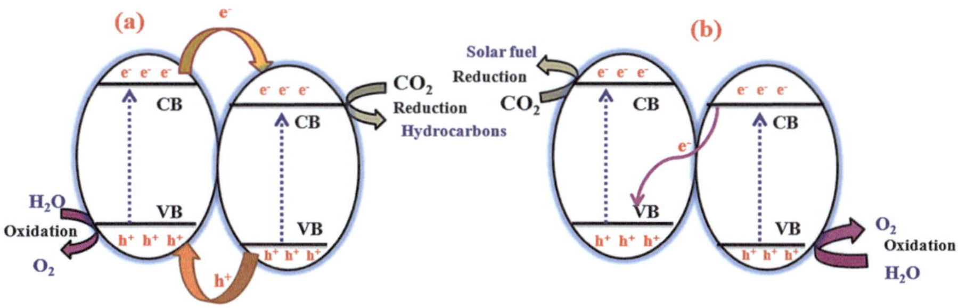

Fig. 6.7 Hybridization of semiconductor on the photocatalyst surface showing (a) heterojunction charge separation system (conventional) and (b) Z-scheme charge carrier separation mechanism. Note: CB conduction band, VB valence band, e^- electrons, h^+ holes, CO_2 carbon dioxide, H_2O water, and O_2 oxygen

6.4.3 Hybridization with Another Semiconductor Material

Coupling with another semiconductor material is an attractive approach for improving photocatalytic performance as schematically illustrated in Fig. 6.7 (Miranda et al. 2013; Jo and Selvam 2017). The key advantages are:

- Extended visible light harvesting efficiency
- Effective photogenerated charge carrier separation by shifting the electrons to the lower conduction band from the upper conduction band and holes from the higher valence band to the lower valence band
- Prevents the photocorrosion behavior of semiconductor material by coupling another stable semiconductor material with unstable material

The hybrid structures of rutile titanium dioxide nanoparticle and anatase titanium dioxide displayed improved carbon dioxide photoconversion activity for the formation of methane as compared with anatase titanium dioxide nanorods, attributed mainly to the shifting of electrons to rutile from anatase with the exposure of ultraviolet light irradiation. Nevertheless, the heterojunction-type charge separation process has the drawback on redox capability of photoinduced electron and hole. Z-scheme system is proposed to overcome the disadvantages associated with the heterojunction-type photocatalytic system. In the Z-scheme, the photogenerated lower conduction band electrons from semiconductor 2 would combine with photogenerated holes in the semiconductor 1 leaving the stronger redox ability of electron holes of semiconductor 1 and semiconductor 2. For instance, $Fe_2V_4O_{13}$/CdS heterostructure enhanced the methane production three times than that of bare $Fe_2V_4O_{13}$, which is attributed to the efficient separation of charge carrier between the $Fe_2V_4O_{13}$ and CdS heterostructures through double charge transfer mechanism. Solid-state Z-scheme system of $Fe_2V_4O_{13}$/reduced graphene oxide (RGO)/CdS has further increased the photocatalytic activity by 30%. The presence of RGO can facilitate the improved charge transfer rate due to more efficient charge separation and high electronic mobility and subsequently enhanced carbon dioxide photoconversion efficiency (Li et al. 2015b). Moreover, the introduction of reduced graphene oxide would enhance cadmium sulfide stability against hole oxidation process (Dougan et al. 2014). Also, silver phosphate on graphitic carbon nitride heterojunction can effectively separate the photogenerated charge carrier by Z-scheme mechanism where the formation of silver nanoparticle acts as the interconnector between the silver phosphates on graphitic carbon nitride (He et al. 2015).

6.5 Effect Operating Parameters on Carbon Dioxide Reduction

Carbon dioxide photoconversion reaction depends on different experimental conditions including reaction medium, pH, wavelength and light intensity, catalyst dosage, temperature, pressure, and particle size. Selection of optimum value of operating parameters is crucial in determining the product formation. Various operating parameters are discussed in detail here.

6.5.1 Reaction Medium

During the carbon dioxide conversion process, equal quantity of photogenerated electrons and holes is utilized. Notably, the holes if not scavenged could have an undesirable role in the photocatalytic reaction, thereby enhancing the possibility of recombination of charge carrier and generating heat. The holes are scavenged by artificial electron donors in aqueous solution (Pastor et al. 2014). A number of artificial donors such as water, sodium hydroxide, sodium bicarbonate, water, acetonitrile, 2-propanol, dichloromethane, sodium sulfite, triethanolamine medium, tetramethylammonium chloride, N,N-dimethylformamide, and 2,5-dihydrofuran have been tested for carbon dioxide photoconversion (Fujiwara et al. 1998; Liu et al. 1998; Mao et al. 2013; Reli et al. 2013; Kuriki et al. 2015). Among all, water is the commonly used electron donor owing to its availability and cost. However, utilizing water is not effective owing to the low carbon dioxide solubility and the formation of hydrogen and hydrogen peroxide from CO_2 photoconversion process, which consumes electrons and proton. The solubility of carbon dioxide is more in hydroxide solution. Furthermore, the presence of hydroxide ion could scavenge the holes and convert hydroxyl radicals, thus suppressing the charge carrier recombination rate, and excited electrons would positively enable the carbon dioxide conversion. Different reaction media influences the yield and selectivity of product formation.

6.5.2 pH

pH of aqueous solution influences the carbon dioxide solubility as well as the formation of carbonate ion species. For instance, carbonic acid species are more at acidic pH (4.0), bicarbonate ions are predominant at pH 7.0 (pKa = 6.4), and carbon dioxide mostly exists in the form of carbonate ions (pKa = 10.3) at alkaline pH =11 (Mele et al. 2015). The proton concentration is more at low pH and may reduce the theoretical carbon dioxide reduction potential by assisting the transfer of electron toward carbon dioxide with protonation of the negative species. Notably, the reduction of carbonate and bicarbonate ions is more difficult than carbon dioxide. Besides, carbonate ion can scavenge the holes and gets oxidized. Pastrana-Martı'nez et al. (2016) reported that methanol and ethanol formation were higher at pH 4 and pH 11, respectively.

6.5.3 Wavelength and Light Intensity

The product yield is greatly affected by the wavelength of light source. It is supposed that smaller light wavelength can enable the generation of charge carriers and thus improve the catalyst efficiency (Kočí et al. 2011). In particular, many wideband materials would absorb in the ultraviolet region. Light intensity influences the excited electrons in the conduction band. The carbon dioxide conversion rate rises with increasing intensity of light source. The rate of product formation at higher light intensity is an independent process. The number of photoactivation sites remains the same at higher light intensity. Hence, the product formation rate reaches only a certain level even when increasing the intensity of light source (Mahmodi et al. 2013; Tahir and Amin 2013b).

6.5.4 Amount of Catalyst

The photocatalyst quantity is a significant parameter for carbon dioxide photoconversion process. Described that an increase in catalyst loading promotes the amount of photon adsorbed on the catalyst surface, thus leading to higher conversion rate of carbon dioxide. In contrast, a higher loading of the catalyst can increase the turbidity level and reduces the light penetration into the reactor, thereby diminishing the conversion efficiency of carbon dioxide (Zuas et al. 2014). So, it is necessary to determine the optimum amount of photocatalyst to reduce the unnecessary usage of the catalyst. Also, the total absorption of the photon for efficient carbon dioxide photoconversion has to be ensured (Herrmann 1999).

6.5.5 Particle Size

The particle size of the photocatalyst is a significant factor in carbon dioxide conversion. The particle size greatly affects the range of the energy band gap. The lesser particle size possesses a larger band gap and alters the oxidation and reduction potential values, thus changing the redox power of the semiconductor. Besides, lowering particle size of photocatalyst produces more surface area and active sites for carbon dioxide adsorption and absorbs more photon energy (Biswa 2002; Nikokavoura and Trapalis 2017). Also, the smaller particle size photocatalyst has the benefit of a smaller transfer pathway for photogenerated charge carriers to reach its surface. Described that the porous graphitic carbon nitride material shows poor carbon dioxide photoconversion than pristine graphitic carbon nitride. Because porousification induces the band gap improvement, the difficulty in electronic excitation and the presence of abundant defects will affect the photoconversion efficiency (Dong and Zhang 2012).

6.5.6 Pressure

Increasing the carbon dioxide concentration and pressure in an aqueous medium will improve the product selectivity in the aqueous medium. For instance, Kočí et al. 2008 described that the yield of methanol in aqueous phase increased with increasing carbon dioxide pressure to 110 kPa to 130 kPa. Moreover, the methanol formation is increased with the increase of carbon dioxide pressure up to 125 kPa but diminishes with raise in pressure level (Tseng et al. 2002).

6.5.7 Temperature

Generally, carbon dioxide photoconversion is studied at room temperature because the increase in temperature declines the carbon dioxide solubility in aqueous suspension. At a temperature range of 293–353 K, the photocatalytic activity is increased with decreasing activity happening beyond this temperature range (Herrmann 1999). This is because frequent collision occurs between the semiconductor and the substrate occurred at higher temperatures (Kočí et al. 2008). Variations in reaction temperature will affect the selectivity of product formation. Fox and Dulay (1993) demonstrated that small differences in temperature do not affect the photocatalytic reactions.

6.6 Photoreactors for Carbon Dioxide Photoconversion

The configuration of the photoreactor is another significant factor that can enhance the carbon dioxide reduction efficiency. The photocatalysts, photons, and reactants combine in the photoreactor vessel generating reaction products. A photoreactor requires improved light harvesting system, controlled photon loss, and enhanced separation of products, thus promoting the reactive surface area as well as reduced recombination rate of photogenerated charge carrier of a photocatalyst. An ideal photoreactor should distribute the light uniformly in the entire system in order to attain the best results. Based on type and mode of operation, the photoreactor system can be divided into two types including fluidized bed reactor and fixed bed reactor (Li et al. 2014a, b).

6.6.1 Fluidized Bed Reactor

The photocatalysts are dispersed in the liquid medium providing uniform distribution of photocatalysts with agitation using a magnetic stirrer to inhibit catalyst particle sedimentation (Fig. 6.8). The reaction rate is determined by catalyst efficiency, light intensity, and absorption properties of reactants in liquid solution. In this system, light sources are placed inside or top of the reactor, and the carbon dioxide gas was filled into the tightly sealed reactor before irradiation. The sample is withdrawn at different time intervals using online automatic sampling or gas tight syringe.

Notably, the reaction medium, size of the reactor, types of light source, and photocatalyst may differ in different photoreaction systems (Ola and Maroto-valer

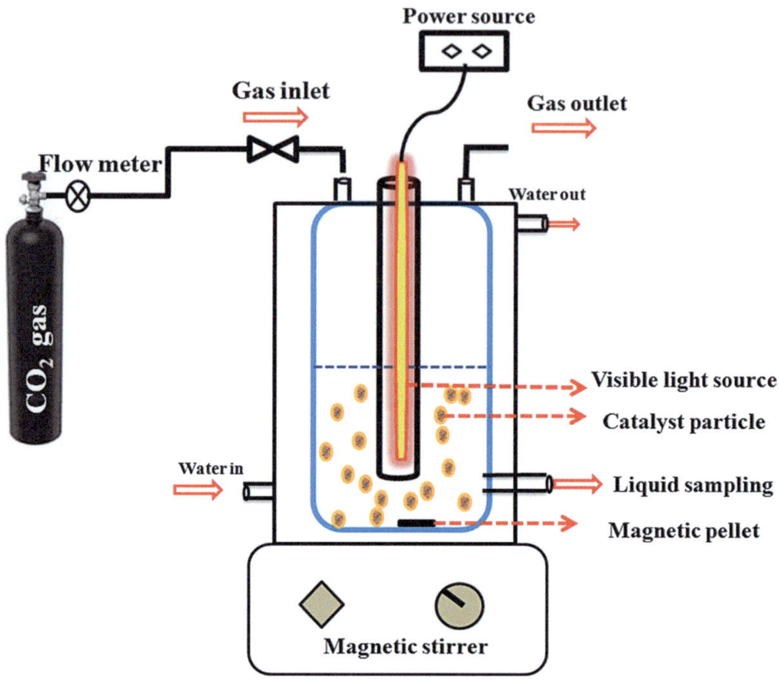

Fig. 6.8 Slurry-type photoreactor and its various parts for photocatalytic conversion of carbon dioxide into hydrocarbons. (Reprinted from Murugesan et al. 2019)

2015). Altering the lamp position can improve the light utilization efficacy and minimizes the light loss by reflection. The key advantages of the fluidized bed reactor are:

- The entire system was illuminated uniformly during the process.
- More loading of catalyst and simple construction.

The major disadvantages of fluidized bed reactors are:

- The photocatalyst particle separation from the reaction mixture is very difficult.
- Overloading of catalyst particles can scatter the light and limit the penetration of light into the suspension.
- Issues related to carbon dioxide solubility in aqueous solution.

6.6.2 Fixed Bed Reactor

In this system, the catalyst particles are uniformly dispersed into the bottom of the reactor, or the catalyst particle is immobilized on the fixed supports (plate, stainless steel webnets, beads, monolith optical fibers) in carbon dioxide and water vapor

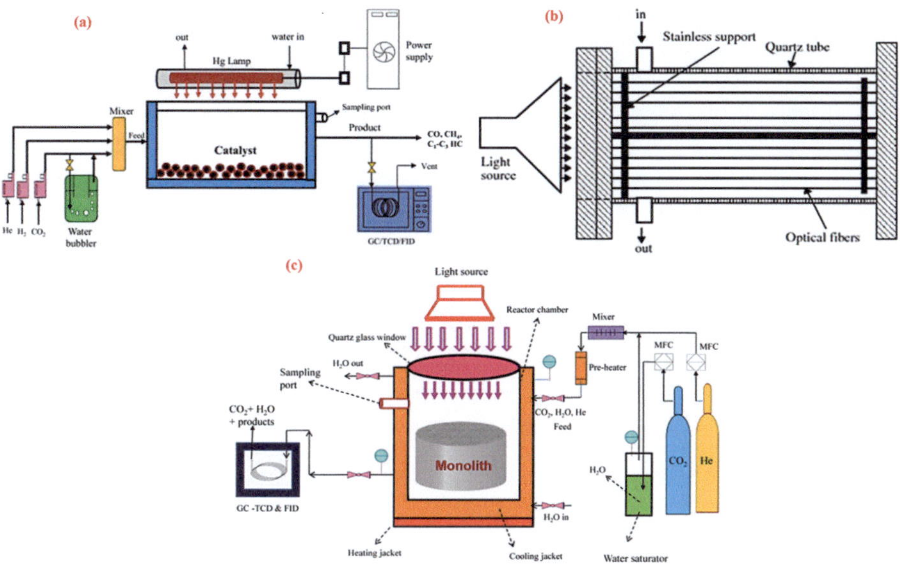

Fig. 6.9 Different types of fixed bed reactor for photoconversion of carbon dioxide (**a**) gas phase photoreactor. (Reprinted from Tahir et al. 2016), (**b**) optical-fiber photoreactor. (Reprinted from Wu and Taichi 2008), and (**c**) monolith photoreactor. (Reprinted from Tahir and Amin 2013a)

atmosphere. Immobilization can avoid the issues of catalyst separation. In some cases, the carbon dioxide gas can be simply generated from in situ neutral reactions of sodium bicarbonate and sulfuric acid or hydrochloric acid (Yu et al. 2014; Li et al. 2017). Moreover, in situ generation can be extensively used to suppress the solubility issue. The most important products for gas phase systems are carbon monoxide and methane. The limiting factor of this system is the light distribution which is affected by light source geometry and gap between the photocatalyst and light source. The fixed bed provides more flexibility than fluidized bed systems. More irradiation area, consumption of photon energy consumption, and volume of the reactor influence the quantum efficiency of the system. The fixed bed reactor type as illustrated in Fig. 6.9.

6.7 Conclusions

Photocatalytic conversion of carbon dioxide is a fascinating strategy in order to reduce the emission of carbon dioxide and encounter energy demand. Several novel photocatalysts have been established for the carbon dioxide photoconversion process. Formerly, the carbon dioxide photoconversion reaction was conducted using simple semiconductor material in aqueous solution. Subsequently, various surface modification techniques including doping, sensitization, and hybridization are amended for enhancing the carbon dioxide conversion efficiency of photocatalysts. Those nanocomposites display better charge separation and improved catalysis. Additionally, the nanoscale structure offers more surface area and improves separation efficiency of charge carrier and electron transfer. Notably, the formation of reaction products depends upon the physicochemical properties of photocatalyst and reaction conditions. This chapter summarizes the progress of carbon dioxide reduction over modified and unmodified semiconductor-based photocatalysts using different photocatalytic reactors. The appropriate selection of operating parameters is essential in order to attain the preferred hydrocarbon products.

References

Alper E, Yuksel Orhan O (2017) CO_2 utilization: developments in conversion processes. Petroleum 3:109–126. https://doi.org/10.1016/j.petlm.2016.11.003

Anandan S, Ikuma Y, Niwa K (2010) An overview of semi-conductor photocatalysis: modification of TiO_2 nanomaterials. Solid State Phenom 162:239–260. https://doi.org/10.4028/www.scientific.net/SSP.162.239

Benson EE, Kubiak CP, Sathrum AJ, Smieja JM (2009) Electrocatalytic and homogeneous approaches to conversion of CO_2 to liquid fuels. Chem Soc Rev 38:89–99. https://doi.org/10.1039/b804323j

Biswa CBA (2002) Role of synthesis method and particle size of nanostructured TiO_2 on its photoactivity. J Catal 211:145–156. https://doi.org/10.1006/jcat.2002.3783

Bukhtiyarova MV (2019) A review on effect of synthesis conditions on the formation of layered double hydroxides. J Solid State Chem 269:494–506. https://doi.org/10.1016/j.jssc.2018.10.018

Chang X, Wang T, Gong J (2016) CO_2 photo-reduction: insights into CO_2 activation and reaction on surfaces of photocatalysts. Energy Environ Sci 9:2177–2196. https://doi.org/10.1039/c6ee00383d

Chen S, Pan B, Zeng L et al (2017) La_2Sn_2O7enhanced photocatalytic CO_2 reduction with H_2O by deposition of Au co-catalyst. RSC Adv 7:14186–14191. https://doi.org/10.1039/c7ra00765e

Cheng C, Yang S, Zhong J (2014) ZnSe/CdS/CdSe triple-sensitized ZnO nanowire arrays for multi-bandgap photoelectrochemical hydrogen generation. RSC Adv 3:22922–22926. https://doi.org/10.1039/C4RA08335K

Dimitrijevic NM, Vijayan BK, Poluektov OG, Rajh T, Gray KA, He H, Zapol P (2011) Role of water and carbonates in photocatalytic transformation of CO_2 to CH_4 on titania. J Am Chem Soc 133:3964–3971. https://doi.org/10.1021/ja108791u

Dindi A, Quang DV, Vega LF et al (2019) Applications of fly ash for CO_2 capture, utilization, and storage. J CO2 Util 29:82–102. https://doi.org/10.1016/j.jcou.2018.11.011

Dong G, Zhang L (2012) Porous structure dependent photoreactivity of graphitic carbon nitride under visible light. J Mater Chem 22:1160–1166. https://doi.org/10.1039/C1JM14312C

Dong F, Zhao Z, Xiong T et al (2013) In situ construction of g-C_3N_4/g-C_3N_4 metal-free heterojunction for enhanced visible-light photocatalysis. Appl Mater Interfaces 5:11392–11401. https://doi.org/10.1021/am403653a

Dougan L, Tych KM, Hughes ML (2014) A single molecule approach to investigate the role of hydrogen bond strength on protein mechanical compliance and unfolding history. Chemcomm 51:800–803. https://doi.org/10.1039/b000000x

Fox MA, Dulay MT (1993) Heterogeneous photocatalysis. Chem Rev 93:341–357. https://doi.org/10.1021/cr00017a016

Fujiwara H, Hosokawa H, Murakoshi K, Wada Y, Yanagida S (1998) Surface characteristics of ZnS Nanocrystallites relating to their Photocatalysis for CO_2 reduction. Longmuir 7463:5154–5159. https://doi.org/10.1021/la9801561

Gaya UI (2014) Heterogeneous photocatalysis using inorganic semiconductor solids. Springer. https://doi.org/10.1007/978-94-007-7775-0

Go S, IshiTani O (2015) Efficient photocatalysts for CO_2 reduction. Inorg Chem 54:5906–5104. https://doi.org/10.1021/ic502675a

Gurudayal BJ, Srankó DF, Towle CM, Lum Y, Hettick M, Scott MC, Javey A, Ager J (2017) Efficient solar-driven electrochemical CO_2 reduction to hydrocarbons and oxygenates. Energy Environ Sci 10:2222–2230. https://doi.org/10.1039/c7ee01764b

Gwak GH, Kim MK, Oh JM (2016) Nanocomposites of magnetite and layered double hydroxide for recyclable chromate removal. J Nanomater 2016:1–10. https://doi.org/10.1155/2016/8032615

Habisreutinger SN, Schmidt-Mende L, Stolarczyk JK (2013) Photocatalytic reduction of CO_2 on TiO_2 and other semiconductors. Angew Chemie-Int Ed 52:7372–7408. https://doi.org/10.1002/anie.201207199

He Y, Zhang L, Teng B, Fan M (2015) New application of Z-scheme Ag_3PO_4/g-C_3N_4 composite in converting CO_2 to fuel. Environ Sci Technol 49:649–656. https://doi.org/10.1021/es5046309

Herrmann J (1999) Heterogeneous photocatalysis: fundamentals and applications to the removal of various types of aqueous pollutants. Catal Today 53:115–129. https://doi.org/10.1016/S0920-5861(99)00107-8

Herzog HJ (2011) Scaling up carbon dioxide capture and storage: from megatons to gigatons. Energy Econ 33:597–604. https://doi.org/10.1016/j.eneco.2010.11.004

Hong J, Zhang W, Ren J, Xu R (2013) Photocatalytic reduction of CO_2: a brief review on product analysis and systematic methods. Anal Methods 5:1086–1097. https://doi.org/10.1039/b000000x

Hsu HC, Shown I, Wei HY et al (2013) Graphene oxide as a promising photocatalyst for CO_2 to methanol conversion. Nanoscale 5:262–268. https://doi.org/10.1039/c2nr31718d

Huang C, Chen S, Fei X et al (2015) Catalytic hydrogenation of CO_2 to methanol: study of synergistic effect on adsorption properties of CO_2 and H_2 in CuO/ZnO/ZrO_2 system. Catalysts 5:1846–1861. https://doi.org/10.3390/catal5041846

Inoue T, Fujishima A, Konishi S, Honda K (1979) Photoelectrocatalytic reduction of carbon dioxide in aqueous suspensions of semiconductor powders. Nature 277:637–638. https://doi.org/10.1038/277637a0

Ioelovich M (2015) Recent findings and the energetic potential of plant biomass as a renewable source of biofuels – a review. Bioresources 10:1879–1914. https://doi.org/10.15376/biores.10.1.1879-1914

Jenisha Barnabas M, Parambadath S, Ha CS (2017) Amino modified core–shell mesoporous silica based layered double hydroxide (MS-LDH) for drug delivery. J Ind Eng Chem 53:392–403. https://doi.org/10.1016/j.jiec.2017.05.011

Jo W, Selvam NCS (2017) Z-scheme CdS/g-C_3N_4 composites with RGO as an electron mediator for efficient photocatalytic H_2 production and pollutant degradation. Chem Eng J 317:913–924. https://doi.org/10.1016/j.cej.2017.02.129

Kalamaras E, Maroto-Valer MM, Shao M, Xuan J, Wang H (2018) Solar carbon fuel via photoelectrochemistry. Catal Today 317:56–75. https://doi.org/10.1016/j.cattod.2018.02.045

Khoo HH, Bu J, Wong RL et al (2011) Carbon capture and utilization: preliminary life cycle CO_2, energy, and cost results of potential mineral carbonation. Energy Procedia 4:2494–2501. https://doi.org/10.1016/j.egypro.2011.02.145

Kiesgen R, Richter D, Ming T, Caillol S (2013) Fighting global warming by photocatalytic reduction of CO_2 using giant photocatalytic reactors. Renew Sust Energ Rev 19:82–106. https://doi.org/10.1016/j.rser.2012.10.026

Koca A, Musa S (2002) Photocatalytic hydrogen production by direct sun light from sulfide/sulfite solution. Int J Hydrog Energy 27:363–367. https://doi.org/10.1016/S0360-3199(01)00133-1

Kočí K, Obalová L, Placha D, Lacnty Z (2008) Effect of temperature, pressure and volume of reacting phase on photocatalytic CO_2 reduction on suspended Nanocrystalline TiO_2. Collect Czechoslov Chem Commun 73:1192–1204. https://doi.org/10.1135/cccc20081192

Kočí K, Zatloukalová K, Obalová L et al (2011) Wavelength effect on photocatalytic reduction of CO_2 by Ag/TiO_2 catalyst. Chinese J Catal 32:812–815. https://doi.org/10.1016/S18722067(10)60199-4

Kothandaraman J, Goeppert A, Czaun M et al (2016) Conversion of CO_2 from air into methanol using a polyamine and a homogeneous ruthenium catalyst. J Am Chem Soc 138:778–781. https://doi.org/10.1021/jacs.5b12354

Kuriki R, Sekizawa K, Ishitani O, Maeda K (2015) Visible-light-driven CO_2 reduction with carbon nitride : enhancing the activity of ruthenium catalysts. Angew Chem Int Ed 54:1–5. https://doi.org/10.1002/anie.201411170

Lee YY, Jung HS, Kang YT (2017) A review : effect of nanostructures on photocatalytic CO_2 conversion over metal oxides and compound semiconductors. J CO2 Util 20:163–177. https://doi.org/10.1016/j.jcou.2017.05.019

Li K, An X, Park KH, Khraisheh M, Tanga J (2014a) A critical review of CO_2 photoconversion: catalysts and reactors. Catal Today 224:3–12. https://doi.org/10.1016/j.cattod.2013.12.006

Li X, Wen J, Low J, Fang Y, Yu J (2014b) Design and fabrication of semiconductor photocatalyst for photocatalytic reduction of CO_2 to solar fuel.Sci. China Mater 57:70–100. https://doi.org/10.1007/s40843-014-0003-1

Li M, Zhang L, Fan X, Zhou Y, Wu M, Shi J (2015a) Highly selective CO_2 photoreduction to CO over g-C_3N_4/Bi_2WO_6 composites under visible light. J Mater Chem A 3:5189–5196. https://doi.org/10.1039/C4TA06295G

Li P, Zhou Y, Li H, Xu Q, Meng X, Wang X, Xiao M, Zou Z (2015b) All-solid-state Z-scheme system arrays of $Fe_2V_4O_{13}$/RGO/CdS for visible light-driving photocatalytic CO_2 reduction into renewable hydrocarbon fuel. Chem Commun 51:800–803. https://doi.org/10.1039/b000000x

Li K, Peng B, Peng T (2016) Recent advances in heterogeneous photocatalytic CO_2 conversion to solar fuels. ACS Catal 6:7485–7527. https://doi.org/10.1021/acscatal.6b02089

Li K, Peng B, Jin J, Zan L, Peng T (2017) Carbon nitride nanodots decorated brookite TiO_2 quasi nanocubes for enhanced activity and selectivity of visible-light-driven CO_2 reduction. Appl Catal B Environ 203:910–916. https://doi.org/10.1016/j.apcatb.2016.11.001

Lin J, Pan Z, Wang X (2014) Photochemical reduction of CO_2 by graphitic carbon nitride polymers. ACS Sustain Chem Eng 2:353–358. https://doi.org/10.1021/sc4004295

Liu BJ, Torimoto T, Yoneyama H (1998) Photocatalytic reduction of CO_2 using surface-modified CdS photocatalysts in organic solvents. J Photochem Photobiol A Chem 113:93–97. https://doi.org/10.1016/S1010-6030(97)00318-3

Liu L, Zhao H, Andino JM, Li Y (2012) Photocatalytic CO_2 reduction with H_2O on TiO_2 nanocrystals: comparison of anatase, rutile, and brookite polymorphs and exploration of surface chemistry. ACS Catal 2:1817–1828. https://doi.org/10.1021/cs300273q

Liu X, Ye L, Liu S, Li Y, Ji X (2016) Photocatalytic reduction of CO_2 by ZnO micro/nanomaterials with different morphologies and ratios of {0001} facets. Sci Rep 6:1–9. https://doi.org/10.1038/srep38474

Low J, Cheng B, Yu J, Jaroniec M (2016) Carbon-based two-dimensional layered materials for photocatalytic CO_2 reduction to solar fuels. Energy Storage Mater 3:24–35. https://doi.org/10.1016/j.ensm.2015.12.003

Maeda K, Sekizawa K, Ishitani O (2013) A polymeric-semiconductor–metal-complex hybrid photocatalyst for visible-light CO_2 reduction. Chem Commun 49:10127. https://doi.org/10.1039/c3cc45532g

Mahmodi G, Sharifnia S, Rahimpour F, Hosseini SN (2013) Photocatalytic conversion of CO_2 and CH_4 using ZnO coated mesh: effect of operational parameters and optimization. Sol Energy Mater Sol Cells 111:31–40. https://doi.org/10.1016/j.solmat.2012.12.017

Mao J, Peng T, Zhang X, Li K, Ye L, Zan L (2013) Effect of graphitic carbon nitride microstructures on the activity and selectivity of photocatalytic CO_2 reduction under visible light. Cat Sci Technol 3:1253. https://doi.org/10.1039/c3cy20822b

Martha S, Sahoo PC, Parida KM (2015) An overview on visible light responsive metal oxide based photocatalysts for hydrogen energy production. RSC Adv 5:61535–61553. https://doi.org/10.1039/C5RA11682A

Mele G, Annese C, Accolti LD et al (2015) Photoreduction of carbon dioxide to formic acid in aqueous suspension: a comparison between phthalocyanine/TiO_2 and porphyrin/TiO_2 catalysed processes. Molecules 20:396–415. https://doi.org/10.3390/molecules20010396

Miranda C, Mansilla H, Yánez J, Obregón S, Colón G (2013) Improved photocatalytic activity of g-C_3N_4/TiO_2 composites prepared by a simple impregnation method. J Photochem Photobiol A Chem 253:16–21. https://doi.org/10.1016/j.jphotochem.2012.12.014

Miyano M, Zhang H, Yoshiba M, Izumi Y (2017) Selective Photoconversion of carbon dioxide into methanol using layered double hydroxides at 0. 40 MPa. Energy Technol 5:892–900. https://doi.org/10.1002/ente.201600578

Murugesan P, Narayanan S, Matheswaran M (2019) Visible light photocatalytic conversion of CO_2 in aqueous solution using 2D-structured carbon-based catalyst-coated γ,β -AgI nanocomposite. J Mater Sci. https://doi.org/10.1007/s10853-019-03389-9

Nie L, Zhang Q (2017) Recent progress in crystalline metal chalcogenides as efficient photocatalysts for organic pollutant degradation. Inorg Chem Front 4:1953–1962. https://doi.org/10.1039/C7QI00651A

Nikokavoura A, Trapalis C (2017) Alternative photocatalysts to TiO_2 for the photocatalytic reduction of. Appl Surf Sci 391:149–174. https://doi.org/10.1016/j.apsusc.2016.06.172

Noji T, Jin T, Nango M, Kamiya N, Amao Y (2017) CO_2 photoreduction by formate dehydrogenase and a ru-complex in a nanoporous glass reactor. ACS Appl Mater Interfaces 9:3260–3265. https://doi.org/10.1021/acsami.6b12744

Ola O, Maroto-valer MM (2015) Review of material design and reactor engineering on TiO_2 photocatalysis for CO_2 reduction. J Photochem Photobiol C: Photochem Rev 24:16–42. https://doi.org/10.1016/j.jphotochemrev.2015.06.001

Owusu PA, Asumadu-Sarkodie S (2016) A review of renewable energy sources, sustainability issues and climate change mitigation. Cogent Eng 3:1–14. https://doi.org/10.1080/23311916.2016.1167990

Pang R, Teramura K, Asakura H, Hosokawa S, Tanaka T (2017) Highly selective photocatalytic conversion of CO2 by water over Ag-loaded $SrNb_2O_6$ nanorods. Appl Catal B Environ 218:770–778. https://doi.org/10.1016/j.apcatb.2017.06.052

Pastor E, Pesci FM, Reynal A, Handoko AD, Guo M, An X, Cowan AJ, Klug DR, Durrant JR, Tang J (2014) Visible light driven CO_2 reduction catalyst. Phys Chem Chem Phys 16:5922–5926. https://doi.org/10.1039/c4cp00102h

Pastrana-Martı'nez LM, AMT S, NNC F, Vaz JR, Figueiredo JL, Faria JL (2016) Photocatalytic reduction of CO_2 with water into methanol and ethanol using graphene derivative – TiO_2 composites : effect of pH and copper (I) oxide. Top Catal 59:1279–1291. https://doi.org/10.1007/s11244-016-0655-2

Ping J, Zhou Y, Wu Y, Papper V, Boujday S, Marks RS, Steele TWJ (2015) Recent advances in aptasensors based on graphene and graphene-like nanomaterials. Biosens Bioelectron 64:373–385. https://doi.org/10.1016/j.bios.2014.08.090

Reli M, Šihor M, Kočí K, Praus P (2013) Influence of reaction medium on CO_2 photocatalytic reduction yields over Zns-MMT. Geosci Eng 58:34–42. https://doi.org/10.2478/v10205-011-0011-5

Rezaei B, Mokhtarianpour M, Ensafi AA, Hadadzadeh H, Shakeri J (2015) Electrocatalytic reduction of CO_2 using the dinuclear rhenium(I) complex [ReCl(CO)$_3$(μ-tptzH)Re(CO)$_3$]. Polyhedron 101:160–164. https://doi.org/10.1016/j.poly.2015.08.014

Salehi-Khojin A, Jhong HRM, Rosen BA et al (2013) Nanoparticle silver catalysts that show enhanced activity for carbon dioxide electrolysis. J Phys Chem C 117:1627–1632. https://doi.org/10.1021/jp310509z

Serpone N, Emeline AV (2012) Semiconductor photocatalysis – past, present, and future outlook. J Phys Chem Lett 3:673–677. https://doi.org/10.1021/jz300071j

Shi H, Wang T, Chen J (2011) Photoreduction of carbon dioxide over $NaNbO_3$ nanostructured. Catal Lett 141:525–530. https://doi.org/10.1007/s10562-010-0482-1

Sun Z, Wang H, Wu Z, Wang L (2018) g-C_3N_4 based composite photocatalysts for photocatalytic CO_2 reduction. Catal Today 300:160–172. https://doi.org/10.1016/j.cattod.2017.05.033

Tahir M, Amin NS (2013a) Photocatalytic CO_2 reduction and kinetic study over In/TiO_2 nanoparticles supported microchannel monolith photoreactor. Applied Catal A Gen 467:483–496. https://doi.org/10.1016/j.apcata.2013.07.056

Tahir M, Amin NAS (2013b) Photocatalytic reduction of carbon dioxide with water vapors over montmorillonite modified TiO_2 nanocomposites. Appl Catal B Environ 142-143:512–522. https://doi.org/10.1016/j.apcatb.2013.05.054

Tahir M, Tahir B, Aishah N, Muhammad A (2016) Photocatalytic CO_2 methanation over NiO/In_2O_3 promoted TiO_2 nanocatalysts using H_2O and/or H_2 reductants. Energy Convers Manag 119:368–378. https://doi.org/10.1016/j.enconman.2016.04.057

Tan LL, Ong WJ, Chai SP, Mohamed AR (2017) Photocatalytic reduction of CO_2 with H_2O over graphene oxide-supported oxygen-rich TiO_2 hybrid photocatalyst under visible light irradiation: process and kinetic studies. Chem Eng J 308:248–255. https://doi.org/10.1016/j.cej.2016.09.050

Tian J, Leng Y, Zhao Z, Xia Y, Sang Y, Zhan PHJ, Li M, Liu H (2015) Carbon quantum dots/hydrogenated TiO_2 nanobelt heterostructures and their broad spectrum photocatalytic properties under UV, visible, and near-infrared irradiation. Nano Energy 11:419–427. https://doi.org/10.1016/j.nanoen.2014.10.025

Tseng I, Chang W, Wu JCS (2002) Photoreduction of CO_2 using sol-gel derived Titania and titania-supported copper catalysts. Appl Catal B 37:37–48. https://doi.org/10.1016/S0926-3373(01)00322-8

Wang C, Thompson RL, Baltrus J, Matranga C (2010) Visible light photoreduction of CO_2 using CdSe/Pt/TiO_2 heterostructured catalysts. J Phys Chem Lett 1:48–53. https://doi.org/10.1021/jz9000032

Wang Z, Yu J, Jin H, Gui R, Jin H, Xia Y (2015) Carbon nanomaterials-based electrochemical aptasensors. Biosens Bioelectron 79:136–149. https://doi.org/10.1016/j.bios.2015.11.093

Wang J, Yao HG, Fan ZU, Zhang L, Wang J, Zang SQ, Li ZJ (2016a) Indirect Z-scheme BiOI/g-C_3N_4 photocatalysts with enhanced photoreduction CO_2 activity under visible light irradiation. ACS Appl Mater Interfaces 8:3765–3775. https://doi.org/10.1021/acsami.5b09901

Wang H, Sun Z, Li Q, Tang Q, Wua Z (2016b) Surprisingly advanced CO_2 photocatalytic conversion over thiourea derived g-$C_3N_4$with water vapor while introducing 200-420 nm UV light. J CO2 Util 14:143–151. https://doi.org/10.1016/j.jcou.2016.04.006

Wu JCS, Taichi (2008) Application of optical-fiber photoreactor for CO_2 photocatalytic reduction. Top Catal 47:131–136. https://doi.org/10.1007/s11244-007-9022-7

Xiao J, Yang S, Feng F, Xue HG, Guo SP (2017) A review of the structural chemistry and physical properties of metal chalcogenide halides. Coord Chem Rev 347:23–47. https://doi.org/10.1016/j.ccr.2017.06.010

Xie X, Liu X, Li X, Liu XZ, Li X, Xie X (2018) Facile synthesis of NiAl-LDHs with tunable establishment of acid-base activity sites. Mater Chem Phys 211:72–78. https://doi.org/10.1016/j.matchemphys.2018.02.015

Yaipimai W, Subjalearndee N, Tumcharern G (2015) Multifunctional metal and metal oxide hybrid nanomaterials for solar light photocatalyst and antibacterial applications. J Mater Sci 50:7681–7697. https://doi.org/10.1007/s10853-015-9333-1

Yang Z, Wei J, Zeng G, Zhang H, Tan X, Ma C, Lia X, Lia Z, Zhang C (2019) A review on strategies to LDH-based materials to improve adsorption capacity and photoreduction efficiency for CO_2. Coord Chem Rev 386:154–182. https://doi.org/10.1016/j.ccr.2019.01.018

Yu J, Wang K, Xiao W, Cheng B (2014) Photocatalytic reduction of CO_2 into hydrocarbon solar fuels over g-C_3N_4–Pt nanocomposite photocatalysts. Phys Chem Chem Phys 16:11492. https://doi.org/10.1039/c4cp00133h

Yu J, Chen Z, Zeng L, Ma Y, Feng Z, Wu Y, Lin H, Zhao L, He Y (2018) Synthesis of carbon-doped $KNbO_3$photocatalyst with excellent performance for photocatalytic hydrogen production. Sol Energy Mater Sol Cells 179:45–56. https://doi.org/10.1016/j.solmat.2018.01.043

Zhang W, Hu Y, Ma L, Wang Y, Xue X, Chen R, Yang S, Jin Z (2018) Progress and perspective of electrocatalytic CO_2 reduction for renewable carbonaceous fuels and chemicals. Adv Sci 5. https://doi.org/10.1002/advs.201700275

Zhou J, Chen W, Sun C, Chen M, Wang X, Wang E, Su Z (2017) Oxidative polyoxometalates modified graphitic carbon nitride for visible-light CO_2 reduction. ACS Appl Mater Interfaces 9:11689–11695. https://doi.org/10.1021/acsami.7b01721

Zhou B, Song J, Xie C, Chen C, Qian Q, Han B (2018a) Mo-Bi-Cd ternary metal chalcogenides: highly efficient photocatalyst for CO_2 reduction to formic acid under visible light Mo-bi-cd ternary metal chalcogenides : highly efficient photocatalyst for CO_2 reduction to formic acid under visible light. ACS Sustain Chem Eng 6:5754–5759. https://doi.org/10.1021/acssuschemeng.8b00956

Zhou L, Luo W, Xu L et al (2018b) Exceptional co-catalyst free photocatalytic activities of B and Fe co-doped $SrTiO_3$ for CO_2 conversion and H_2 evolution. Nano Res 11:6391–6404. https://doi.org/10.1007/s12274-018-2164-z

Zuas O, Kim JS, Gunlazuardi J (2014) Influence of operational parameters on the photocatalytic activity of powdered TiO_2 for the reduction of CO_2. Indo J Chem 14:122–130. https://doi.org/10.22146/ijc.21248

Zubair M, Daud M, McKay G, Shehzad F, Harthi MAA (2017) Recent progress in layered double hydroxides (LDH)-containing hybrids as adsorbents for water remediation. Appl Clay Sci 143:279–292. https://doi.org/10.1016/j.clay.2017.04.002

Chapter 7
Electrocatalytic Production of Methanol from Carbon Dioxide

Esperanza Ruiz Martínez and José María Sánchez Hervás

Contents

7.1 Introduction 167
7.2 Liquid Phase Electrocatalytic Production of Methanol from Carbon Dioxide 170
7.2.1 Electrocatalysts 170
7.2.2 Electrolytes 176
7.2.3 Electrode Structure 179
7.2.4 Electrochemical Cell Configuration 180
7.2.5 Operation Parameters 185
7.3 Gaseous Phase Electrocatalytic Production of Methanol from Carbon Dioxide 188
7.3.1 Electrocatalysts 189
7.3.2 Electrolytes 190
7.3.3 Electrochemical Cell Configuration 192
7.3.4 Operation Parameters 197
7.4 Conclusions 199
References 202

Abstract The increasing CO_2 atmospheric levels, associated with adverse climate change, and the decreasing fossil fuel reserves resulting from extensive power generation are of major environmental and sustainability concerns. The electrocatalytic production of methanol from CO_2 is a promising technology for CO_2 emission reduction and renewable energy storage as sustainable platform chemicals and combustibles. However, commercial application of the technology requires stable electrocatalysts that exhibit current densities above 0.1 A cm^{-2}, overpotentials below 200 mV, and Faradaic efficiencies close to 100%. This chapter highlights different recent approaches to fulfil these challenges and reviews the effect of applied catalytic and electrolyte material, electrode and electrochemical cell structure, and utilized operational parameters on process performance. Mo, Ru, and Cu are the most active catalysts. Strategies to improve electrocatalysts performance included alloying with other metals; oxidation and mixing with other metal

---

E. Ruiz Martínez (✉) · J. M. Sánchez Hervás
Unit for Sustainable Thermochemical Valorisation, CIEMAT, Madrid, Spain
e-mail: esperanza.ruiz@ciemat.es; josemaria.sanchez@ciemat.es

Inamuddin et al. (eds.), *Conversion of Carbon Dioxide into Hydrocarbons Vol. 1 Catalysis*, Environmental Chemistry for a Sustainable World 40,
https://doi.org/10.1007/978-3-030-28622-4_7

oxides; being part of metal nanostructures, complexes, or organic frameworks; and depositing onto metal or porous carbonaceous supports. Finally, best catalysts were incorporated into practical structures, such as gas diffusion electrodes or thin films deposited on solid oxide electrolytes that operate on gaseous CO_2 and in a continuous mode in polymer electrolyte membrane or solid oxide electrolyte cells, respectively.

Keywords CO_2 recycling · Renewable energy storage · Methanol production · CO_2 electrocatalytic reduction · Electrocatalysts · Gas diffusion electrodes · Polymer electrolyte membranes · Solid oxide electrolyzers · Electrochemical promotion · Faradaic efficiency

Abbreviations

[Bmim]BF4	1-butyl-3-methylimidazolium tetrafluoroborate
2HNN	2-hydroxil-1-nitrosonaphthalene
3DPN	Co(II) 3,6-disulphonatocobal(II)
4TBP	4-tert-butylpyridinium
APCF	aquo-pentacyanoferrate (II): $Na_3[Fe(CN)_5(H_2O)$
BDD	boron-doped diamond
CNF	carbon nanofiber
CNT	carbon nanotube
CuAdeAce	$[Cu_2$(m3-adeninate$)_2$(m2-OOC$(CH_3)_2)]$n
DTA	bis-bidentate dithiooxamidate
ES	Everitt's Salt: K_2Fe(II)[Fe(II)$(CN)_6$
GDE	gas diffusion electrode
H_2/CO_2	H_2 to CO_2 molar ratio
M	molarity in mol L^{-1}
MEA	membrane electrode assembly
MeCN	acetonitrile
MOA	mesoporous metal-organic aerogel
MOF	metal-organic framework
MWCNTs	multiwalled carbon nanotubes
NG	nitrogen-doped graphene
n-GaAs	n-gallium arsenide
NP	nanoparticle
Pan	polyaniline
PEM	polymer electrolyte membrane
p-GaP	p-gallium phosphide
p-InP	p-indium phosphide
PRB	Prussian blue: KFe(III)[Fe(II)$(CN)_6$
PYD	4-(3-Phenoxy-2,2-bis(phenoxymethyl)propoxy)pyridine
SPE	solid polymer electrolyte
TRPP	tetraphenylporphyrin

YSZ	yttria-stabilized-zirconia
ΔCO_2 conversion	increment in overall CO_2 conversion in number of times
ΔSelectivity	ncrement in methanol selectivity in number of times

7.1 Introduction

Energy sources have diversified in recent decades. However, the production of energy, and also chemicals, from fossil sources is still playing an important role (Goeppert et al. 2014). Owing to the growth in carbon dioxide atmospheric concentrations and the decreasing fossil fuel reserves resulting from extensive power generation, conversion of released carbon dioxide into valuable fuels and platform chemicals is regarded as a supplementary approach to capture and store for an efficient quantifiable decrease of carbon dioxide emissions, permitting carbon dioxide reusing and, consequently, a more sustainable usage of resources (Centi and Perathoner 2009).

Each year about 1×10^{12} kilos of CO_2 are emitted to the atmosphere (Goeppert et al. 2014). The CO_2 atmospheric concentration has increased by about 1.5 times from the preindustrial value of around 280 ppm up to above 400 ppm today. CO_2 is a powerful greenhouse gas which warms the earth by trapping the infrared radiation of the sun and is considered as the main contributor to the adverse global warning and climate change effects (Ganesh 2014; Goeppert et al. 2014).

Recycling of wasted CO_2 to hydrocarbon fuels and chemicals can be carried out as long as any attainable renewable power source, such as solar, wind, or hydraulic, is utilized for the conversion of carbon dioxide. In this manner, CO_2 can become from a damaging greenhouse gas producing global warning, into a rewarding, renewable, carbon neutral and endless combustibles and chemicals source for the forthcoming generations (Centi and Perathoner 2009; Ganesh 2014; Goeppert et al. 2014). Electrocatalytic conversion of CO_2 to fuels provides also a way for storing the surplus of intermittent renewable energy, which cannot be discharged into the electricity grid, in the form of carbon neutral, safe, tested, and socially accepted hydrocarbon fuels, which storage, distribution, and energy generation technologies are fully developed. These fuels could subsequently be used for power generation or transport, contributing to the decarbonization of both sectors (Centi and Perathoner 2011).

The development and potential technology transfer of processes for the electrocatalytic conversion of CO_2 to fuels would help to overcome the limitations to renewable energy utilization, to reduce the dependence on fossil reserves, and to ensure power supply. The electrocatalytic conversion of CO_2 to fuels is a potential way to enable a proper economic growth and social welfare while allowing some reduction of greenhouse gas emission and contributing also to the fulfilment of the energy, environmental, and social objectives established at local and global levels (Centi and Perathoner 2009; Ganesh 2014; Goeppert et al. 2014).

Among the possible hydrocarbons products, methanol has high energy density, of 15.6 MJ·dm^{-3}, and octane rating, of about 110, stable storage properties, easeness of transport, and established use. Methanol can be used for energy generation directly or indirectly, after conversion into gasoline or hydrogen, respectively, in internal combustion engines and in fuel cells. Methanol is also an important intermediate for the fabrication of many chemicals, such as paint and polymers (Ganesh 2014; Goeppert et al. 2014).

Among the available technologies, electrocatalytic conversion of carbon dioxide to methanol offers a simple, selectively controlled and environmental friendly process to produce methanol from waste CO_2 (Albo et al. 2015a, b; Durst et al. 2015; Whipple and Kenis 2010). This technological alternative utilizes electricity, including from renewable sources, to convert CO_2 into methanol under milder temperature and pressure than the CO_2 thermocatalytic conversion counterpart, which uses H_2 generated by a separated water electrolyzer to hydrogenate CO_2. On the contrary, the electrocatalytic approach combines water electrolysis with CO_2 hydrogenation to methanol, at the anode and cathode, respectively, of an electrochemical cell. Therefore, the electrocatalytic process allows CO_2 recycling and chemical storage of discontinuous renewable energy as sustainable fuels and platform chemicals in a single step.

The electrocatalytic production of methanol from CO_2 requires the input of electricity to create a potential difference between two or three electrodes: the cathode or catalyst-working electrode and the anode or counter and reference electrodes. Electrolytes can be liquids, such as aqueous, nonaqueous, and ionic liquids, or solids, ion-conducting polymers or oxides. The process can proceed in gaseous and liquid phases at different cell and electrode structures and under different operating conditions (Albo et al. 2015a; Whipple and Kenis 2010).

The electrochemical aqueous reduction of carbon dioxide to methanol is thermodynamically feasible. However the reduction potential of CO_2 is very close to that of hydrogen generation by proton reduction; as a result, there is a competition between both reactions. In addition to the hydrogen evolution reaction, there are various competing hydrocarbon synthesis reactions which also consume the electrical current supplied to the electrochemical cell, reducing the percentage of electrons that ends up in methanol formation, which is defined as Faradaic efficiency of the process (Albo et al. 2015a; Jia et al. 2014).

The reduction of CO_2 into CH_3OH implies six electrons, so, the process has a slow kinetics. The slow kinetics of the CO_2 reduction to CH_3OH reaction results in high cathode overpotentials, understood as the difference between the applied electrode potential and the thermodynamic potential, which lead also to low conversions and energy efficiencies (Albo et al. 2015a; Jia et al. 2014; Jones et al. 2014). In addition, according to the Faraday's law, the rate of an electrochemical reaction per electrode area is proportional to the electrical current density established by the applied potential. Therefore, for low current densities, large electrode surfaces will be required, leading to high investment and operational costs (Jones et al. 2014).

High selectivity to methanol, typically defined as the Faradaic efficiency, and high methanol production rates, usually defined as current density, can only be attained by finding electrocatalysts that are active enough to reduce CO_2 to methanol efficiently and at high rate but inactive for the above commented competing reactions (Albo et al. 2015a; Jones et al. 2014). Unfortunately, the production of methanol via CO_2 electrocatalytic conversion remains below the level required for economic viability due to limited selectivity, lack of stability, and low efficiency of present-day electrocatalysts. The later implies the use of high overpotentials, resulting also in high energy requirements (Albo et al. 2015a; Kuhl et al. 2014; Qiao et al. 2014).

Many catalyst-electrode materials, such as metals, alloys, metallic complexes and metallic oxides, and chalcogenides, have been employed for generating methanol from CO_2 (Hirunsit et al. 2015; Irfan Malik et al. 2016; Qiao et al. 2014; Sun et al. 2016; Zhang et al. 2018b). The use of these electrocatalysts is reviewed in this chapter.

Among the available metals, the most active for electrocatalytically converting CO_2 into methanol were Mo, Ru, Ti, Fe, Ga, In, Pd, Pt, and, in particular, Cu. Cu has been used in the form of metal foils, Cu electrochemically deposited on carbon-based materials, chemically deposited Cu films on solid electrolytes, Cu alloys, and Cu-containing perovskite-type electrocatalysts (Irfan Malik et al. 2016; Schwartz et al. 1993).

This chapter reviews meaningful work performed in the field of electrocatalytic conversion of carbon dioxide to methanol, taking into consideration the impact of the applied catalytic and electrolyte materials, the electrode and electrochemical cell structure, as well as the utilized working conditions. This review attempts also to introduce the progress of diverse recent process approaches, such as the utilization of solid polymer or solid oxide electrochemical membrane reactors for the gaseous phase electrocatalytic conversion of CO_2 into methanol, as well as the electrochemically promoted catalytic CO_2 hydrogenation to methanol in solid oxide electrolyte cells. Therefore, this chapter reviews the present-day status, pending challenges, and prospect opportunities for electrocatalytic conversion of CO_2 to methanol.

Given the enormous amount of work published on electrocatalytic production of methanol from CO_2, most of the reported results are presented in tabular form. Different tables within the chapter summarize the best performance data reported in terms of Faradaic efficiency or selectivity and production yields, current densities, or total charge transferred, for methanol formation. Tables include cathode potentials versus different reference electrodes reported in the literature, so the tabulated results are not fully comparable. The tables also compile information about the reaction medium, operating conditions, and catalyst-electrode and cell configuration utilized for the electrocatalytic production of methanol from CO_2.

7.2 Liquid Phase Electrocatalytic Production of Methanol from Carbon Dioxide

7.2.1 Electrocatalysts

Many catalysts, such as metals, metal alloys, metal complexes and metal oxides, and chalcogenides, have been developed for producing methanol from carbon dioxide (Hirunsit et al. 2015; Irfan Malik et al. 2016; Qiao et al. 2014; Zhang et al. 2018b). However, electrocatalysts that simultaneously exhibit low overpotentials and high Faradaic efficiencies and current densities needed for commercial applications are still missing (Jhong et al. 2013). Among the available metals, the most active for the electrocatalytic conversion of CO_2 to methanol are Mo, Ru, Ti, Fe, Ga, In, Pd, Pt, and, in particular, Cu. Copper has been used in the form of metal foils, Cu electrochemically or chemically deposited on carbon-based materials and solid electrolytes, Cu oxides, Cu alloys, Cu metal-organic porous materials, and Cu-containing perovskite-type electrocatalysts (Albo et al. 2017; Irfan Malik et al. 2016; Schwartz et al. 1993). The use of these electrocatalysts for CO_2 electrocatalytic conversion to methanol is discussed in this section.

The formation of methanol was observed but, generally, with high overpotentials and low selectivities, at various Cu-based electrocatalysts, such as Cu, electrodeposited Cu, and Cu-coated electrodes, at ambient temperature and pressure in aqueous solutions (Albo et al. 2015a; Schwartz et al. 1993). Substantial efforts have been made to enhance energy efficiency for methanol production by electrocatalytic CO_2 conversion. The different approaches include altering surface structures, morphologies, and compositions, resulting in the development of diverse types of Cu-based catalysts, which use is reviewed in the following.

The electrocatalytic reduction of carbon dioxide to methanol occurred at various metallic and oxidized copper electrodes, such as Cu and Cu oxide foils and Cu oxides electrodeposited on Ti foils (Frese 1991), Cu_2O-catalyzed carbon clothes (Albo et al. 2015a, b; Albo and Irabien 2016; Chang et al. 2009), electrodeposited cuprous oxide thin films electrodes (Le et al. 2011), and Cu(core)-CuO(shell) catalysts (Lan et al. 2014). The presence of cuprous oxide plays a decisive role in electrocatalytic activity and selectivity for methanol formation. However, cuprous oxide-based catalysts suffer from a lack of stability ascribable to some reduction of the copper oxide (Albo et al. 2015a).

Yadav and Purkait (2015) investigated the selective electroreduction of CO_2 to methanol over Cu_2O coated on a graphite plate. The maximum Faradaic efficiency for methanol production was about 8% (Yadav and Purkait 2015). The electroreduction of CO_2 to methanol also took place on carbon paper supported Cu_2O and Cu_2O and ZnO. The presence of ZnO improves stability of the system (Albo and Irabien 2016; Albo et al. 2015b). Irfan Malik et al. (2016) also studied the performance of multiwalled carbon nanotubes impregnated with Cu_2O for CO_2 electroreduction to methanol. The supported system exhibited higher activity, selectivity, and stability than the unsupported Cu_2O (Irfan Malik et al. 2016).

Table 7.1 Best performance data for aqueous electroreduction of carbon dioxide to methanol on copper and copper oxides

Electrode	Electrolyte/cell	Potential (V)	Current density ($mA \cdot cm^{-2}$)	Faradaic efficiency (%)	References
Cu_2O(1)-carbon paper	0.5 M $KHCO_3$/GDE filter-press cell (Nafion 117)	−1.30	6.93	45.7	Albo et al. (2015b)
Cu_2O + ZnO(1/1)-carbon paper	0.5 M $KHCO_3$/GDE filter-press cell (Nafion 117)	−1.30	10.64	17.7	Albo et al. (2015b)
MWCNTs impregnated with Cu_2O (30%)-Cu foil	0.5 M $NaHCO_3$/H-type cell (Nafion)-three electrodes	−0.8	7.5	38	Irfan Malik et al. (2016)
Pre-oxidized Cu foil	0.5 M $KHCO_3$/undivided	−0.05	0.069	240	Frese (1991)
Cu_2O-carbon paper	0.1 M $KHCO_3$/GDE filter-press cell (Nafion 117)	−1.39	10	45	Albo and Irabien (2016)
Cu_2O and ZnO-carbon paper	0.1 M $KHCO_3$/GDE filter-press cell (Nafion 117)	−1.16	10	26	Albo and Irabien (2016)
Cu_2O-stainless steel	$KHCO_3$/undivided-three electrodes	−1.10	5	38	Le et al. (2011)
Cu_2O-graphite plate	0.5 M $KHCO_3$/undivided -two electrodes	3.5	8	8.3	Yadav and Purkait (2015)

This table summarizes maximum Faradaic efficiencies and current densities for methanol formation over metallic copper and copper oxide-based catalyst-electrodes. Potential values, which should be as low as possible, are also shown as an indicative of process energy consumption. The presence of Cu_2O enhances current density and Faradaic efficiency for methanol formation, whereas that of ZnO improves stability. Supported systems exhibited higher current density, Faradaic efficiency, and stability than the unsupported Cu_2O

MWCNTs multiwalled carbon nanotubes, *GDE* gas diffusion electrode, *M* molarity in mol L^{-1}

Best performance data obtained for Cu metal and Cu oxide-based catalysts are shown in Table 7.1.

Alloying copper with various metals, such as lead, nickel, zinc, tin, cadmium, and silver, enhanced electrocatalytic activity and CH_3OH selectivity (Lim et al. 2014; Qiao et al. 2014; Schizodimou and Kyriacou 2012; Watanabe et al. 1991a, b). Electroreduction of CO_2 at nanostructured Cu-Au alloys revealed that methanol Faradaic efficiencies depend on alloy nanostructure and composition (Jia et al. 2014). A summary of the best performance results obtained on Cu alloys are listed in Table 7.2.

Yang et al. (2016) studied the electrocatalytic conversion of CO_2 to methanol over a highly stable and recyclable catalyst, [4-(3-phenoxy-2,2-bis(phenoxymethyl) propoxy)pyridine]@Cu-Pt composite (Yang et al. 2016). More recently, Lu et al.

Table 7.2 Best performance data for aqueous electroreduction of CO_2 to methanol on copper alloys

Electrode	Electrolyte/cell	Potential (V)	Current density ($mA \cdot cm^{-2}$)	Faradaic efficiency (%)	References
$Cu_{63.9}Au_{36.1}$ on Cu films	0.5 M $KHCO_3$/H-type cell (Nafion 117)	−1	0.85	15.9	Jia et al. (2014)
Cu-Ni alloy (Cu/Ni = 60/40)	0.5 M $KHCO_3$/H-type cell (Nafion 117)	−0.9	–	10	Watanabe et al. (1991b)
Cu-Cd alloy (Cu/Cd = 38/62)	0.5 M $KHCO_3$/H-type cell (Nafion 117)	−1.75	–	5	Watanabe et al. (1991a)
$Cu_{88}Sn_6Pb_6$ alloy foil	1.5 M HCl-0.08 M $LaCl_3$/H-type cell (Nafion 117)	−0.7	0.68	35.7	Schizodimou and Kyriacou (2012)
[PYD]@Cu-Pt	0.5 M KCl/H-type cell (Nafion 117)	−0.6	22	37	Yang et al. (2016)
$Pd_{83}Cu_{17}$ aerogel on carbon paper	25 mol% [Bmim]BF4 and 75 mol% water/H-type cell (Nafion 117)	−2.1	31.8	80	Lu et al. (2018)

This table summarizes maximum Faradaic efficiencies and current densities for methanol formation obtained over different copper alloys. Alloying copper with other metals enhances current density and Faradaic efficiency for methanol formation. Supported nanostructured Cu alloys exhibited higher current density and Faradaic efficiency at relatively low potentials
PYD 4-(3-phenoxy-2,2-bis(phenoxymethyl)propoxy)pyridine, *[Bmim]BF4* 1-butyl-3-methylimidazolium tetrafluoroborate, *M* molarity in mol L^{-1}

(2018) investigated the electrocatalytic conversion of CO_2 into methanol over Pd-Cu bimetallic aerogels using an aqueous solution of 1-butyl-3-methylimidazolium tetrafluoroborate as electrolyte. The Faradaic efficiency and the current density for methanol formation simultaneously reached values as high as 80% and 31.8 mA cm^{-2}, respectively, at overpotentials as low as 0.24 V, which is one of the best results reported to date (Lu et al. 2018).

The use of modified electrodes with various surface structures and morphologies, such as Cu nanostructures and organic frameworks, resulted in more effective, selective, and stable Cu-based materials for the electrocatalytic conversion of carbon dioxide into methanol in aqueous solutions. More specifically, the electrochemical CO_2 reduction to methanol proceeded at Cu-transition-metal-based perovskite oxides (Brown Bourzutschky et al. 1990; Qiao et al. 2014; Schwartz et al. 1993). nanostructured Cu on ZnO (Andrews et al. 2013), oxide-derived Cu (Lee et al. 2015), nanostructured Cu (Kim et al. 2017; Kuhl et al. 2012, 2014; Loiudice et al. 2016). Cu and CuO nanoparticles supported on carbon nanotubes (Safdar Hossain et al. 2014) and TiO_2 (Yuan et al. 2018), as well as Cu-based metal-organic porous materials (Albo et al. 2017). Results for the best performing systems are displayed in Table 7.3.

Table 7.3 Best performance data for aqueous electroreduction of CO_2 to methanol on copper nanostructures and organic frameworks

Electrode	Electrolyte/cell	Potential (V)	Current density ($mA \cdot cm^{-2}$)	Faradaic efficiency (%)	References
MOA with DTA and Cu and Zn oxides (CuZnDTA)	0.5 M $KHCO_3$/ GDE filter-press cell (Nafion 117)	−1.41	10	3.4	Albo et al. (2017)
Perovskite ($La_{1.8}Sr_{0.2}CuO_4$)-based carbon GDE	0.5 M KOH/H-type cell (Nafion 117)	−2.30−−2.60	180	2	Schwartz et al. (1993)
Cu nanocluster-ZnO	0.1 M $KHCO_3$/ two compartments (fritted glass)	−1.45	12	2.8	Andrews et al. (2013)
Cu_2O-Cu NPs	0.1 M KCl/ H-type cell (Nafion 115)	−1.4	10.5	5	Lee et al. (2015)
Cu nanocrystal-glassy carbon	0.1 M $KHCO_3$/ H-type cell (Selemion)	−1.1	6	8.1	Loiudice et al. (2016)
20% Cu-CNT	–	−1.7	–	38.4	Safdar Hossain et al. (2014)
Cu-TiO_2 NPs-NG	0.1 M $KHCO_3$	−0.2	0.1	19.5	Yuan et al. (2018)

This table summarizes maximum Faradaic efficiencies and current densities for methanol formation obtained over copper electrodes with different surface structures and morphologies. The use of copper-based nanostructures and organic frameworks improved not only current density and Faradaic efficiency for methanol formation but also the stability of copper-based materials
MOA mesoporous metal-organic aerogel, *DTA* bis-bidentate dithiooxamidate, *GDE* gas diffusion electrode, *M* molarity in mol L^{-1}, *NP* nanoparticle, *CNT* carbon nanotube, *NG* nitrogen-doped graphene

Other transition metal-based catalysts were successfully applied for aqueous electroreduction of CO_2 to CH_3OH. Best performing results obtained are shown in Table 7.4.

The formation of CH_3OH by CO_2 electroreduction in aqueous solution took place not only on Ni- (Barton Cole et al. 2010), Mo- (Summers et al. 1986), or MoO_2-based catalysts (Bandi 1990) but also over Mo-Bi alloy chalcogenides (Sun et al. 2016) and MoS_2 rods supported on TiO_2 nanotubes (Li et al. 2014).

Methanol formation also occurred over Ru-based electrodes such as Teflon-supported Ru (Frese and Leach 1985); mixtures of RuO_2 and other conductive oxides, such as TiO_2, MoO_2, SnO_2, Co_3O_4, and Rh_2O_3, deposited on Ti foils (Bandi 1990); mixed Ru and Ti materials covered with Cu (Bandi and Kuhne 1992); Ru metal; copper and cadmium-modified Ru metal; ruthenium and iridium

Table 7.4 Best results for aqueous electroreduction of CO_2 to CH_3OH on other transition metals

Electrode	Electrolyte/cell	Potential (V)	Current density (mA·cm^{-2})	Faradaic efficiency (%)	References
Mo foil-Cu	0.2 M Na_2SO_4/ undivided	−0.80	0.12	84	Summers et al. (1986)
RuO_2 + TiO_2 (35/65)-Ti	0.2 M Na_2SO_4/two compartments	−1.48	0.06	76	Bandi (1990)
Mo-Bi alloy chalcogenide-carbon paper	0.5 M [bmin]BF4 in MeCN	−0.7	12.1	71.2	Sun et al. (2016)
MoS_2-rods-TiO_2 nanotubes	0.1 M $KHCO_3$/ undivided, three electrodes	−1.30	0.75	44.9	Li et al. (2014)
Electroplated Ru on Cu foil	Na_2SO_4/two compartments (agar bridge)	−0.54	0.08	42	Frese and Leach (1985)
TiO_2 + RuO_2 (75/25)-Ti	0.5 M $KHCO_3$/ undivided-rotating-disk electrode	−1.00	5	29.8	Bandi and Kuhne (1992)
RuOx-Cu-Ti	0.5 M $NaHCO_3$/ undivided-three electrodes	−0.80	2	18.2–41.3	Popić et al. (1997)
RuO_2-diamon (BDD)	0.4 M Britton-Robinson solution/two compartments	−0.80	5	8.12	Spataru et al. (2003)
RuO_2-TiO_2 nanotubes-Pt	0.5 M $NaHCO_3$/H-type cell (Nafion 117)	−0.80	1.2	60.5	Qu et al. (2005)
Co(II) TRPP-Pt plate	APCF-0.1 M KCl-CH_3OH/ undivided, three electrodes	−0.5	0.07	32.9	Ogura and Yoshida (1988)
ES supported-Pt	0.1 M KCl-10 mM APCF-15 mM CH_3OH/ two compartments (glass frit)	−0.7	–	56.3–45.3	Ogura and Fujita (1987)
ES coated-Pt	APCF-CH_3OH/two compartments (glass frit)	−0.9	–	15.5	Ogura and Takamagari (1986)
Fe-C-Pan-PRB-Pt	0.5 M KCl/H-type cell	−0.8	–	12.2	Ogura et al. (1995)
Ni foil	0.1 M $KHCO_3$/H-type cell (Selemion)	−1	4.9	2.3	Kuhl et al. (2014)

This table shows the improvements in current density and Faradaic efficiency for CH_3OH formation attained through altering the configuration, surface structure, morphology, and composition of different transition metals, such as Ni-, Co-, Mo-, Ru-, Fe-, and Ti-based electrocatalysts

BDD boron-doped diamond, *TRPP* tetraphenylporphyrin, *ES* Everitt's Salt, *Pan* polyaniline, *PRB* Prussian blue, *[Bmim]BF4* 1-butyl-3-methylimidazolium tetrafluoroborate, *MeCN* acetonitrile, *APCF* aquopentacyanoferrate (II), *M* molarity in mol L^{-1}

Table 7.5 Best results for aqueous electroreduction of CO_2 to CH_3OH on post-transition metals

Electrode	Electrolyte/cell	Potential (V)	Current density ($mA \cdot cm^{-2}$)	Faradaic efficiency (%)	References
p-GaP-pyridine	0.5 M KCl,10 mM pyridinium /two compartments-three electrodes	−0.06	0.27	90	Jones et al. (2014), Barton et al. (2008)
n-GaAs-crystal-(111)As	0.2 M Na_2SO_4/two compartments (agar bridge)	−1.2−−1.4	0.16–0.2	100	Canfield and Frese (1983), Frese and Canfield (1984)
n-GaAs-crystal-(111)Ga	0.2 M Na_2SO_4/two compartments (agar bridge)	−1.2−−1.4	0.34	30–80	Canfield and Frese (1983), Frese and Canfield (1984)
p-InP	0.2 M Na_2SO_4/two compartments (agar bridge)	−1.2−−1.4	0.1–0.06	44–80	Canfield and Frese (1983)

This table summarizes maximum Faradaic efficiencies and current densities for methanol formation obtained over post-transition metal, such as Ga- and In-based photoelectrodes. These systems show high Faradaic efficiencies but at low current densities and high overpotentials
p-GaP p-gallium phosphide, *n-GaAs* n-gallium arsenide, *p-InP* p-indium phosphide, *M* molarity in mol L^{-1}

oxide mixtures modified with copper and cadmium (Popić et al. 1997); RuO_2-coated diamond (Spataru et al. 2003); and Pt modified with RuO_2 and TiO_2 nanotubes and nanoparticles (Qu et al. 2005). The use of TiO_2 with other metals or oxides deposited on Ti gave rise to enhanced CO_2 reduction to CH_3OH (Bandi 1990; Bandi and Kuhne 1992; Koudelka et al. 1984; Li et al. 2014; Monnier et al. 1980; Ogura and Yoshida 1988; Spataru et al. 2003).

Table 7.5 shows results of CO_2 electrocatalytic conversion to CH_3OH over the best performing post-transition metal-based catalytic systems.

The photoelectrochemical conversion of carbon dioxide to methanol in water solutions happened over p-gallium phosphide (Aurian-Blajeni et al. 1983; Barton et al. 2008; Halmann 1978; Jones et al. 2014), p-gallium arsenide (Aurian-Blajeni et al. 1983), n-gallium arsenide (Canfield and Frese 1983; Frese and Canfield 1984), and p-indium phosphide. Post-transition metal-based photo-electrocatalysts exhibited high selectivity for CH_3OH formation but only at high overpotentials (Canfield and Frese 1983).

In addition, CO_2 reduction in aqueous solutions on various Pt (Barton Cole et al. 2010; Eggins and McNeill 1983; Ogura and Fujita 1987; Sun et al. 2016) and Pd (Barton Cole et al. 2010; Ohkawa et al. 1993; Ohkawa et al. 1994; Podlovchenko et al. 1994; Seshadri et al. 1994; Sun et al. 2016)-based catalyst-electrodes resulted in the production of CH_3OH (Albo et al. 2015a). Best results obtained for both noble metal-based systems are listed in Table 7.6.

Table 7.6 Best performance data for aqueous electroreduction of CO_2 to CH_3OH on noble metals

Electrode	Electrolyte/cell	Potential (V)	Current density ($mA \cdot cm^{-2}$)	Faradaic efficiency (%)	References
Indigo-graphite-Pt	0.1 M KCl-10 mM APCF-15 mM CH_3OH/ two compartments (glass frit)	−0.7	–	70.2−37.2	Ogura and Fujita (1987)
Alizarin-graphite-Pt	0.1 M KCl-10 mM APCF-15 mM CH_3OH/ two compartments (glass frit)	−0.5 −−0.7	–	44.3–69.4	Ogura and Fujita (1987)
Pd disk (HNO_3 and NaOH treated)	0.5 M $NaClO_4$-10 mM pyridine/two compartments (glass frit)	–	0.04	30	Ohkawa et al. (1993)
Pt- or Pd-pyridine	0.5 M KCl-10 mM pyridinium /two compartments-three electrodes	–	0.05	22	Sun et al. (2016), Barton Cole et al. (2010)
Pt- or Pd-pyridine	0.5 M KCl-10 mM 4TBP/ two compartments-three electrodes	–	0.05	14	Barton Cole et al. (2010)

This table summarizes maximum Faradaic efficiencies and current densities for methanol formation obtained over noble metal-, such as Pt and Pt, based catalysts. High Faradaic efficiencies, but with very low current densities, were obtained for noble metal-based catalytic systems
APCF aquopentacyanoferrate (II), *4TBP* 4-tert-butylpyridinium, *M* molarity in mol L^{-1}

7.2.2 Electrolytes

In liquid CO_2 electrocatalytic reduction, electrolytes enable a medium to transfer protons and electrons. The nature and composition of the electrolytes influence catalyst activity and selectivity. In addition, trace impurities incoming from the electrolyte may cause catalyst deactivation, affecting long-term catalytic performance. More specifically, even for the same catalyst-electrode, the electrolyte choice has an acute influence on current density, selectivity and energy efficiency of electrocatalytic conversion of CO_2 to methanol (Albo et al. 2015a; Jhong et al. 2013).

Most reported works generally carried out the electrocatalytic conversion of CO_2 to methanol in aqueous solutions (Albo et al. 2015a). However, the CO_2 solubility in water at atmospheric pressure and room temperature is rather low, which hence retards carbon dioxide adsorption over the electrode and, therefore, the reduction process (Zhang et al. 2018a). The use of nonaqueous solutions or ionic liquids instead of aqueous solutions may enhance CO_2 electrocatalytic reduction performance by improving the solubility of CO_2 and reducing mass transfer limitations (Jhong et al. 2013).

Table 7.7 Influence of aqueous electrolyte type on electrocatalytic reduction of carbon dioxide to methanol

Electrode	Electrolyte/cell	Potential (V)	Current density ($mA{\cdot}cm^{-2}$)	Faradaic efficiency (%)	References
$Cu_{88}Sn_6Pb_6$ alloy	2 M HCl/H-type cell (Nafion)	−0.65	0.24	34.3	Schizodimou and Kyriacou (2012)
$Cu_{88}Sn_6Pb_6$ alloy	0.5 M NaCl–1.5 M HCl/H-type cell (Nafion)	−0.7	0.36	28.2	Schizodimou and Kyriacou (2012)
$Cu_{88}Sn_6Pb_6$ alloy	0.17 M $MgCl_2$–1.5 M HCl/H-type cell (Nafion)	−0.7	0.38	34.1	Schizodimou and Kyriacou (2012)
$Cu_{88}Sn_6Pb_6$ alloy	0.17 M $CaCl_2$–1.5 M HCl/H-type cell (Nafion)	−0.7	0.39	29.6	Schizodimou and Kyriacou (2012)
$Cu_{88}Sn_6Pb_6$ alloy	0.17 M $BaCl_2$–1.5 M HCl/H-type cell (Nafion)	−0.7	0.41	36.3	Schizodimou and Kyriacou (2012)
$Cu_{88}Sn_6Pb_6$ alloy	0.08 M $AlCl_3$–1.5 M HCl/H-type cell (Nafion)	−0.7	0.43	17.8	Schizodimou and Kyriacou (2012)
$Cu_{88}Sn_6Pb_6$ alloy	0.33 M $ZrCl_4$–1.5 M HCl/H-type cell (Nafion)	−0.7	0.45	23.7	Schizodimou and Kyriacou (2012)
$Cu_{88}Sn_6Pb_6$ alloy	0.08 M $NdCl_3$–1.5 M HCl/H-type cell (Nafion)	−0.7	0.61	34.6	Schizodimou and Kyriacou (2012)
$Cu_{88}Sn_6Pb_6$ alloy	0.08 M $LaCl_3$–1.5 M HCl/H-type cell (Nafion)	−0.7	0.68	35.7	Schizodimou and Kyriacou (2012)

This table shows the influence of the utilized ion conductive water solution on Faradaic efficiencies and current densities for methanol formation over a $Cu_{88}Sn_6Pb_6$ alloy foil. Current density increases with surface charge of the cationic species. Faradaic efficiency varies as a function of the utilized cathionic and anionic species, which also affects the pH of the reaction medium

M molarity in mol L^{-1}

Aqueous electrolytes commonly employed in electrochemical reduction of CO_2 consisted of solutions of alkali cations, such as Na^+ and K^+, and various anions, such as Cl^-, HCO_3^- and OH^-, which exhibit great conductivity and are able to supply the needed hydrogen ions (Jhong et al. 2013; Sun et al. 2017; Zhang et al. 2018a).

At the same applied potential, Faradaic efficiency for CH_3OH production depends on the utilized ion conductive solution (Frese et al. 1986). As an example, Table 7.7 shows the effect of electrolyte composition on the CO_2 electrocatalytic conversion to CH_3OH over a $Cu_{88}Sn_6Pb_6$ alloy foil (Schizodimou and Kyriacou 2012). The CO_2 conversion rate increased with surface charge of the cationic species. The selectivity of the process can be also tuned via changing the utilized cation and the pH of the aqueous solution (Schizodimou and Kyriacou 2012).

The utilized anion species also affects the pH in the proximity of the catalyst-electrode surface and therefore the type and selectivity of the formed compounds. Alcohol formation at Cu electrodes was enhanced in KCl, K_2SO_4, $KClO_4$, and diluted $KHCO_3$ solutions at high local pH which, in turn, inhibited H_2 evolution reaction (Hori et al. 1989). As commented above, the CO_2 electroreduction and the H_2 formation reactions compete for the electrocatalytic active sites, protons and electrons, leading to poorer carbon dioxide reduction rates and Faradaic efficiencies toward the preferred compound(s) (Higgins et al. 2019).

Nevertheless, the utilization of water solutions in CO_2 electroreduction processes brings a number of shortcomings, such as small selectivity for CO_2 conversion, due to the competing hydrogen evolution reaction; slow reaction kinetics, which results in increased overpotentials and costs; complicated and expensive products separation and recovery; low solubility of CO_2; and electrode surface contamination by electrolyte impurities, which results in lower yields and catalyst deactivation (Albo et al. 2015a).

Nonaqueous organics solvents, such as methanol, propylene carbonate, and dimethylformamide, present a substantially higher CO_2 solubility than aqueous solvents.

Different nonaqueous electrolytes have been used to improve the solubility of carbon dioxide in the reaction medium and to retard hydrogen evolution, with subsequent enhancement in Faradaic efficiency. However, the main product in nonaqueous organic systems appears to be CO regardless of the utilized electrode materials, due to the absence of protons. Sacrificial proton donors, including trace amounts of water, can be added to supply protons and enable CO_2 reduction to other carbonaceous compounds (Endrődi et al. 2017; Sun et al. 2017). However, none of these studies deal with the potential application of nonaqueous electrolytes in the electroreduction of CO_2 to methanol.

Ionic liquids are nitrogen-containing organic salts that stay as liquids around ambient temperature and have an extraordinary CO_2 capture capacity via coordination with molecular CO_2 (Jones et al. 2014; Sun et al. 2017; Zhang et al. 2018a).

Replacing the conventional aqueous electrolytes with ionic liquids results in the following advantages: suppression of the competing hydrogen evolution reaction, enhancement of electrolyte ionic conductivity, and higher CO_2 solubility at a wider temperature rangeability, thus reducing mass transfer restrictions and facilitating CO_2 electroreduction at lower overpotentials (Albo et al. 2015a; Endrődi et al. 2017; Jones et al. 2014; Rosen et al. 2011; Sun et al. 2017; Welton 1999; Zhang et al. 2018a). In addition, ionic liquids exhibit high thermal and chemical stability, high viscosity, and extremely low vapor pressure and can be recycled or used continuously in a flow system (Jones et al. 2014). On the contrary, ionic liquids are much more costly than water (Jones et al. 2014).

Some recent studies pointed out the potential application of ionic liquids in CO_2 electroreduction (Barrosse-Antle and Compton 2009; Rosen et al. 2011), but few studies deal with the conversion of carbon dioxide into methanol (Sun et al. 2016). In one of these studies, Mo-Bi bimetallic chalcogenide supported on carbon paper was used as electrocatalyst and 1-butyl-3-methylimidazolium tetrafluoroborate in acetonitrile as reaction medium for methanol production by electrocatalytic reduction of carbon dioxide. Simultaneous high values of Faradaic efficiency and current density were obtained, which outperform previous results reported to date (see Table 7.4).

A solid polymer electrolyte is an ion-conducting polymer able to transport charged species. The operation of most of the solid polymer electrolytes is restricted to temperatures below 393 K. The usage of solid polymer electrolytes can overcome the problems imposed by the use of aqueous electrolytes such as high resistance loss, potential reactor leakage, enlarged hydrogen evolution, and transport constraint of absorbed carbon dioxide from electrolyte to catalyst surface. Moreover, the usage of anionic solid polymer electrolytes can contribute to suppress proton reduction (Bandi and Kuhne 1992; Endrődi et al. 2017).

Additional potential advantages of solid polymer electrolytes over liquid electrolytes are easiness of manipulation, enabling the manufacture of thin films with small resistance; simple separation of CO_2 reduction products; low convection, which reduces electrode erosion increasing electrode lifetime; and appropriate tensile strength (Aeshala et al. 2013).

Depending on the conductive nature, two main types of solid polymer electrolytes are distinguished: cationic, mostly H^+ conductors such as CMI-7000, Nafion, and SPEEK, and anionic, mainly OH^-conductors such as SustainionTM, CMI-7001, 1% Amberlyst-SPEEK, and alkali doped PVA (Aeshala et al. 2012, 2013; Endrődi et al. 2017). Most recently, Weekes et al. (2018) reported the development of bipolar membranes which are able to drive protons and hydroxyl ions to the cathodic and anodic electrode, correspondingly, facilitating water splitting (Weekes et al. 2018).

The properties of the polymeric ion-exchange membrane, such as chemical structure, thickness, and wetting properties, affect the performance of the cell, making possible to tailor the physical and chemical characteristics of the polymer to carbon dioxide electroreduction or to a particular compound production. Moreover, durability and pressure handling of the polymeric membrane should be also considered (Endrődi et al. 2017).

Many studies reported the use of solid polymer electrolytes to divide catholyte and anolyte chambers in two-compartment cells (Aeshala et al. 2013). Some papers stated the employment of solid polymer electrolytes in flow cells, where carbon dioxide was continuously delivered to the reactor in the form of CO_2 saturated electrolyte solutions (Aeshala et al. 2013; Endrődi et al. 2017).

7.2.3 Electrode Structure

Optimization of both electrode and cell performance involves improving all of the mass transfer steps which are markedly influenced by electrode structure.

Metallic electrodes applied in CO_2 electrocatalytic reduction can be classified as follows: bulk metal electrodes; metals deposited on a metallic or a carbon-based support, such as multiwalled carbon nanotubes (Irfan Malik et al. 2016), graphite (Yadav and Purkait 2015), and glassy carbon (Hernández et al. 1999; Loiudice et al. 2016); and metals supported on gas diffusion layers, the so-called gas diffusion electrodes (Ikeda et al. 2009).

The carbon dioxide electrocatalytic conversion into methanol occurred at several copper foils, namely, metallic, anodized, and air-oxidized Cu foils (Kuhl et al. 2012; Le et al. 2011; Frese 1991). However, in general, the small electrode superficial area and the low carbon dioxide surface concentration at the cathode, resulting from the reduced CO_2 solubility in water solutions, gave rise to a poor performance of bulk metal electrodes (Hori et al. 1994).

Electrochemical conversion of carbon dioxide into methanol took place at diverse metal- (Bandi 1990; Bandi and Kuhne 1992; Canfield and Frese 1983; Frese and Leach 1985; Le et al. 2011; Ogura et al. 1995; Ogura and Fujita 1987; Ogura and Takamagari 1986; Ogura and Yoshida 1986, 1988; Popić et al. 1997; Qu et al. 2005; Seshadri et al. 1994; Summers et al. 1986; Frese 1991) and carbon-based supported electrodes (Hernández et al. 1999; Irfan Malik et al. 2016; Loiudice et al. 2016; Yadav and Purkait 2015). As an example, Le et al. (2011) studied the CO_2 electrochemical conversion to CH_3OH over a cuprous oxide thin film electrodeposited on stainless steel. Faradaic efficiencies of about 38% were attained, being noticeably higher than those obtained for air-oxidized or anodized copper electrodes (Le et al. 2011).

In a gas diffusion electrode, the catalytic material, usually polymer-bonded catalyst particles, is dispersed by different methods, such as hand and spray painting, onto a porous structure, typically a carbon-based material, support, the so-called gas diffusion layer (Merino-Garcia et al. 2016).

The use of gas diffusion electrodes led to a considerable improvement in the performance of carbon dioxide electroreduction to methanol in liquid phases by improving the mass transfer of CO_2 (Albo and Irabien 2016; Albo et al. 2015b, 2017; Chang et al. 2009; Lan et al. 2014).

7.2.4 *Electrochemical Cell Configuration*

Different electrochemical cell configurations have been developed in order to accelerate the reaction rate, decrease the energy consumption, and increase the energy efficiency of the electrocatalytic conversion of CO_2 (Jhong et al. 2013). Electrochemical cell design has a pronounced effect on mass transport.

Liquid phase CO_2 electroreduction implies the use of electrochemical cell configurations able to work with both cathodic and anodic electrolytes in liquid state and may include a solid polymer electrolyte separation membrane. These cell configurations also enable direct reduction of the CO_2 captured in the electrolyte solutions. In this regard, CO_2 capture capacity can be tuned by electrolyte composition and pH adjustment.

Table 7.8 Electroreduction of CO_2 to methanol in undivided three-electrode cells

Electrode	Electrolyte/ cell	Potential (V)	Current density ($mA \cdot cm^{-2}$)	Faradaic efficiency (%)	References
Pre-oxidized Cu foil (17 h, 403 K)	0.5 M $KHCO_3$	−1.55	7.1	-	Frese (1991)
Anodized Cu foil	0.5 M $KHCO_3$	−1.25	1.4	120	Frese (1991)
Pre-oxidized Cu-TiO_x (50 min., 573 K)	0.5 M $KHCO_3$	−1.06	0.74	180	Frese (1991)
Pre-oxidized Cu-TiO_x (45 min., 773 K)	0.5 M $KHCO_3$	−0.45	0.30	30	Frese (1991)
Air-furnace oxidized Cu	0.5 M $KHCO_3$	−1.50	10	2	Le et al. (2011)
Mo foil on Cu wire	0.2 M Na_2SO_4	−0.80	0.05	55	Summers et al. (1986)
Mo foil on Cu wire	0.05 M H_2SO_4	−0.69	0.31	46	Summers et al. (1986)
RuO_x thermally deposited-Ti	0.5 M $KHCO_3$	−0.80	2	30.5–17.2	Popić et al. (1997)
RuO_x-Cd thermally deposited-Ti	0.5 M $KHCO_3$	−0.80	2	20.4–38.2	Popić et al. (1997)

This table shows that in a classical undivided three-electrode electrochemical cell configuration, modest Faradaic efficiencies and current densities for methanol formation at relatively high potentials are usually obtained due to mass transfer restrictions and CO_2 reduction products reoxidation at the anode electrode

M molarity in mol L^{-1}

There is no standardized procedure or experimental setup for CO_2 electroreduction in liquid phase (Chen et al. 2018; Sun et al. 2017). Various cell arrangements utilized so far are reviewed in the following.

Undivided electrochemical cells consist of three dense electrodes immersed in a common anode and cathode liquid electrolyte (Bandi and Kuhne 1992; Chang et al. 2009; Le et al. 2011; Li et al. 2014; Popić et al. 1997; Summers et al. 1986; Frese 1991). In the undivided electrochemical cell reaction configuration, recovery of product is difficult and involves subsequent separation stages, making the process more expensive (Jhong et al. 2013). There are some studies in literature on electrochemical reduction of carbon dioxide to methanol in undivided cells (Bandi and Kuhne 1992; Le et al. 2011; Li et al. 2014; Popić et al. 1997; Summers et al. 1986; Frese 1991; Yadav and Purkait 2015). Some results for the best performing undivided cell systems are displayed in Table 7.8.

Most of the studies on liquid phase electrochemical conversion of CO_2 to methanol used a two-compartment electrochemical cell configuration. This system consists of two compartments and three electrodes, the catalyst, the counter, and the reference electrodes. In this cell setup, the cathode and anode compartments are separated by an anion or a proton-conducting polymeric membrane, in H-type electrochemical cells (Albo et al. 2015b; Andrews et al. 2013; Bandi 1990; Irfan

Malik et al. 2016; Jia et al. 2014; Kuhl et al. 2012; Lee et al. 2015; Loiudice et al. 2016; Ogura et al. 1995; Ogura and Yoshida 1988; Qu et al. 2005; Schizodimou and Kyriacou 2012; Schwartz et al. 1993; Spataru et al. 2003; Sun et al. 2017; Watanabe et al. 1991a, b; Yang et al. 2016); an agar bridge (Canfield and Frese 1983; Frese and Leach 1985), or a glass frit (Andrews et al. 2013; Ogura and Fujita 1987; Ogura and Takamagari 1986; Ogura and Yoshida 1986; Seshadri et al. 1994). Two-compartment cell configuration enables ionic conductivity while keeping the electrodes in close proximity and prevents the transport of CO_2 reduction products to the anodic chamber, avoiding reoxidation reactions and improving separation of products (Sun et al. 2017). Results for the best performing two-compartment systems are shown in Table 7.9.

Continuous flow cells have some benefits compared with batch cell equivalents, such as improving blending of distinct phases, heat transmission, and mass transport, as well as the control of temperature and residence time in the reactor (Endrődi et al. 2017; Rosen et al. 2011).

Some studies on electrocatalytic conversion of CO_2 to methanol employed flow cells where CO_2 was delivered to the cathode as a CO_2 saturated electrolyte solution (Albo and Irabien 2016; Albo et al. 2015b, 2017; Lan et al. 2014). Some reported results are summarized in Table 7.10. For instance, the performance of Cu_2O and Cu_2O and ZnO deposited on carbon papers toward the electrocatalytic reduction of CO_2 to methanol has been studied in a continuous flow filter-press reactor. The ZnO-containing electrodes were stable, whereas those including only Cu_2O showed long-term deactivation (Albo and Irabien 2016; Albo et al. 2015b).

The parent setup of a continuous flow system comprises an anode and a cathode channel for the flow of the corresponding electrolyte. An ion-conducting polymer membrane enables the separation between the two channels. In continuous flow cells, the electrolyte is supplied to the corresponding catalytic electrode, typically in a gas diffusion electrode configuration, providing a catalyst-liquid electrolyte interface. Gaseous carbon dioxide is supplied on one side of the cathode electrocatalyst, whereas the catholyte flows on the other side.

Removing the membrane from the parent continuous flow cell configuration leads to a typical microfluidic flow cell, where the two electrodes are separated by a common liquid electrolyte flow channel, enabling the study of both anode and cathode performance by means of an external reference electrode (Endrődi et al. 2017; Higgins et al. 2019; Rosen et al. 2011; Sun et al. 2017; Weekes et al. 2018).

Another derived configuration is the membrane electrode assembly continuous flow cell, where the two gas diffusion electrodes are pressed together on both sides of an ion-conducting polymer separation membrane with no flow channels between electrodes, enhancing the contact and the transport of species between electrodes (Endrődi et al. 2017; Merino-Garcia et al. 2016; Weekes et al. 2018). The anolyte and dissolved CO_2 catholyte are fed to the corresponding compartment. Typical membrane electrode assembly designs comprise parallel plate and filter-press cell configurations (Endrődi et al. 2017; Sun et al. 2017).

Table 7.9 Electroreduction of CO_2 to CH_3OH in two-compartment cells

Electrode	Electrolyte/cell	Potential (V)	Current density ($mA \cdot cm^{-2}$)	Faradaic efficiency (%)	References
$RuO_2 + TiO_2$ (35/65)-Ti	0.05 M H_2SO_4/two compartments	−0.55	5	24	Bandi (1990)
$RuO_2 + MoO_2 + TiO_2$ (25/30/45)-Ti	0.05 M H_2SO_4/two compartments	−0.55	5	12	Bandi (1990)
$RuO_2 + Co_3O_4 + SnO_2 + TiO_2$ (20/10/8/62)-Ti	0.05 M H_2SO_4/two compartments	−0.55	5	7	Bandi (1990)
$RuO_2 + TiO_2$ (35/65)-Ti	Phosphate buffer 0.2 M/two compartments	−1.44	0.08	35	Bandi (1990)
$RuO_2 + Co_3O_4 + SnO_2 + TiO_2$ (20/10/8/62)-Ti	0.2 M Na_2SO_4/two compartments	−1.49	0.05	53	Bandi (1990)
RuO_2-TiO_2 NPs-Pt	0.5 M $NaHCO_3$/H-type cell (Nafion)	−0.80	1.2	40.2	Qu et al. (2005)
Fe(II) TRPP-Pt	2HNN-3DPNCo(II) in 0.1 M KCl-CH_3OH/H-type cell (H^+ membrane)	−0.5	0.36	12.2	Ogura and Yoshida (1988)
Co(II) TRPP-Pt	2HNN-3DPNCo(II) in 0.1 M KCl-CH_3OH/H-type cell (H^+ membrane)	−0.5	0.39	15.1	Ogura and Yoshida (1988)
Ni(II) TRPP-Pt	2HNN-3DPNCo(II) in 0.1 M KCl-CH_3OH/H-type cell (H^+ membrane)	−0.5	0.15	14.4	Ogura and Yoshida (1988)
Cr(III) TRPP-Cl-Pt	2HNN-3DPNCo(II) l in 0.1 M KCl-CH_3OH/H-type cell (H^+ membrane)	−0.5	0.1	11.6	Ogura and Yoshida (1988)
Fe(III) TRPP-Cl-Pt	2-HNN-3DPNCo in 0.1 M KCl-CH_3OH/H-type cell (H^+ membrane)	−0.5	0.37	11.1	Ogura and Yoshida (1988)
Indigo-graphite-Fe	0.1 M KCl-10 mM APCF-15 mM CH_3OH/two compartments (glass frit)	−0.7		42.8	Ogura and Fujita (1987)
2-Aminoanthraquinone-graphite-Fe	0.1 M KCl-10 mM APCF-15 mM CH_3OH/two compartments (glass frit)	−0.7		31	Ogura and Fujita (1987)
Alizarin-graphite-Fe	0.1 M KCl-10 mM APCF-15 mM CH_3OH/two compartments (glass frit)	−0.7		30.8	Ogura and Fujita (1987)
p-Benzoquinone-graphite-Fe	0.1 M KCl-10 mM APCF-15 mM CH_3OH/two compartments (glass frit)	−0.7		36	Ogura and Fujita (1987)

(continued)

Table 7.9 (continued)

Electrode	Electrolyte/cell	Potential (V)	Current density (mA·cm^{-2})	Faradaic efficiency (%)	References
ES coated-Pt	$[Fe(C_6H_2(OH)_2(SO_3)_2)_2]$-$CH_3OH$/two compartments (glass frit)	−0.90	–	14.5	Ogura and Takamagari (1986)
ES coated-Pt	$K[Cr(C_2O_4)_2(H_2O)_2)]$-CH_3OH/two compartments (glass frit)	−0.9	–	14.5	Ogura and Takamagari (1986)
Fe-S-Pan-PRB-Pt	0.5 M KCl/H-type cell	−0.8	–	6.8	Ogura et al. (1995)
Fe-T-Pan-PRB-Pt	0.5 M KCl/H-type cell	−0.8	–	8.3	Ogura et al. (1995)
Fe-C-Pan-PRB-Pt-self assembled	0.2 M KCl/H-type cell	−0.8	–	10.1	Ogura et al. (1995)
n-GaAs-crystal-(110)Ga	0.2 M Na_2SO_4/two compartments (agar bridge)	−1.2	0.13	14	Canfield and Frese (1983)

This table summarizes maximum Faradaic efficiencies and current densities for methanol formation obtained in a typical two-compartment electrochemical cell configuration. This cell configuration enables ionic conductivity while keeping the electrodes in close proximity and prevents the transport of CO_2 reduction products to the anodic chamber, avoiding reoxidation reactions and improving separation of products

TRPP tetraphenylporphyrin, *ES* Everitt's Salt, *Pan* polyaniline, *PRB* Prussian blue, *APCF* aquopentacyanoferrate (II), *n-GaAs* n-gallium arsenide, *NP* nanoparticle, *2HNN* 2-hydroxil-1-nitrosonaphthalene, *3DPN Co(II)* 3,6-disulphonatocobal(II), *M* molarity in mol L^{-1}

Table 7.10 Electroreduction of CO_2 to methanol in membrane electrode assembly flow cells

Electrode	Electrolyte/cell	Potential (V)	Current density ($mA \cdot cm^{-2}$)	Faradaic efficiency (%)	References
Benchmark MOF HKUST-1	0.5 M $KHCO_3$/GDE filter-press cell (Nafion 117)	−0.9	10	5.6	Albo et al. (2017)
CuAdeAce	0.5 M $KHCO_3$/GDE filter-press cell (Nafion 117)	−1.75	10	0.7	Albo et al. (2017)
CuMOA with CuDTA	0.5 M $KHCO_3$/GDE filter-press cell (Nafion 117)	−1.41	10	1.9	Albo et al. (2017)
Carbon nanopowder supported CuO and Cu	1 M $KHCO_3$/GDE two compartments MEA flow cell (Nafion 117)	−1.40	17.3	2.5	Lan et al. (2014)

This table summarizes maximum Faradaic efficiencies and current densities for methanol formation obtained in membrane electrode assembly continuous flow cell configurations, where two porous electrodes are pressed together on both sides of an ion-conducting polymer separation membrane, being the anolyte and dissolved CO_2 catholyte fed to the corresponding compartment. This cell configuration reduces mass transfer restrictions and enhances ionic transference between electrodes, which resulted in higher current densities with respect to undivided and two-compartment cell configurations

MOF metal-organic framework, *MOA* mesoporous metal-organic aerogel, *DTA* bis-bidentate dithiooxamidate, *GDE* gas diffusion electrode, *CuAdeAce* [Cu_2(m3-adeninate)$_2$(m2-OOC(CH_3)$_2$)] n, *MEA* membrane electrode assembly, *M* molarity in mol L^{-1}

7.2.5 Operation Parameters

Although operation parameters, such as applied potential, reaction temperature and pressure, and electrolyte pH, may highly influence product yields and selectivities, only a small number of works dealt with the influence of these parameters on the liquid phase CO_2 electroreduction performance (Gattrell et al. 2006; Riyanto and Afiati 2018).

The external applied potential has a different effect on the selectivity of CO_2 electrocatalytic reduction, especially in relation to the competitive hydrogen evolution reaction. In addition, the applied potential also affects the transfer of electrons and the energy barrier during the process, resulting in different power consumption.

As a rule, the selectivity for CO_2 electrocatalytic reduction shows a typical volcano-type dependence on applied potential increase. Therefore, there is an optimum potential which maximizes selectivity to CO_2 reduction for a specific electrocatalyst system and under a given operating conditions, which should be identified for each particular case (Zhang et al. 2018a).

The electrolyte pH significantly impacts the electrocatalytic activity and selectivity for aqueous-phase CO_2 electroreduction. In this regard, one of the problems to be faced is that the pH of the electrolyte is restricted to neutral or slightly acid values by the acid-base CO_2, bicarbonate and carbonate buffer equilibria (Higgins et al. 2019; Zhang et al. 2018a). In fact, dissolved molecular CO_2 can be hydrated to carbonic acid or dissociated into carbonate or bicarbonate.

The ratios between the different species which can adsorb on the electrode surface, and, thus, the reaction pathways and final products, are dictated by the pH of the solvent (Higgins et al. 2019; Zhang et al. 2018a).

Moreover, electrocatalytic reduction of CO_2 is thermodynamically favored at high proton concentrations or low pHs. However, the competitive hydrogen evolution reaction is thermodynamically favored over CO_2 reduction with decreasing the pH and rendering low selectivity and Faradaic efficiency for CO_2 reduction (Zhang et al. 2018a).

In addition, OH^- is formed during the CO_2 reduction to methanol, leading to an increased pH on the surface of the electrode which retarded CO_2 reduction. In this regard, electrolytes with buffering properties, which are able to supply anions, are normally utilized in order to minimize the local pH change (Albo et al. 2015a).

A number of works reported the influence of pH on the electroreduction of CO_2 to CH_3OH (Frese and Leach 1985; Frese and Canfield 1984; Schizodimou and Kyriacou 2012; Spataru et al. 2003; Frese et al. 1986).

More specifically, Schizodimou and Kyriacou (2012) analyzed the influence of the electrolyte acidity on methanol formation by electrocatalytic conversion of carbon dioxide over a copper alloy. The obtained results indicated that the CO_2 reduction rate and current density increase, whereas the Faradaic efficiency decreases on decreasing the pH of the reaction medium. Best performing results in terms of simultaneous high values of Faradaic efficiency and current density for methanol production by electrocatalytic conversion of carbon dioxide were obtained at a pH of 4.14 (Schizodimou and Kyriacou 2012).

Studies of methanol formation by carbon dioxide reduction over Ru-based electrodes revealed also that the Faradaic efficiency increased on decreasing acidity of the electrolyte solution (Frese and Leach 1985; Spataru et al. 2003). The highest current density and Faradaic efficiency for methanol production were reached at a pH of 5.9 over a RuO_2-coated diamond electrode (Spataru et al. 2003).

In the same way, the current density and Faradaic efficiency for methanol production by electrocatalytic conversion of carbon dioxide at Mo-based electrodes slightly improve on decreasing the pH of the reaction medium from 4.2 to 1.5 under similar potential and temperature conditions (Frese et al. 1986).

Results obtained for the effect of pH of the reaction medium on the electrocatalytic production of CH_3OH from carbon dioxide over the different above commented electrocatalyst systems are listed in Table 7.11.

In CO_2 aqueous electrocatalytic reduction, operation temperature considerably affects product yield and selectivity as a result of changes in reaction rates (Gattrell et al. 2006). Moreover, temperature also influences adsorption equilibria, altering product selectivity (Zhang et al. 2018a).

Table 7.11 Effect of pH on aqueous electroreduction of CO_2 to CH_3OH

Electrode	Electrolyte/cell	pH	Faradaic efficiency (%)	Potential (V)	Charge (C)	References
$Cu_{88}Sn_6Pb_6$ alloy	0.01 M CH_3COOH, 0.01 M CH_3COONa_2/H-type cell (Nafion)	4.53	10.2	−0.75	14.82	Schizodimou and Kyriacou (2012)
$Cu_{88}Sn_6Pb_6$ alloy	0.01 M CH_3COOH, 0.003 M $BaCl_2$ /H-type cell (Nafion)	4.56	7.9	−0.75	19.56	Schizodimou and Kyriacou (2012)
$Cu_{88}Sn_6Pb_6$ alloy	0.01 M CH_3COOH, 0.0016 M $LaCl_3$/H-type cell (Nafion)	4.14	7.1	−0.75	48.36	Schizodimou and Kyriacou (2012)
$Cu_{88}Sn_6Pb_6$ alloy	0.01 M CH_3COOH, 0.05 M $LaCl_3$/H-type cell (Nafion)	4.02	6.7	−0.75	49.11	Schizodimou and Kyriacou (2012)
Electroplated Ru on Cu foil	0.2 M Na_2SO_4/two compartments (agar bridge)	4.2	2.9	−0.65	98.5	Frese and Leach (1985)
Electroplated Ru on Cu foil	0.1 M H_2SO_4/two compartments (agar bridge)	1.4	Traces	−0.35	114	Frese and Leach (1985)
RuO_2-diamon (BDD)	0.4 M Britton-Robinson solution/ two compartments	2.4	5.93	−0.5	50.9	Spataru et al. (2003)
RuO_2-diamon (BDD)	0.4 M Britton-Robinson solution/ two compartments	3.9	7.73	−0.6	37.9	Spataru et al. (2003)
RuO_2-diamon (BDD)	0.4 M Britton-Robinson solution/ two compartments	5.9	8.12	−0.8	51.3	Spataru et al. (2003)
HCl pretreated Mo foil	0.2 M Na_2SO_4/two compartments (agar bridge)	4.2	42	−0.7	11.8	Frese et al. (1986)
HCl pretreated Mo foil	0.05 M H_2SO_4/two compartments (agar bridge)	1.5	46	−0.68	18.7	Frese et al. (1986)

This table shows the influence of the pH of the reaction medium on Faradaic efficiency and current density, expressed as electric charge, for methanol formation by aqueous-phase CO_2 electroreduction. In general, current density increases, whereas Faradaic efficiency decreases on decreasing the pH of the reaction medium, because higher pH enhances CO_2 solubility and reaction rate while hinders methanol synthesis in favor of hydrogen evolution reaction, leading to lower methanol selectivity

BDD boron-doped diamond, *M* molarity in mol L^{-1}

On the other hand, competitive hydrogen formation significantly lessens with decreasing temperature. Lower temperature increases not only CO_2 solubility, which induces a pH variation, but also diffusivity and reaction ability, which generally results in a subsequent increase in production. Moreover, current density increases with operating temperatures (Endrődi et al. 2017; Frese and Leach 1985; Kim et al. 2017; Frese et al. 1986; Zhang et al. 2018a).

Temperature exhibited an important effect on CO_2 electrocatalytic conversion to methanol when an electroplated Ru electrode in CO_2 saturated 0.2 M Na_2SO_4 electrolytes was used (Frese and Leach 1985). The Faradaic efficiency reached a maximum value as high as 42% at 60 °C with a simultaneous enhancement in current (Frese and Leach 1985).

The effect of temperature on CO_2 electrocatalytic conversion to methanol was also confirmed on studying electrocatalytic reduction of CO_2 at Mo-based electrodes in 0.05 M H_2SO_4 solutions (Frese et al. 1986). The Faradaic efficiency for methanol production increases from 21% to 46%, whereas the current density declines from 590 to 310 mA cm^{-2} on decreasing temperature from 52 to 22 °C (Frese et al. 1986).

The low carbon dioxide solubility in aqueous medium can restrict CO_2 electroreduction. Increased pressure enhances CO_2 solubility and enables the use of higher operation temperatures, endowing higher CO_2 conversion rate and current density (Endrődi et al. 2017; Zhang et al. 2018a).

High CO_2 pressure also favors the CO_2 reduction versus proton reduction. The flow of carbon dioxide from the electrolyte solution toward the cathode also increases at higher pressures. Therefore, the use of pressurized CO_2 accelerates the CO_2 reduction process and often leads to appreciable changes in product selectivity, current density, and Faradaic efficiencies (Hara et al. 1994; Zhang et al. 2018a).

The study of the effect of pressure on methanol formation via CO_2 liquid phase electroreduction is almost an unexplored territory. A few works analyzed the influence of CO_2 pressure on Faradaic efficiency and current density for methanol formation over p-gallium phosphide-based photoelectrodes (Aurian-Blajeni et al. 1983; Barton et al. 2008). Results obtained in these studies are tabulated in Table 7.12.

7.3 Gaseous Phase Electrocatalytic Production of Methanol from Carbon Dioxide

In recent years, the discovery of novel electrocatalysts and acquirement of a basic knowledge on electrocatalytic production of methanol from CO_2 substantially progressed as a result of the intensive research carried out in liquid phase electrochemical CO_2 reduction processes. However, these studies provided limited practical information which is required for the potential application of the technology (Higgins et al. 2019).

Table 7.12 Effect of pressure on aqueous electroreduction of CO_2 to CH_3OH

Electrode	Electrolyte/cell	Pressure (Atm.)	Faradaic efficiency (%)	Potential (V)	Charge (C)	References
p-GaP-pyridine	0.5 M KCl,10 mM Pyridinium /two compartments-three electrodes	10	90	−0.06	0.27[a]	Barton et al. (2008)
p-GaP	0.5 M Na_2CO_3/two compartments-three electrodes (BDH)	3.5	0.5	−1	312	Aurian-Blajeni et al. (1983)
p-GaP	0.5 M $NaHCO_3$/two compartments-three electrodes (BDH)	7.5	3.3	−1.4	370	Aurian-Blajeni et al. (1983)

This table shows the influence of operation pressure on Faradaic efficiency and current density, expressed as electric charge, for methanol formation by aqueous-phase CO_2 electroreduction. In general, current density and Faradaic efficiency increases with pressure, because higher pressure improves CO_2 solubility and reaction rate and favors methanol synthesis over hydrogen evolution reaction

a current density mA cm^{-2}, *p-GaP* p-gallium phosphide, *M* molarity in mol L^{-1}

7.3.1 Electrocatalysts

Liquid phase CO_2 electroreduction studies resulted in an improved understanding of basic aspects, such as the effect of configuration of electrocatalyst and electrochemical cell, pH, temperature, and pressure on catalyst activity and selectivity, which led to the discovery of more active and selective catalyst morphologies and compositions.

Currently, the research efforts aims to transfer the acquired understanding to the development of easily scalable electrocatalytic reactor designs, by incorporating best performing identified catalysts into feasible catalyst-electrode structures based on solid electrolytes, such as gas diffusion electrodes or thin electrocatalyst films deposited on solid oxide electrolytes, able to work employing gaseous CO_2 as reagent.

In CO_2 gas-fed electrocatalytic systems, the intrinsic limitations of solution-based systems toward achieving high yields and low energy costs in the preferred conversion route can be overcome. In fact, by delivering gaseous CO_2 to the cathodic electrocatalyst, the local partial pressure of CO_2 on the catalyst surface can be decoupled from CO_2 solubility in water, enabling strategies to tune surface coverage of reaction intermediates, and, thus, selectivity, by controlling reactant transport. The use of CO_2 gas-based electrocatalytic systems allows also minimizing mass transfer limitations associated with solution-based systems by close interaction at the triple interface between CO_2 gas, the catalytic electrode, and the solid or liquid electrolyte (Chen et al. 2018; Higgins et al. 2019).

Gaseous CO_2-fed gas diffusion electrodes can be improved to attain enhanced yields and current densities. The proper kind of catalytic material and preparation procedure should be carefully chosen to obtain the maximum catalytic surface area. Micro- and nanoscale electrode structures should be designed to maximize the simultaneous transport of reactive species, such as electrons, CO_2, and protons, as well as products. The porosity, pore structure, hydrophilicity, and thickness of the diffusion layer must be also optimized for an appropriate electrode performance (Ampelli et al. 2016; Chen et al. 2018; Endrődi et al. 2017; Higgins et al. 2019; Weng et al. 2018).

The technique utilized for catalyst immobilization on the gas diffusion layer has a profound effect on gas diffusion electrode performance, determining current density, product distribution, and cell stability (Endrődi et al. 2017). The gas diffusion electrodes are typically prepared by airbrushing of dispersions of catalyst nanopowders and ionomers onto the carbon-based diffusion layer (Ma et al. 2016), by electrodeposition over a carbon paper (Hoang et al. 2017), via incorporation into a carbon-based material (Abdelwahab et al. 2017) or by deposition of carbon pastes on metal mesh using rolling out procedures (Wang et al. 2014).

Other aspect to consider is the presence and content of an ionomer, such as Nafion, as binder material, as the addition of an ionomer to the catalyst precursor dispersion may result in an enhancement of the electronic and ionic conductivity, active surface area, and hydrophilicity of the catalyst layer (Endrődi et al. 2017).

The design of gaseous CO_2-fed gas diffusion electrodes is more complex than that of dissolved CO_2-fed ones. The formers require the use of multiple layers of porous cathode to control the rate of reactants delivery. The layout of catalyst within carbon layers can also considerably affect the gas diffusion electrode performance (Zhuang et al. 2018).

7.3.2 *Electrolytes*

Solid electrolytes are dense membranes where the transfer of ionic species is driven by electrical potential difference. In gas phase electrocatalytic conversion of carbon dioxide, depending on the nature of the membrane, two types of solid electrolytes can be distinguished: solid polymer electrolytes and ceramic solid oxide electrolytes.

The solid polymer electrolytes applied for liquid phase CO_2 electroreduction and described above can also be used for gas phase electrocatalytic production of methanol from CO_2 (Aeshala et al. 2012, 2013).

Both the current density and product distribution vary depending on the chemical characteristics of the polymer electrolyte, mainly due to the different ionic conductivity and ion-exchange capacity (Aeshala et al. 2012, 2013).

Aeshala et al. (2012) concluded that the use of solid polymer electrolytes relieved the restrictions to carbon dioxide transport in the electrochemical reactor (Aeshala et al. 2012). In aqueous-phase CO_2 electroreduction superficial water content on the catalyst is greater than in a typical ion-exchange polymeric membrane, where some humidity is needed to ensure proper ionic conductivity, resulting in an almost suppression of the hydrogen evolution reaction (Higgins et al. 2019).

Solid oxide electrolytes can be classified based on electrolyte conductive nature (Vernoux et al. 2013) into:

- Alkaline ion, such as Na^+, K^+, and Li^+, conductors, of which the most typical ones are β-alumina and NASICON.
- Anionic ion, such as O^{2-}, Cl^-, and F^-, conductors, of which the most common ones are those conducing O^{2-} ions, represented by two main solid oxide electrolytes: fluorite-type and perovskite-structured oxides.
- Protonic conductors, where two different types can be distinguished depending on the operation temperature: intermediate temperature protonic electrolytes, including both solid acid electrolytes, such as phosphates, arsenates, sulfates, and seleniates, which operate between 473 and 573 K, and perovskites, which work at temperatures between 723 and 873 K, and high temperature, typically above 923 K, protonic conductors such as $La_xSr_{1-x}CoO_{3-\delta}$, $La_xSr_{1-x}MnO_{3-\delta}$, and $La_xSr_{1-x}Co_yFe_{1-y}O_{3-\delta}$.
- Mixed conductors, which conduct both ions and electrons. There are two main types of mixed conductors: CeO_2-based and perovskite-based electrolytes.

Anionic, mainly yttria-stabilized-zirconia, and alkaline, such as β-alumina and NASICON, solid oxide electrolytes are well known for being effective materials in electrochemical promotion of catalytic CO_2 hydrogenation, whereas in the case of protonic electrolytes, promoter species, H^+, can take part not only in enhancing the catalytic rate through electrochemical promotion of catalysis but also in oxidizing H_2O and reducing CO_2 reactant gases implicated in the electrochemical reaction itself (Vernoux et al. 2013).

Oxygen- and proton-conducting solid oxide electrolytes have been also used for co-electrolysis of gaseous carbon dioxide and steam in double-chamber solid oxide electrolyte cells. Electrolytes should meet a series of requirements to be applied in these systems, such as very low electron conduction, high ion conduction, stability under different applied potentials and reaction environments, and easiness of conforming into layers. Regarding the later, the electrolyte layer should be thin and hard enough as to make the ionic transport resistance low and to avoid the mixing of gases between the anodic and cathodic chambers, correspondingly. In addition, low costly and widely available electrolytes are preferred. Yttria-stabilized-zirconia meets applicability requirements at high temperature, being the most commonly used electrolyte. On the contrary, the oxygen ion conductivity of yttria-stabilized-zirconia considerably decreases with temperature, limiting the application of yttria-stabilized-zirconia at intermediate-low temperature ranges (Zhang et al. 2017).

Intermediate temperature protonic electrolytes offer some advantages over oxygen ion conductors, such as higher ionic conductivity and good chemical compatibility with electrodes, mainly with Ni. The most intensively studied materials during the past years are ABO_3 perovskite-type oxides, where A = Ba, Sr and B = Ce, Zr, since perovskite-type oxides have the best protonic conductivity. Among the perovskite-type oxides, $BaZr_{0.1}Ce_{0.7}Y_{0.2}O_{3-\delta}$ is considered as the best material because it combines both large bulk conductivity and good chemical stability (Nguyen and Blum 2015).

7.3.3 Electrochemical Cell Configuration

Solution phase electrochemical cells are useful for comparing intrinsic activities of different electrocatalysts but impractical for the electrocatalytic production of methanol from carbon dioxide on a more developed scale. Moreover, the rate of CO_2 reduction is restricted due to the relatively low solvability of CO_2 in water (Albo et al. 2015a). In order to overcome these challenges, it is necessary to move toward more practical reactor designs based on delivering gaseous CO_2 at the cathodic catalyst-electrode.

In CO_2 gas-fed electrochemical cells, the electrolyte can be a liquid, as in the above commented microfluidic flow reactors, or a solid ion-conducting polymer, acid, or oxide material. The CO_2 reduction takes place at the triple boundary between the gas, the catalyst-electrode, and the electrolyte. Depending on the type of solid electrolyte, two different configurations can be distinguished (Chen et al. 2018; Endrődi et al. 2017; Higgins et al. 2019): polymer electrolyte membrane cell and solid oxide electrolyte cell.

A polymer electrolyte membrane cell is basically a membrane electrode assembly flow cell where gaseous CO_2 is supplied into the cathode chamber. In a polymer electrolyte membrane system, a solid polymer electrolyte provides protonic or anionic conductivity, separation of gaseous products, and electric insulation between electrodes. In a typical polymer electrolyte membrane cell configuration, a concentrated CO_2 gaseous stream is supplied to the cathodic gas diffusion electrode, where the carbon dioxide electroreduction occurs at the gas, catalyst, and polymer electrolyte interface, whereas water is supplied to the anodic electrode.

The polymer electrolyte membrane cell setup can surpass liquid phase mass transport restrictions, enable products separation, and alleviate liquid electrolyte associated degradation issues, resulting in higher current density, energy efficiency, and durability (Chen et al. 2018; Higgins et al. 2019; Merino-Garcia et al. 2016). However, the application of polymer electrolyte membrane cells to diluted CO_2 residual streams is still challenging, given that the change in feed composition can harmfully impact the catalytic performance, polymer electrolyte lifetime and local pH at the catalyst (Higgins et al. 2019).

The available literature in gas phase electrocatalytic production of methanol from CO_2 is slightly scarce (Aeshala et al. 2012, 2013; Centi and Perathoner 2009, 2010; Gangeri et al. 2009; Genovese et al. 2013; Gutiérrez-Guerra et al. 2016; Jiménez et al. 2017; Merino-Garcia et al. 2018; Sebastián et al. 2017; Shironita et al. 2013a, b). The main results obtained in these studies for the gaseous CO_2 electroreduction to methanol in polymer electrolyte membrane cells are summarized in Table 7.13.

Table 7.13 Gas phase electrocatalytic conversion of CO_2 to methanol in polymer electrolyte membrane cells

Electrode	Electrolyte/ cell	Potential (V)	Current density ($mA \cdot cm^{-2}$)	Faradaic efficiency (%)	References
Cu_2O-carbon paper	SPE (CMI-7000)	-2	2.4	20	Aeshala et al. (2013)
Cu_2O-carbon paper	SPE (AMI-7001)	-2	3.7	5	Aeshala et al. (2013)
Cu-carbon paper	SPE (Nafion)	–	11.1	0.54	Aeshala et al. (2012)
Cu-carbon paper	SPE (SPEEK)	–	8.9	0.13	Aeshala et al. (2012)
Pt-C-carbon paper	SPE (Nafion 117)	−0.35	–	40	Shironita et al. (2013b)
Pt-C-carbon paper	SPE (Nafion 117)	−0.4	20	35	Shironita et al. (2013a)
Pt-Ru-C-carbon paper	SPE (Nafion 117)	−0.45	15	75	Shironita et al. (2013a)
Pt-CNTs-carbon cloth	SPE (Nafion)	–	16	1.9	Jiménez et al. (2017)
Pt-Ru-C-carbon cloth	SPE (Nafion 115)	1.25	1	0.33	Sebastián et al. (2017)
Ru-C-carbon cloth	SPE (Nafion 115)	1.25	0.5	0.91	Sebastián et al. (2017)
Cu-graphite GDE	SPE (Sterion)	−1.7	30	80[a]	Gutiérrez-Guerra et al. (2016)
Cu-activated carbon GDE	SPE (Sterion)	−3	30	45[a]	Gutiérrez-Guerra et al. (2016)
Cu-CNFs GDE	SPE (Sterion)	−2.5	30	5[a]	Gutiérrez-Guerra et al. (2016)

(continued)

Table 7.13 (continued)

Electrode	Electrolyte/ cell	Potential (V)	Current density ($mA \cdot cm^{-2}$)	Faradaic efficiency (%)	References
Cu-activated carbon GDE	SPE (Sterion)	–	–	0.013[b]	Lucas-Consuegra et al. (2018)
Pt-CNTs-carbon cloth	SPE (Nafion 117)	–	–	0.0005[c]	Gangeri et al. (2009), Centi and Perathoner (2010), Genovese et al. (2013), Centi and Perathoner (2009)
Fe-CNTs-carbon cloth	SPE (Nafion 117)	–	–	0.0005[c]	Gangeri et al. (2009), Centi and Perathoner (2010), Genovese et al. (2013), Centi and Perathoner (2009)

This table summarizes maximum Faradaic efficiencies and current densities for methanol formation obtained in a more practical polymer electrolyte membrane cell configuration, which is basically a membrane electrode assembly flow cell where gaseous CO_2 is supplied into the cathode chamber. This cell setup overcomes mass transport restrictions, enables products separation, and avoids electrocatalyst deactivation by electrolyte components, resulting in higher current density, Faradaic efficiency, and durability

a selectivity, *b* μmol/h mg Cu, *c* μmol/h cm^2, *SPE* solid polymer electrolyte, *GDE* gas diffusion electrode, *CNF* carbon nanofiber, *CNT* carbon nanotube, *PEM* polymer electrolyte membrane

A couple of studies investigated the conversion of gaseous CO_2 to CH_3OH, with modest current efficiencies, at gas diffusion electrodes consisting of Cu electroplated on carbon paper over various cationic and anionic solid polymer electrolytes (Aeshala et al. 2012, 2013).

The gas phase electrocatalytic production of methanol from carbon dioxide also took place, with Faradaic efficiencies and current densities up to about 75% and 20 mA cm^{-2}, respectively, at carbon supported Pt (Shironita et al. 2013b) and Pt-Ru (Shironita et al. 2013a) gas diffusion electrodes. Ru addition efficiently enhances CO_2 reduction to CH_3OH. In these systems, temperature also affected the CO_2 equilibrium adsorption.

Shironita et al. (2013b) also observed on a carbon-supported Pt-based membrane electrode assembly of a reversible fuel cell that the higher the temperature, in the 299–363 K range, the higher the Faradaic efficiency for methanol formation and the easier the methanol desorption from the electrocatalytic surface (Shironita et al. 2013b).

Nanostructured carbon-based electrodes, such as carbon nanotubes, are reported to help electronic and ionic conduction and prevent chemical and mechanic deterioration of the catalyst. Recently, Jiménez et al. (2017) studied the gas phase electrocatalytic conversion of CO_2 to methanol using carbon nanotubes supported Pt catalysts (Jiménez et al. 2017). Sebastián et al. (2017) also studied the co-electrolysis of steam and carbon dioxide in a similar system but using $IrRuO_x$ as anode (Sebastián et al. 2017). The main products were methanol on a carbon nanotubes supported Pt-Ru catalyst and a mixture of alcohols, including methanol, on a carbon nanotubes supported Ru catalyst.

Other carbon nanotube-supported Fe, Pt, and Co metal catalysts were also explored for electrocatalytic conversion of CO_2 to methanol. The carbon nanotubes supported iron catalysts exhibited superior activity but inferior stability than the carbon nanotubes supported platinum catalysts. The carbon nanotubes supported bimetallic Fe-Co systems were the most stable (Centi and Perathoner 2009, 2010; Gangeri et al. 2009; Genovese et al. 2013). More recently, Cu supported on activated carbon, graphite, and carbon nanofibers was employed to catalyze CO_2 conversion to methanol. In these systems IrO_2 was used as anodic electrode (Gutiérrez-Guerra et al. 2016; Lucas-Consuegra et al. 2018).

Solid oxide electrolyte cells have been used as electrochemical membrane reactors to carry out, among other heterogeneous catalytic reactions, gaseous CO_2 hydrogenation. A solid oxide electrolyte cell consists of a dense solid oxide electrolyte membrane and two or three porous electrodes. In closed circuit operation, the conducting ions travel from one electrode to the other where the conducting ions react with the gaseous content of the corresponding chamber.

The typical configuration of a solid oxide electrolyte cell is the double-chamber cell also known as solid oxide electrolyzer. In solid oxide electrolyzers, the anode and cathode are separated so that each electrode can be exposed to different gaseous mixtures.

Solid oxide electrolyzers can exploit available exhaust heat streams to maximize electrical efficiency and operate at temperatures typically higher than 923 K, offering both thermodynamic and kinetic advantages over low temperature CO_2 and H_2O co-electrolysis, leading to higher electrolysis efficiency (Uhm and Kim 2014; Wu et al. 2013; Yan et al. 2014). At this temperature level, carbon dioxide is more easily activated on the catalyst-electrode surface and mass transport ceases to be the limiting step on the reaction rate, leading to higher current densities for CO_2 electrocatalytic reduction (Zhang et al. 2017). Moreover, solid oxide electrolyzers enable the use of inexpensive oxide-derived electrodes, instead of the precious metals typically utilized in solution-based electrolysis cells (Yan et al. 2014).

Despite the anticipated advantages of solid oxide electrolyzers, the potential practical application of this technology, making use of discontinuous renewable energy sources, requires the development of cells able to sustain multiple start-stop cycles (Jones et al. 2014).

Ideally, in a solid oxide electrolyzer employing oxygen ion conductors, carbon dioxide and water accept electrons and dissociate at the cathodic electrode to produce oxygen ions, which pass through the electrolyte membrane to arrive at the anode, where oxygen ions drop electrons being oxidized to gaseous oxygen. Simultaneously, at the cathode, chemisorbed carbonaceous and hydrogen species may react to yield different products (Zhang et al. 2017).

In the case of solid oxide electrolyzers utilizing proton-conducting electrolytes, H_2O is oxidized to gaseous oxygen at the anodic electrode, simultaneously producing protons, which pass through the protonic electrolyte and react with CO_2 chemisorbed on the cathodic electrode to form different products (Wu et al. 2013).

Most of the work in the area of CO_2 conversion to fuels in solid oxide electrolyzers has focused on trying to co-electrolyze carbon dioxide and steam to produce a syngas, consisting mainly of carbon monoxide and hydrogen, which is processed downstream to produce hydrocarbons in a separated reactor (Jones et al. 2014; Lei et al. 2017; Frese et al. 1986). However, the need of long gas transport networks for connecting both processes leads to additional cost and huge energy dissipation. Therefore, the integration of CO_2 and H_2O co-electrolysis, favored at high temperature, and Fischer-Tropsch or other synthesis reactions, favored at low temperature, in a single unit can potentially reduce the investment, due to the resulting more compact equipment, and increase the energy efficiency, by better heat integration, of the combined system.

In the future, to obtain methanol or other valuable compounds, the solid electrolyte and the cathodic and anodic catalyst-electrodes should be elaborated to enable operation of the electrolyzer at temperatures between 473 and 873 K, which allows the formation of various products with high current densities (Zhang et al. 2017). The integration of CO_2 and H_2O co-electrolysis and the synthesis reaction can be realized by utilizing intermediate temperature solid electrolytes operating at temperatures compatible with methane synthesis, such as perovskites (Xie et al. 2011b), or with methanol production, such as solid acids (Jones et al. 2014). The integration of CO_2 and H_2O co-electrolysis and the synthesis reaction can be also realized by establishing a temperature gradient in the cell, which allows one part to be controlled at high temperature for CO_2 and H_2O co-electrolysis, while the other is at reduced temperature for the synthesis reaction, resulting in a significant enhancement in hydrocarbon yield (Xie et al. 2011a, b).

The second solid electrolyte cell configuration is the single-chamber cell, where the solid oxide electrolyte only acts as an ionically conductive catalyst support and both electrodes are exposed to the same reacting mixture with two or one electrodes, the reference and/or the counter electrode(s), being inert for the reaction of interest. The single-chamber solid oxide electrolyte cell can be used to electrochemically promote catalytic reaction rates by pumping promoter ions, such as O^{2-} and K^+, to or from the catalyst-electrode surface during reaction, causing a dramatic change in catalytic activity and/or selectivity and, therefore, in product yield (Marnellos and Stoukides 2004).

In electrochemical promotion systems, the application of small potentials between a working-catalyst-electrode and an inert gold electrode results in movement of promoting ions incoming or toward the catalyst surface, resulting in a change in the chemisorptive properties of coexisting reagent, intermediates or product species that enables to increase reaction rate, and, thus, catalyst to be operated at moderated conditions, lowering associated energy costs; selectivity tuning toward the preferred compound formation, improving energetic efficiency in compound production; enhancing catalyst sufferance to harmful feed constituents, electrochemically hindering poisons fixation on catalytic active sites; increasing the durability of the catalytic system via on-site electrochemical reactivation; and making possible to electrochemically monitor and control the reaction by an attached sensing device (Ruiz et al. 2013, 2014a, b; 2016).

Electrochemically promoted CO_2 hydrogenation to, among other compounds, methanol has been carried out in single-chamber solid oxide electrolyte cells which consist of thin films of Pt (Ruiz et al. 2013, 2014a), Cu (Ruiz et al. 2014b), or TiO_2-supported Fe (Ruiz et al. 2016), acting as catalyst-working electrodes, deposited on ion-conducting solid electrolyte supports, such as YSZ (yttria-stabilized-zirconia), an O^{2-} conductor (Ruiz et al. 2014a, 2016), or K-βAl_2O_3, a K^+ conductor (Ruiz et al. 2013, 2014b). These works studied the effect of metallic catalyst, Pt, Pd, Cu, or Fe; solid oxide electrolyte, YSZ or K-βAl_2O_3; film preparation procedure, "paint-brushing," "dip-coating," or "electroless"; and operating conditions, potential, temperature, and H_2 to CO_2 molar ratio, on CO_2 conversion, on methanol selectivity, and on the level of electrochemical promotion of the catalyst (Ruiz et al. 2013, 2014a, b, 2016).

In these studies, the electrochemical promotion of CO_2 hydrogenation was investigated at bench scale, on a tubular electrochemical catalyst configuration and using concentrated CO_2 gases, simulating carbon dioxide capture streams, as well as varying hydrogen to carbon dioxide molar ratios between 1 and 4, to mimic the intermittency of hydrogen production by water electrolysis using intermittent renewable power. Results revealed that the catalytic CO_2 hydrogenation reaction can be electrochemically enhanced and the selectivity to methanol can be tuned by varying the cell voltage (Ruiz et al. 2013, 2014a, b, 2016).

7.3.4 Operation Parameters

In the case of carbon dioxide electrocatalytic conversion to methanol in gas phase, the evaluation of the effect of operating conditions on product yield and selectivity is practically unexplored yet. The influence of different operation conditions, such as potential, temperature, H_2 to CO_2 molar ratio, and flowrate, on the electropromoted carbon dioxide conversion to methanol over different systems is summarized in Table 7.14.

In electrochemically promoted CO_2 hydrogenation to methanol, applied potential significantly affects CO_2 conversion and product distribution, as a result of the effect of potential on the strength of the interaction of reactants, mainly H_2 and CO_2, and reaction intermediates with the catalyst surface.

Formation of methanol was promoted under conditions where the superficial coverage of both reactants, carbon dioxide and hydrogen, onto the catalyst-electrode is comparable, such as at small potassium or oxygen ion surface coverages, which match with positive potentials and with slightly positive or negative potentials for K-βAl_2O_3 and YSZ, correspondgly (Ruiz et al. 2013, 2014a, b, 2016).

In general, CO_2 does not adsorb on clean Pt (Ruiz et al. 2013, 2014a), Cu (Ruiz et al. 2014b), or Fe (Ruiz et al. 2016) surfaces, i.e., in the absence of promoting ions, at highly positive potentials for K^+ or at nearly open circuit potentials for O^{2-}. At the same time, metals, Pt in higher extent than Cu, rather catalyze hydrogen evolution at highly negative voltages. Consequently, a strong competitive adsorption among CO_2

Table 7.14 Electrochemically promoted catalytic CO_2 hydrogenation to methanol in single-chamber solid oxide electrolyte membrane reactors

Electrode	Electrolyte/cell	Potential (V)	CO_2 conversion (%) / ΔCO_2 conversion	Selectivity (%)/Δ selectivity	References
Dip-coated Pt	K-βAl_2O_3/tubular-single chamber, $H_2/CO_2 = 3$	1	≈1/–	1.5/27	Ruiz et al. (2013)
Paint-brushed Pt	YSZ/tubular-single chamber, $H_2/CO_2 = 2$	0.5	24/3.2	8/800	Ruiz et al. (2014a)
Electroless-coated Cu	K-βAl_2O_3/tubular-single chamber, $H_2/CO_2 = 2/4$	2.5/1	25/4.3[a]	55.4/34	Ruiz et al. (2014b)
Dip-coated Fe-TiO_2	YSZ/tubular-single chamber, $H_2/CO_2 = 3$	−0.5	15/3.7	50/50	Ruiz et al. (2016)

This table summarizes maximum carbon dioxide conversion and methanol selectivity and improvement ratio attained by electrochemical promotion of gas phase catalytic CO_2 hydrogenation in single-chamber solid oxide electrolyte membrane reactors. The selectivity and efficiency of the carbon dioxide conversion process can be electrochemically enhanced, in several orders of magnitude, by varying the cell voltage

YSZ yttria-stabilized-zirconia, *ΔCO_2 conversion* increment in overall CO_2 conversion in number of times, *ΔSelectivity* increment in methanol selectivity in number of times, H_2/CO_2 H_2 to CO_2 molar ratio, *a* obtained at $H_2/CO_2=4$ and 1 V

and H_2 can be anticipated, being methanol production restricted by the chemisorption of both reactants.

Therefore, depending on operation temperature and gas composition, certain values of voltage or promoting ion surface concentration maximizes catalytic activity and methanol selectivity.

The effect of temperature on electrochemically promoted carbon dioxide hydrogenation to methanol was analyzed on Cu-K-βAl_2O_3 (Ruiz et al. 2014b) and Fe-TiO_2-YSZ (Ruiz et al. 2016). For stoichiometric H_2 to CO_2 molar ratios of three, which thermodynamically favors methanol formation by CO_2 hydrogenation, both CO_2 conversion and selectivity to CH_3OH exhibit a maximum at a given temperature independently of the utilized potential value (Ruiz et al. 2014b, 2016). The dependence of CH_3OH yield on temperature is in agreement with the statement that formation of methanol by CO_2 hydrogenation is thermodynamically disfavored with increasing reaction temperature. In addition, methanol is transformed into dimethyl ether at 523–573 K, and the two oxygenates are transformed into hydrocarbons at temperatures around 673 K, where both reverse water-gas shift and methanation reactions are as well favored. Decreasing the applied potential around open circuit conditions resulted in a significant increment in the maximum value of selectivity to methanol, which is obtained at lesser temperature (Ruiz et al. 2014b, 2016).

In general, in electrochemically promoted CO_2 hydrogenation to methanol on Pt-K-βAl_2O_3 (Ruiz et al. 2013), Pt-YSZ (Ruiz et al. 2014a), and Cu-K-βAl_2O_3 (Ruiz et al. 2014b) systems, an increment in the maximum carbon dioxide conversion and a slight decrease in the electropromotion level were observed with the increase of the hydrogen to carbon dioxide molar ratio, supposedly, due to the presence of hydrogen in excess with respect to that stoichiometrically necessary, of about 3, for methanol formation, resulting also in a slight decrease in methanol selectivity (Ruiz et al. 2013, 2014a, b).

On the contrary, selectivity to CH_3OH exhibited a maximum for hydrogen to carbon dioxide ratios of two, under optimum conditions of low promoter surface coverage, which matched with the stoichiometry of CH_3OH formation by hydrogenation of the adsorbed CO deposited by CO_2 decomposition on the catalyst surface. So, methanol formation is favored by thermodynamics over other synthesis reactions (Ruiz et al. 2013, 2014a, b). In contrast, over Fe-TiO_2-YSZ, selective CO_2 conversion to methanol exhibited a maximum at the stoichiometrically required ratio of three (Ruiz et al. 2016).

The influence of flow rate on the electropromoted CO_2 hydrogenation to CH_3OH over Cu-K-βAl_2O_3 (Ruiz et al. 2014b) and Fe-TiO_2-YSZ (Ruiz et al. 2016) has been also analyzed. The increase in gas flow rate at a constant temperature of 598 K and using a stoichiometric hydrogen to carbon dioxide molar ratio led to a decrease in the maximum CH_3OH selectivity, which, in the case of Cu, also moved to more negative voltages (Ruiz et al. 2014b).

7.4 Conclusions

The electrocatalytic production of methanol from CO_2 is one of the most promising technologies for recycling CO_2 as fuels and chemicals. Unfortunately, the technology readiness level remains below that needed for commercial applications which require the availability of stable electrocatalytic systems which concurrently show overpotentials below about 200 mV, current densities higher than 0.1 A cm^{-2}, and Faradaic efficiencies close to 100%.

The electrocatalytic production of methanol from carbon dioxide occurred at different types of electrocatalytic materials, such as metals, metal alloys, and metallic oxides and complexes, among others. However, the poor selectivity, low efficiency, and lack of stability of present-day electrocatalysts limit the potential industrial application of the technology.

Among the catalyst-electrode materials studied in liquid phase carbon dioxide reduction, copper-, ruthenium-, and molybdenum-based materials were the most active for the process. Substantial research work has been carried out in order to enhance the performance of electrocatalysts toward methanol production through altering configuration, surface structure, morphology, and composition of electrocatalysts.

Particularly, Cu oxide electrodes, more specifically Cu (I) species in Cu_2O, are promissory catalysts given the good values of combined current density and methanol selectivity, with Faradaic efficiencies as high as 100%, obtained over these catalysts. However, methanol yields diminish on increasing time on stream, due to a lack of stability. In this regard, the addition of ZnO improves stability of the system. Alloying Cu with other metals improved the reversibility, CO_2 reduction rate, and methanol selectivity at lower overpotentials. Nanostructured copper also showed improved performance for the electrocatalytic production of methanol from CO_2 in terms of Faradaic efficiency, selectivity, current density, stability, and overpotential.

An excellent alternative to improve catalyst performance for aqueous CO_2 reduction is the usage of gas diffusion electrodes with both, protonic and anionic, solid polymer electrolytes. In this way, Cu-containing nanostructured or metal-organic porous materials supported on porous carbon-based gas diffusion layers promoted CO_2 conversion to methanol, reaching simultaneous values of Faradaic efficiency and current density as high as 80% and 31.8 mA cm^{-2}, respectively, at overpotentials as low as 240 mV, which is one of the preeminent outcomes reported to date.

Currently, the research efforts focus on incorporating best performing electrocatalysts identified in liquid phase electrochemical CO_2 reduction into practical catalyst-electrode structures based on solid electrolytes, such as gas diffusion electrodes or thin electrocatalyst films deposited on solid oxide electrolytes that operate using gaseous CO_2 instead of CO_2 saturated electrolyte solutions.

Different electrochemical cell setups have been developed in order to improve the performance of the CO_2 electrocatalytic conversion process and to advance in the potential practical application of the technology, such as the utilization of electrochemical membrane reactors, based on solid polymer or solid oxide electrolytes.

Liquid-liquid solid polymer two-compartment cells are the most studied for liquid phase CO_2 electroreduction. Little work has been performed on continuous flow membrane electrode assembly cells, even though the continuous flow membrane electrode assembly cell might be the unique scalable configuration for liquid phase CO_2 electroreduction. Works based on feeding gaseous CO_2 at the cathodic catalyst-electrode are coming up, aiming to overpass mass transport restrictions.

Gas phase co-electrolysis of carbon dioxide and water to methanol has been carried out in solid polymer electrochemical membrane reactors, also known as polymer electrolyte membrane cells, usually with low efficiencies, but relatively high values of current densities and Faradaic efficiencies, up to about 15 mA cm^{-2} and 75%, respectively, were recently obtained at carbon supported Pt-Ru-based gas diffusion electrodes.

Solid oxide electrochemical membrane reactors can also provide gas phase co-electrolysis of CO_2 and H_2O to methanol, but the efficiencies are still very low, and further developments are still needed. In addition, electrochemical promotion of CO_2 hydrogenation to methanol in solid oxide electrolyte cells is proposed as a promising alternative to improve, in several orders of magnitude, the selectivity and efficiency of the carbon dioxide conversion process.

Although the electrocatalytic production of methanol from CO_2 shows a great potential, substantial technical progresses are necessary for the process to be feasible and commercially applicable. The main challenge for the advancement of the electrocatalytic production of methanol from CO_2 is to decrease the energy cost for methanol production. The tendency is to maximize methanol yield with a minimum energy input. In this regard, low energy cost is attained through high selectivity, measured as Faradaic efficiency, for methanol formation at low overpotentials. On the other hand, reaction rate, measured as current density, defines the size of the reactor and the investment cost, whereas reaction conditions govern the operating costs of the process.

Despite the many advances made in the electrocatalytic conversion of CO_2 in water solutions, some challenges are still to be overcome, such as the high overpotential required; the low carbon dioxide solubility in aqueous medium at ambient temperature and atmospheric pressure; the challenging product separation, which implies a cost; and the plugging and deactivation of the electrodes by electrolyte impurities.

Therefore, there are still opportunities for improvement, such as optimization of reaction conditions to improve reaction rates, which allow increasing the concentration of CO_2 in the electrolyte water solution, i.e., by decreasing temperature and increasing pressure; the use of new reaction media, such as nonaqueous solutions and ionic liquids, which enhance the solubility of CO_2 and suppress the formation of hydrogen; improvement of the activity, selectivity, and stability of the electrocatalysts, by exploring innovative composite and nanostructured materials; optimization of the design of electrode, reactor, and system, by utilizing gas diffusion electrodes and electrolyte membrane-based electrochemical reactors, which enable reduction of internal resistance of the cell and enhanced mass transfer and, thus, improve reaction rate or current density; and development and testing in continuous mode of easily scalable flow reactors.

One way to enhance CO_2 conversion, to decrease overpotentials, and to increase current density is by raising the operating temperature. Solid oxide electrolyte cells, which conduct ions such as O^{2-}, K^+, or H^+, allow the conversion of CO_2 at temperatures higher than 673 K. Although the electrochemical promotion of catalytic CO_2 hydrogenation to methanol, among other products, and the co-electrolysis of carbon dioxide and steam to carbon monoxide and hydrogen or to CH_4 in solid oxide electrolyte cells have undergone significant advances, additional research is still necessary to enhance the efficiency, selectivity, stability, and lifetime of the electrocatalysts; to lower the operation temperature of the system, with the subsequent increase in energy efficiency and catalyst durability; to cheapen the process, by material cost minimization and compact reactor designs; and to obtain greater simplicity and scalability in the preparation of electrocatalyst and solid electrolyte materials, as well as in reactor design.

In the future, the solid electrolytes and the cathodic and anodic catalyst-electrodes will have to be developed to enable electrolyzers to work at temperatures between 473 and 573 K, in order to also produce methanol in a single step with high current densities and in a continuous mode of operation. In this regard, one of the most promising options is the development of solid acid-based electrolytes, such as phosphates, arsenates, sulfates, and seleniates, with improved ionic conductivity and thermal stability at the temperature range of interest. Other potential approach is to optimize electrolyzer design in order to provide a temperature gradient for co-electrolysis, to CO and H_2, and methanol synthesis reactions to consecutively proceed.

References

Abdelwahab A, Castelo-Quibén J, Pérez-Cadenas M, Elmouwahidi A, Maldonado-Hódar FJ, Carrasco-Marín F, Perez-Cadenas A (2017) Cobalt-doped carbon gels as electro-catalysts for the reduction of CO_2 to hydrocarbons. Catalysts 7. https://doi.org/10.3390/catal7010025

Aeshala LM, Rahman SU, Verma A (2012) Effect of solid polymer electrolyte on electrochemical reduction of CO_2. Sep Purif Technol 94:131–137. https://doi.org/10.1016/j.seppur.2011.12.030

Aeshala LM, Uppaluri RG, Verma A (2013) Effect of cationic and anionic solid polymer electrolyte on direct electrochemical reduction of gaseous CO_2 to fuel. J CO2 Util 3-4:49–55. https://doi.org/10.1016/j.jcou.2013.09.004

Albo J, Irabien A (2016) Cu_2O-loaded gas diffusion electrodes for the continuous electrochemical reduction of CO_2 to methanol. J Catal 343:232–239. https://doi.org/10.1016/j.jcat.2015.11.014

Albo J, Alvarez-Guerra M, Castaño P, Irabien A (2015a) Towards the electrochemical conversion of carbon dioxide into methanol. Green Chem 17:2304–2324. https://doi.org/10.1039/C4GC02453B

Albo J, Sáez A, Solla-Gullón J, Montiel V, Irabien A (2015b) Production of methanol from CO_2 electroreduction at Cu_2O and Cu_2O/ZnO-based electrodes in aqueous solution. Appl Catal B Environ 176–177:709–717. https://doi.org/10.1016/j.apcatb.2015.04.055

Albo J, Vallejo D, Beobide G, Castillo O, Castaño P, Irabien A (2017) Copper-based metal–organic porous materials for CO_2 electrocatalytic reduction to alcohols. ChemSusChem 10:1100–1109. https://doi.org/10.1002/cssc.201600693

Ampelli C, Centi G, Passalacqua R, Perathoner S (2016) Electrolyte-less design of PEC cells for solar fuels: prospects and open issues in the development of cells and related catalytic electrodes. Catal Today 259:246–258. https://doi.org/10.1016/j.cattod.2015.07.020

Andrews E, Ren M, Wang F, Zhang Z, Sprunger PT, Kurtz R, Flake J (2013) Electrochemical reduction of CO_2 at cu nanocluster / (100) ZnO electrodes. J Electrochem Soc 160:H841–H846. https://doi.org/10.1149/2.105311jes

Aurian-Blajeni B, Halmann M, Manassen J (1983) Electrochemical measurement on the photoelectrochemical reduction of aqueous carbon dioxide on p-gallium phosphide and p-gallium arsenide semiconductor electrodes. Solar Energy Mater 8:425–440. https://doi.org/10.1016/0165-1633(83)90007-2

Bandi A (1990) Electrochemical reduction of carbon dioxide on conductive metallic oxides. J Electrochem Soc 137:2157–2160. https://doi.org/10.1149/1.2086903

Bandi A, Kuhne HM (1992) Electrochemical reduction of carbon-dioxide in water-analysis of reaction-mechanism on ruthenium-titanium-oxide. J Electrochem Soc 139:1605–1610. https://doi.org/10.1149/1.2069464

Barrosse-Antle LE, Compton RG (2009) Reduction of carbon dioxide in 1-butyl-3-methylimidazolium acetate. Chem Commun 25:3744–3746. https://doi.org/10.1039/B906320J

Barton Cole E, Lakkaraju PS, Rampulla DM, Morris AJ, Abelev E, Bocarsly AB (2010) Using a one-electron shuttle for the multielectron reduction of CO_2 to methanol: kinetic, mechanistic, and structural insights. J Am Chem Soc 132:11539–11551. https://doi.org/10.1021/ja1023496

Barton EE, Rampulla DM, Bocarsly AB (2008) Selective solar-driven reduction of CO_2 to methanol using a catalyzed p-GaP based Photoelectrochemical cell. J Am Chem Soc 130:6342–6344. https://doi.org/10.1021/ja0776327

Brown Bourzutschky JA, Homs N, Bell AT (1990) Hydrogenation of CO_2 and CO_2/CO mixtures over copper-containing catalysts. J Catal 124:73–85. https://doi.org/10.1016/0021-9517(90)90104-R

Canfield D, Frese KW (1983) Reduction of carbon dioxide to methanol on n - and p - GaAs and p - InP . Effect of crystal face, electrolyte and current density. J Electrochem Soc 130:1772–1773. https://doi.org/10.1149/1.2120090

Centi G, Perathoner S (2009) Opportunities and prospects in the chemical recycling of carbon dioxide to fuels. Catal Today 148:191–205. https://doi.org/10.1016/j.cattod.2009.07.075

Centi G, Perathoner S (2010) Problems and perspectives in nanostructured carbon-based electrodes for clean and sustainable energy. Catal Today 150:151–162. https://doi.org/10.1016/j.cattod.2009.09.009

Centi G, Perathoner S (2011) CO_2-based energy vectors for the storage of solar energy. Greenhouse Gases: Sci Technol 1:21–35. https://doi.org/10.1002/ghg3.3

Chang T-Y, Liang R-M, Wu P-W, Chen J-Y, Hsieh Y-C (2009) Electrochemical reduction of CO_2 by Cu_2O-catalyzed carbon clothes. Mater Lett 63:1001–1003. https://doi.org/10.1016/j.matlet.2009.01.067

Chen C, Khosrowabadi Kotyk J, Sheehan S (2018) Progress toward commercial application of electrochemical carbon dioxide reduction. Chem 4:2571–2586. https://doi.org/10.1016/j.chempr.2018.08.019

Durst J et al (2015) Electrochemical CO_2 reduction-a critical view on fundamentals, materials and applications. CHIMIA Int J Chem 69:769–776. https://doi.org/10.2533/chimia.2015.769

Eggins BR, McNeill J (1983) Voltammetry of carbon dioxide: part I. A general survey of voltammetry at different electrode materials in different solvents. J Electroanal Chem Interf Electrochem 148:17–24. https://doi.org/10.1016/S0022-0728(83)80127-2

Endrődi B, Bencsik G, Darvas F, Jones R, Rajeshwar K, Janáky C (2017) Continuous-flow electroreduction of carbon dioxide. Prog Energy Combust Sci 62:133–154. https://doi.org/10.1016/j.pecs.2017.05.005

Frese KW (1991) Electrochemical reduction of CO_2 at intentionally oxidized copper electrodes. J Electrochem Soc 138:3338–3344. https://doi.org/10.1149/1.2085411

Frese KW, Canfield D (1984) Reduction of CO_2 on n - GaAs electrodes and selective methanol synthesis. J Electrochem Soc 131:2518–2522. https://doi.org/10.1149/1.2115351

Frese KW, Leach S (1985) Electrochemical reduction of carbon dioxide to methane, methanol, and CO on Ru electrodes. J Electrochem Soc 132:259–260. https://doi.org/10.1149/1.2113780

Frese KW, Leach SC, Summers DP (1986) Electrochemical reduction of aqueous carbon dioxide to methanol. US patent no. 4,609,441

Ganesh I (2014) Conversion of carbon dioxide into methanol – a potential liquid fuel: fundamental challenges and opportunities (a review). Renew Sust Energ Rev 31:221–257. https://doi.org/10.1016/j.rser.2013.11.045

Gangeri M et al (2009) Fe and Pt carbon nanotubes for the electrocatalytic conversion of carbon dioxide to oxygenates. Catal Today 143:57–63. https://doi.org/10.1016/j.cattod.2008.11.001

Gattrell M, Gupta N, Co A (2006) A review of the aqueous electrochemical reduction of CO_2 to hydrocarbons at copper. J Electroanal Chem 594:1–19. https://doi.org/10.1016/j.jelechem.2006.05.013

Genovese C, Ampelli C, Perathoner S, Centi G (2013) Electrocatalytic conversion of CO_2 to liquid fuels using nanocarbon-based electrodes. J Energy Chem 22:202–213. https://doi.org/10.1016/S2095-4956(13)60026-1

Goeppert A, Czaun M, Jones J-P, Surya Prakash GK, Olah GA (2014) Recycling of carbon dioxide to methanol and derived products – closing the loop. Chem Soc Rev 43:7995–8048. https://doi.org/10.1039/C4CS00122B

Gutiérrez-Guerra N, Moreno-López L, Serrano-Ruiz JC, Valverde JL, de Lucas-Consuegra A (2016) Gas phase electrocatalytic conversion of CO_2 to syn-fuels on cu based catalysts-electrodes. Appl Catal B Environ 188:272–282. https://doi.org/10.1016/j.apcatb.2016.02.010

Halmann M (1978) Photoelectrochemical reduction of aqueous carbon dioxide on p-type gallium phosphide in liquid junction solar cells. Nature 275:115. https://doi.org/10.1038/275115a0

Hara K, Tsuneto A, Kudo A, Sakata T (1994) Electrochemical reduction of CO_2 on a Cu electrode under high pressure: factors that determine the product selectivity. J Electrochem Soc 141:2097–2103. https://doi.org/10.1149/1.2055067

Hernández RM, Márquez J, Márquez OP, Choy M, Ovalles C, Garcia JJ, Scharifker B (1999) Reduction of carbon dioxide on modified glassy carbon electrodes. J Electrochem Soc 146:4131–4136. https://doi.org/10.1149/1.1392603

Higgins D, Hahn C, Xiang C, Jaramillo TF, Weber AZ (2019) Gas-diffusion electrodes for carbon dioxide reduction: a new paradigm. ACS Energy Lett 4:317–324. https://doi.org/10.1021/acsenergylett.8b02035

Hirunsit P, Soodsawang W, Limtrakul J (2015) CO_2 electrochemical reduction to methane and methanol on copper-based alloys: theoretical insight. J Phys Chem C 119:8238–8249. https://doi.org/10.1021/acs.jpcc.5b01574

Hoang TTH, Ma S, Gold JI, Kenis PJA, Gewirth AA (2017) Nanoporous copper films by additive-controlled electrodeposition: CO_2 reduction catalysis. ACS Catal 7:3313–3321. https://doi.org/10.1021/acscatal.6b03613

Hori Y, Murata A, Takahashi R (1989) Formation of hydrocarbons in the electrochemical reduction of carbon dioxide at a copper electrode in aqueous solution. J Chem Soc Faradaic Trans 1 (85):2309–2326. https://doi.org/10.1039/f19898502309

Hori Y, Wakebe H, Tsukamoto T, Koga O (1994) Electrocatalytic process of CO selectivity in electrochemical reduction of CO_2 at metal electrodes in aqueous media. Electrochim Acta 39:1833–1839. https://doi.org/10.1016/0013-4686(94)85172-7

Ikeda S, Ito K, Noda H (2009) Electrochemical reduction of carbon dioxide using gas diffusion electrodes loaded with fine catalysts. AIP Conf Proc 1136:108. https://doi.org/10.1063/1.3160110

Irfan Malik M, Malaibari ZO, Atieh M, Abussaud B (2016) Electrochemical reduction of CO_2 to methanol over MWCNTs impregnated with Cu_2O. Chem Eng Sci 152:468–477. https://doi.org/10.1016/j.ces.2016.06.035

Jhong H-RM, Ma S, Kenis PJA (2013) Electrochemical conversion of CO_2 to useful chemicals: current status, remaining challenges, and future opportunities. Curr Opin Chem Eng 2:191–199. https://doi.org/10.1016/j.coche.2013.03.005

Jia F, Yu X, Zhang L (2014) Enhanced selectivity for the electrochemical reduction of CO_2 to alcohols in aqueous solution with nanostructured Cu–Au alloy as catalyst. J Power Sources 252:85–89. https://doi.org/10.1016/j.jpowsour.2013.12.002

Jiménez C, García J, Camarillo R, Martínez F, Rincón J (2017) Electrochemical CO_2 reduction to fuels using Pt/CNT catalysts synthesized in supercritical medium. Energy Fuel 31:3038–3046. https://doi.org/10.1021/acs.energyfuels.6b03017

Jones J-P, Prakash GKS, Olah GA (2014) Electrochemical CO_2 reduction: recent advances and current trends. Israel J Chem 54:1451–1466. https://doi.org/10.1002/ijch.201400081

Kim D, Kley CS, Li Y, Yang P (2017) Copper nanoparticle ensembles for selective electroreduction of CO_2 to C_2-C_3 products. Proc Natl Acad Sci 114:10560–10565. https://doi.org/10.1073/pnas.1711493114

Koudelka M, Monnier A, Augustynski J (1984) Electrocatalysis of the cathodic reduction of carbon dioxide on platinized titanium dioxide film electrodes. J Electrochem Soc 131:745–750. https://doi.org/10.1149/1.2115692

Kuhl KP, Cave ER, Abram DN, Jaramillo TF (2012) New insights into the electrochemical reduction of carbon dioxide on metallic copper surfaces. Energy Environ Sci 5:7050–7059. https://doi.org/10.1039/C2EE21234J

Kuhl KP, Hatsukade T, Cave ER, Abram DN, Kibsgaard J, Jaramillo TF (2014) Electrocatalytic conversion of carbon dioxide to methane and methanol on transition metal surfaces. J Am Chem Soc 136:14107–14113. https://doi.org/10.1021/ja505791r

Lan Y, Ma S, Lu J, Kenis PJA (2014) Investigation of a Cu(core)/CuO(shell) catalyst for electrochemical reduction of CO_2 in aqueous solution. Int J Electrochem Sci 9:7300–7308. https://doi.org/10.1002/celc.201402182

Le M, Ren M, Zhang Z, Sprunger PT, Kurtz RL, Flake JC (2011) Electrochemical reduction of CO_2 to CH_3OH at copper oxide surfaces. J Electrochem Soc 158:E45–E49. https://doi.org/10.1149/1.3561636

Lee S, Kim D, Lee J (2015) Electrocatalytic production of C_3-C_4 compounds by conversion of CO_2 on a chloride-induced bi-phasic Cu_2O-Cu catalyst. Angew Chem Int Ed 54:14701–14705. https://doi.org/10.1002/anie.201505730

Lei L, Liu T, Fang S, Lemmon JP, Chen F (2017) The co-electrolysis of CO_2–H_2O to methane via a novel micro-tubular electrochemical reactor. J Mater Chem A 5:2904–2910. https://doi.org/10.1039/C6TA10252B

Li P et al (2014) New insights into the photo-enhanced electrocatalytic reduction of carbon dioxide on MoS_2-rods/TiO_2 NTs with unmatched energy band. Appl Catal B Environ 147:912–919. https://doi.org/10.1016/j.apcatb.2013.10.010

Lim RJ, Xie M, Sk MA, Lee J-M, Fisher A, Wang X, Lim KH (2014) A review on the electrochemical reduction of CO_2 in fuel cells, metal electrodes and molecular catalysts. Catal Today 233:169–180. https://doi.org/10.1016/j.cattod.2013.11.037

Loiudice A, Lobaccaro P, Kamali EA, Thao T, Huang BH, Ager JW, Buonsanti R (2016) Tailoring copper nanocrystals towards C_2 products in electrochemical CO_2 reduction. Angew Chem Int Ed 55:5789–5792. https://doi.org/10.1002/anie.201601582

Lu L et al (2018) Highly efficient Electroreduction of CO_2 to methanol on palladium–copper bimetallic aerogels. Angew Chem Int Ed 57:14149–14153. https://doi.org/10.1002/anie.201808964

Lucas-Consuegra A, Serrano J, Gutiérrez-Guerra N, Luis Valverde J (2018) Low-temperature Electrocatalytic conversion of CO_2 to liquid fuels: effect of the Cu particle size. Catalysts 8:340. https://doi.org/10.3390/catal8080340

Ma S, Sadakiyo M, Luo R, Heima M, Yamauchi M, Kenis PJA (2016) One-step electrosynthesis of ethylene and ethanol from CO_2 in an alkaline electrolyzer. J Power Sources 301:219–228. https://doi.org/10.1016/j.jpowsour.2015.09.124

Marnellos G, Stoukides M (2004) Catalytic studies in electrochemical membrane reactors. Solid State Ionics 175:597–603. https://doi.org/10.1016/j.ssi.2004.03.038

Merino-Garcia I, Alvarez-Guerra E, Albo J, Irabien A (2016) Electrochemical membrane reactors for the utilisation of carbon dioxide. Chem Eng J 305:104–120. https://doi.org/10.1016/j.cej.2016.05.032

Merino-Garcia I, Albo J, Irabien A (2018) Tailoring gas-phase CO_2 electroreduction selectivity to hydrocarbons at Cu nanoparticles. Nanotechnology 29:014001. (9pp). https://doi.org/10.1088/1361-6528/aa994e

Monnier A, Augustynski J, Stalder C (1980) On the electrolytic reduction of carbon dioxide at TiO_2 and TiO_2-Ru cathodes. J Electroanal Chem Interfacial Electrochem 112:383–385. https://doi.org/10.1016/S0022-0728(80)80420-7

Nguyen VN, Blum L (2015) Syngas and Synfuels from H_2O and CO_2: current status. Chemie Ingenieur Technik 87:354–375. https://doi.org/10.1002/cite.201400090

Ogura K, Fujita M (1987) Electrocatalytic reduction of carbon dioxide to methanol: part 7. With quinone derivatives immobilized on platinum and stainless steel. J Mol Catal 41:303–311. https://doi.org/10.1016/0304-5102(87)80108-6

Ogura K, Takamagari K (1986) Electrocatalytic reduction of carbon dioxide to methanol. Part 2. Effects of metal complex and primary alcohol. J Chem Soc Dalton Trans 8:1519–1523. https://doi.org/10.1039/DT9860001519

Ogura K, Yoshida I (1986) Electrocatalytic reduction of carbon dioxide to methanol in the presence of 1,2-dihydroxybenzene-3,5-disulphonatoferrate(III) and ethanol. J Mol Catal 34:67–72. https://doi.org/10.1016/0304-5102(86)87039-0

Ogura K, Yoshida I (1988) Electrocatalytic reduction of CO_2 to methanol: part 9: mediation with metal porphyrins. J Mol Catal 47:51–57. https://doi.org/10.1016/0304-5102(88)85072-7

Ogura K, Endo N, Nakayama M, Ootsuka H (1995) Mediated activation and electroreduction of CO_2 on modified electrodes with conducting polymer and inorganic conductor films. J Electrochem Soc 142:4026–4032. https://doi.org/10.1149/1.2048457

Ohkawa K, Noguchi Y, Nakayama S, Hashimoto K, Fujishima A (1993) Electrochemical reduction of carbon dioxide on hydrogen-storing materials.: part II. Copper-modified palladium electrode. J Electroanal Chem 348:459–464. https://doi.org/10.1016/0022-0728(93)80152-8

Ohkawa K, Noguchi Y, Nakayama S, Hashimoto K, Fujishima A (1994) Electrochemical reduction of carbon dioxide on hydrogen-storing materials: part 3. The effect of the absorption of hydrogen on the palladium electrodes modified with copper. J Electroanal Chem 367:165–173. https://doi.org/10.1016/0022-0728(93)03011-D

Podlovchenko BI, Kolyadko EA, Lu S (1994) Electroreduction of carbon dioxide on palladium electrodes at potentials higher than the reversible hydrogen potential. J Electroanal Chem 373:185–187. https://doi.org/10.1016/0022-0728(94)03324-2

Popić JP, Avramov-Ivić ML, Vuković NB (1997) Reduction of carbon dioxide on ruthenium oxide and modified ruthenium oxide electrodes in 0.5 M $NaHCO_3$. J Electroanal Chem 421:105–110. https://doi.org/10.1016/S0022-0728(96)04823-1

Qiao J, Liu Y, Hong F, Zhang J (2014) A review of catalysts for the electroreduction of carbon dioxide to produce low-carbon fuels. Chem Soc Rev 43:631–675. https://doi.org/10.1039/C3CS60323G

Qu J, Zhang X, Wang Y, Xie C (2005) Electrochemical reduction of CO_2 on RuO_2/TiO_2 nanotubes composite modified Pt electrode. Electrochim Acta 50:3576–3580. https://doi.org/10.1016/j.electacta.2004.11.061

Riyanto, Afiati TA (2018) Electrochemical synthesis of ethanol from carbon dioxide using copper and carbon-polyvinyl chloride (C-PVC) electrode. IOP Conf Ser Mater Sci Eng 288:012128. https://doi.org/10.1088/1757-899X/288/1/012128

Rosen BA, Salehi-Khojin A, Thorson MR, Zhu W, Whipple DT, Kenis PJA, Masel RI (2011) Ionic liquid-mediated selective conversion of CO_2 to CO at low overpotentials. Science 334:643–644. https://doi.org/10.1126/science.1209786

Ruiz E, Cillero D, Martínez PJ, Morales Á, Vicente GS, de Diego G, Sánchez JM (2013) Bench scale study of electrochemically promoted catalytic CO_2 hydrogenation to renewable fuels. Catal Today 210:55–66. https://doi.org/10.1016/j.cattod.2012.10.025

Ruiz E, Cillero D, Martínez PJ, Morales Á, Vicente GS, de Diego G, Sánchez JM (2014a) Bench-scale study of electrochemically assisted catalytic CO_2 hydrogenation to hydrocarbon fuels on Pt, Ni and Pd films deposited on YSZ. J CO2 Util 8:1–20. https://doi.org/10.1016/j.jcou.2014.09.001

Ruiz E, Cillero D, Martínez PJ, Morales Á, Vicente GS, de Diego G, Sánchez JM (2014b) Electrochemical synthesis of fuels by CO_2 hydrogenation on cu in a potassium ion conducting membrane reactor at bench scale. Catal Today 236:108–120. https://doi.org/10.1016/j.cattod.2014.01.016

Ruiz E, Martínez PJ, Morales Á, San Vicente G, de Diego G, Sánchez JM (2016) Electrochemically assisted synthesis of fuels by CO_2 hydrogenation over Fe in a bench scale solid electrolyte membrane reactor. Catal Today 268:46–59. https://doi.org/10.1016/j.cattod.2016.02.025

Safdar Hossain S, Rahman S, Ahmed S (2014) Electrochemical reduction of carbon dioxide over CNT-supported nanoscale copper electrocatalysts. J Nanomater 374318, 10 pp. https://doi.org/10.1155/2014/374318

Schizodimou A, Kyriacou G (2012) Acceleration of the reduction of carbon dioxide in the presence of multivalent cations. Electrochim Acta 78:171–176. https://doi.org/10.1016/j.electacta.2012.05.118

Schwartz M, Cook RL, Kehoe VM, MacDuff RC, Patel J, Sammells AF (1993) Carbon dioxide reduction to alcohols using perovskite-type electrocatalysts. J Electrochem Soc 140:614–618. https://doi.org/10.1149/1.2056131

Sebastián D et al (2017) CO_2 reduction to alcohols in a polymer electrolyte membrane CO-electrolysis cell operating at low potentials. Electrochim Acta 241:28–40. https://doi.org/10.1016/j.electacta.2017.04.119

Seshadri G, Lin C, Bocarsly AB (1994) A new homogeneous electrocatalyst for the reduction of carbon dioxide to methanol at low overpotential. J Electroanal Chem 372:145–150. https://doi.org/10.1016/0022-0728(94)03300-5

Shironita S, Karasuda K, Sato K, Umeda M (2013a) Methanol generation by CO_2 reduction at a Pt–Ru/C electrocatalyst using a membrane electrode assembly. J Power Sources 240:404–410. https://doi.org/10.1016/j.jpowsour.2013.04.034

Shironita S, Karasuda K, Sato M, Umeda M (2013b) Feasibility investigation of methanol generation by CO_2 reduction using Pt/C-based membrane electrode assembly for a reversible fuel cell. J Power Sources 228:68–74. https://doi.org/10.1016/j.jpowsour.2012.11.097

Spataru N, Tokuhiro K, Terashima C, Rao NT, Fujishima A (2003) Electrochemical reduction of carbon dioxide at ruthenium dioxide deposited on boron-doped diamond. J Appl Electrochem 33:1205–1210. https://doi.org/10.1023/B:JACH.0000003866.85015.b6

Summers DP, Leach S, Frese KW (1986) The electrochemical reduction of aqueous carbon dioxide to methanol at molybdenum electrodes with low overpotentials. J Electroanal Chem Interfacial Electrochem 205:219–232. https://doi.org/10.1016/0022-0728(86)90233-0

Sun X, Zhu Q, Kang X, Liu H, Qian Q, Zhang Z, Han B (2016) Molybdenum–bismuth bimetallic chalcogenide nanosheets for highly efficient electrocatalytic reduction of carbon dioxide to methanol. Angew Chem Int Ed 55:6771–6775. https://doi.org/10.1002/anie.201603034

Sun Z, Ma T, Tao H, Fan Q, Han B (2017) Fundamentals and challenges of electrochemical CO_2 reduction using two-dimensional materials. Chem 3:560–587. https://doi.org/10.1016/j.chempr.2017.09.009

Uhm S, Kim YD (2014) Electrochemical conversion of carbon dioxide in a solid oxide electrolysis cell. Curr Appl Phys 14:672–679. https://doi.org/10.1016/j.cap.2014.02.013

Vernoux P et al (2013) Ionically conducting ceramics as active catalyst supports. Chem Rev 113:8192–8260. https://doi.org/10.1021/cr4000336

Wang Q, Dong H, Yu H (2014) Development of rolling tin gas diffusion electrode for carbon dioxide electrochemical reduction to produce formate in aqueous electrolyte. J Power Sources 271:278–284. https://doi.org/10.1016/j.jpowsour.2014.08.017

Watanabe M, Shibata M, Kato A, Azuma M, Sakata T (1991a) Design of alloy electrocatalysts for CO_2 reduction: III. The selective and reversible reduction of on Cu alloy electrodes. J Electrochem Soc 138:3382–3389. https://doi.org/10.1149/1.2085417

Watanabe M, Shibata M, Katoh A, Sakata T, Azuma M (1991b) Design of alloy electrocatalysts for CO_2 reduction: improved energy efficiency, selectivity, and reaction rate for the CO_2 electroreduction on Cu alloy electrodes. J Electroanal Chem Interfacial Electrochem 305:319–328. https://doi.org/10.1016/0022-0728(91)85528-W

Weekes DM, Salvatore DA, Reyes A, Huang A, Berlinguette CP (2018) Electrolytic CO_2 reduction in a flow cell. Acc Chem Res 51:910–918. https://doi.org/10.1021/acs.accounts.8b00010

Welton T (1999) Room-temperature ionic liquids. Solvents for synthesis and catalysis. Chem Rev 99:2071–2084. https://doi.org/10.1021/cr980032t

Weng L-C, Bell AT, Weber AZ (2018) Modeling gas-diffusion electrodes for CO_2 reduction. Phys Chem Chem Phys 20:16973–16984. https://doi.org/10.1039/C8CP01319E

Whipple DT, Kenis PJA (2010) Prospects of CO_2 utilization via direct heterogeneous electrochemical reduction. J Phys Chem Lett 1:3451–3458. https://doi.org/10.1021/jz1012627

Wu G, Xie K, Wu Y, Yao W, Zhou J (2013) Electrochemical conversion of H_2O/ CO_2 to fuel in a proton-conducting solid oxide electrolyser. J Power Sources 232:187–192. https://doi.org/10.1016/j.jpowsour.2013.01.039

Xie K, Zhang Y, Meng G, Irvine JTS (2011a) Direct synthesis of methane from CO_2/H_2O in an oxygen-ion conducting solid oxide electrolyser. Energy Environ Sci 4:2218–2222. https://doi.org/10.1039/C1EE01035B

Xie K, Zhang Y, Meng G, Irvine JTS (2011b) Electrochemical reduction of CO_2 in a proton conducting solid oxide electrolyser. J Mater Chem 21:195–198. https://doi.org/10.1039/C0JM02205E

Yadav VSK, Purkait MK (2015) Electrochemical studies for CO_2 reduction using synthesized Co_3O_4 (anode) and Cu_2O (cathode) as electrocatalysts. Energy Fuel 29:6670–6677. https://doi.org/10.1021/acs.energyfuels.5b01656

Yan J, Chen H, Dogdibegovic E, Stevenson JW, Cheng M, Zhou X-D (2014) High-efficiency intermediate temperature solid oxide electrolyzer cells for the conversion of carbon dioxide to fuels. J Power Sources 252:79–84. https://doi.org/10.1016/j.jpowsour.2013.11.047

Yang H-P, Yue Y-N, Qin S, Wang H, Lu J-X (2016) Selective electrochemical reduction of CO_2 to different alcohol products by an organically doped alloy catalyst. Green Chem 18:3216–3220. https://doi.org/10.1039/C6GC00091F

Yuan J, Yang M-P, Hu Q-L, Li S-M, Wang H, Lu J-X (2018) Cu/TiO_2 nanoparticles modified nitrogen-doped graphene as a highly efficient catalyst for the selective electroreduction of CO_2 to different alcohols. J CO2 Util 24:334–340. https://doi.org/10.1016/j.jcou.2018.01.021

Zhang L, Hu S, Zhu X, Yang W (2017) Electrochemical reduction of CO_2 in solid oxide electrolysis cells. J Energy Chem 26:593–601. https://doi.org/10.1016/j.jechem.2017.04.004

Zhang N, Long R, Gao C, Xiong Y (2018a) Recent progress on advanced design for photoelectrochemical reduction of CO_2 to fuels. Sci China Mater 61:771–805. https://doi.org/10.1007/s40843-017-9151-y

Zhang W et al (2018b) Progress and perspective of electrocatalytic CO_2 reduction for renewable carbonaceous fuels and chemicals. Adv Sci 5:1700275. https://doi.org/10.1002/advs.201700275

Zhuang T-T et al (2018) Steering post-C–C coupling selectivity enables high efficiency electroreduction of carbon dioxide to multi-carbon alcohols. Nature Catal 1:421–428. https://doi.org/10.1038/s41929-018-0084-7

Index

A
Abudula, A., 27–38
Aeshala, L.M., 190
Albrecht, M., 51
Alcohols, 2, 5, 6, 8–13, 15, 16, 105, 110, 116, 118, 120, 178, 195
Aljabour, A., 17
Al-Rowaili, F.N., 92–121
Álvarez, C., 51
Alvarez-Guerra, M., 36, 39
An, X., 27–38
Artificial photosynthesis, 140, 141

B
Back, S., 33
Balun, D., 39
Bonin, J., 3
Boreriboon, N., 5
Boudjemaa, A., 65–79
Boudouard reaction, 48, 50

C
Carbon-based materials, 31, 148, 169, 170, 180, 190
Carbon dioxide (CO_2), 17, 27–38, 57, 65, 92–121, 143, 167–201
  conversion, 2–19, 27–38, 54, 65, 93, 94, 96, 111, 120, 135–158, 167–170, 172, 175, 178, 180, 186, 188, 190, 195–200
  electrocatalytic reduction, 94, 96–98, 170, 176, 177, 179, 182, 185, 186, 188, 195
  hydrogenation, 3–9, 11, 15, 19, 46, 47, 50, 51, 57, 168, 169, 191, 195, 197–201
  recycling, 168
Carbon monoxide (CO), 3–5, 7, 8, 10–14, 16, 18, 19, 27, 28, 31, 33, 48, 69–71, 73, 75–77, 79, 97, 98, 101, 103–105, 108, 109, 137–139, 141, 143, 146, 157, 178, 196, 199, 201
Cardoso, J.C., 119
Carrasquillo-flores, R., 51
Catalysts, 3, 27, 46, 66, 94, 137, 168
Catalytic mechanisms, 28, 33–34, 56
CeO_2, *see* Cerium dioxide (CeO_2)
Ceria, 50
Cerium dioxide (CeO_2), 8, 9, 47, 51, 52, 54, 56, 57, 71–73, 149
Charge carriers, 66, 69, 70, 76, 78, 139, 148–155, 158
Chen, X., 51
Cherifi, N., 65–79
CH_3OH, *see* Methanol (CH_3OH)
Choi, S.Y., 39
CO, *see* Carbon monoxide (CO)
CO_2, *see* Carbon dioxide (CO_2)
Current density, 15, 16, 18, 28–32, 36–38, 118, 168–177, 179, 181, 184–190, 192, 194–196, 199–201

D
Dai, B., 51, 52
Das, T., 8
Deo, G., 8

Inamuddin et al. (eds.), *Conversion of Carbon Dioxide into Hydrocarbons Vol. 1 Catalysis*, Environmental Chemistry for a Sustainable World 40,
https://doi.org/10.1007/978-3-030-28622-4

Doping, 70, 77, 79, 112, 120, 142, 149, 150, 157
Dorner, R.W., 11
Dulay, M.T., 155

E
Electrocatalysts, 15–17, 27–32, 34, 38, 95–97, 99–100, 104, 105, 108, 110, 112–114, 116, 121, 169–176, 188–190, 194, 199–201
Electrochemical, 3, 15–19, 27–29, 33, 34, 37, 66, 92–121, 136, 137, 142, 168, 169, 172, 180–182, 188–197, 200, 201
 promotion, 191, 196–198, 200, 201
 reduction, 2, 18, 27, 28, 30–38, 92–121, 177, 181
Electroreduction, 27–33, 35–38, 95–101, 103–106, 110, 116–118, 121, 170, 171, 173–176, 178–181, 183, 185–190, 192, 200

F
Fan, M., 29, 39
Faradaic efficiency (FE), 18, 28–33, 36–38, 75, 97, 109–111, 118, 121, 168–181, 184–189, 194, 199, 200
FE, *see* Faradaic efficiency (FE)
Feaster, J.T., 33
Fischer-Tropsch, 2, 9
Formate, 2, 3, 12, 15, 16, 27–38, 56, 97, 98, 103–105
Formic acid (HCOOH), 2, 3, 15, 16, 19, 27–38, 68, 70, 75, 76, 79, 98, 108, 111, 137, 139, 141, 143
Fox, M.A., 155
Fuels, 2–5, 17, 19, 27, 44–46, 49, 66, 67, 75, 93–95, 121, 135–137, 140–142, 148, 167, 168, 194, 196, 199

G
Gao, S., 17
García, J., 31
Gas diffusion electrodes (GDE), 36, 38, 106, 117, 118, 121, 171, 173, 180, 182, 185, 189, 190, 192, 194, 200, 201
GDE, *see* Gas diffusion electrodes (GDE)
Global warming, 45, 65, 92, 135, 136
Gorimbo, J., 44–58
Guan, G., 27–38

H
HCOOH, *see* Formic acid (HCOOH)
Heterogeneous catalysis, 52, 99, 100
Hildebrandt, D., 44–58
Hori, Y., 39
Hybridization, 149, 151, 152, 157
Hydrocarbons, 16, 27, 49, 65, 97, 104, 105, 151, 167, 168, 196, 198
Hydrogenation, 2, 3, 6, 12, 19, 46, 47, 49, 51, 66, 105, 136, 137, 139, 191, 197–199

I
Innocent, B., 39
Inoue, T., 139
Irfan Malik, M., 170

J
Jamal, A., 92–121
Jiménez, C., 194

K
Kanan, M.W., 32, 39
Khajeh, A., 2–19
Kim, D.H., 51
Kočí, K., 71, 154
Köleli, F., 36, 39
Kyriacou, G., 186

L
Lead (Pb), 15, 28–33, 36, 38, 75, 104, 109
Lee, C.H., 32, 39
Le, M., 180
Liu, L., 71, 146
Liu, Y., 51
Li, Y., 71
Long-chain hydrocarbons, 2, 4–7, 9, 11, 15, 19
Lu, L., 172

M
Machunda, R.L., 36, 39
Martínez, E.R., 167–201
Matejova, L., 71
Matheswaran, M., 135–158
Metal-organic frameworks, 10, 92–121, 185
Methanation reactions, 48, 50, 198
Methanol (CH_3OH), 2, 27, 46, 68, 94, 139, 168

Methanol production, 96–98, 110, 111, 114, 115, 169, 170, 179, 186, 188, 196, 198–200
Mizuno, T., 28, 36, 39
Murugesan, P., 135–158

N
Nanocomposites, 70, 76, 77, 158
Narayanan, S., 135–158
Non-metal-organic frameworks, 92–121

O
Organic acids, 27
Overpotentials, 18, 28–30, 33, 38, 98, 100, 101, 105, 109, 110, 116, 168–170, 172, 175, 178, 199–201
Owen, R.E., 9

P
Park, K.T., 30
Pastrana-Martínez, L.M., 153
PEM, *see* Polymer electrolyte membranes (PEM)
Photocatalysis, 66, 67, 71, 72, 120, 136, 139, 141, 143, 145, 146, 148, 150
Photochemical, 3, 19, 27, 94, 109
Photoreactors, 139, 155, 157
Polymer electrolyte membranes (PEM), 192–194, 200
Porous materials, 75, 76, 119, 170, 172, 200
Purkait, M.K., 32, 38, 39, 170

R
Reduction, 2, 27, 47, 66, 93, 136, 167
Renewable energy storage, 167, 168
Reverse water-gas shift, 3–5, 7, 8, 19, 48, 198
Ronda-Loret, M., 51

S
Sabatier reaction, 47, 48
Sánchez Hervás, J.M., 167–201
Satthawong, R., 6
Schizodimou, A., 186
Sebastián, D., 194
Semiconductors (SCs), 66, 67, 71, 73, 75, 79, 139, 140, 142, 145–147, 149–152, 154, 155, 157
Sensitization, 142, 149–151, 157
Shahbazi, A., 2–19
Shironita, S., 194
Solid oxide electrolysers, 195
Sun, F., 51

T
TiO_2, *see* Titanium dioxide (TiO_2)
Titanium dioxide (TiO_2), 8, 9, 14, 19, 51, 67, 76, 77, 79, 110, 111, 119, 149, 151, 172, 173, 175, 197, 198
Todoroki, M., 28, 37, 39

U
UV light, 67, 71, 72

V
Visible irradiation, 71, 79, 120

W
Wang, J., 28, 39
Wang, L., 51
Wang, Y., 31, 39
Weekes, D.M., 179

X
Xie, Y., 71

Y
Yadav, V.S.K., 32, 38, 39, 170
Yang, H.P., 171
Yang, L., 51
Yang, X., 51
Yan, H., 79
Yoshida, A., 27–38

Z
Zhang, Y., 4
Zhao, H., 76
Zhao, Z., 19

MIX
Papier aus verantwortungsvollen Quellen
Paper from responsible sources
FSC® C105338

If you have any concerns about our products,
you can contact us on
ProductSafety@springernature.com

In case Publisher is established outside the EU,
the EU authorized representative is:
Springer Nature Customer Service Center GmbH
Europaplatz 3, 69115 Heidelberg, Germany

Printed by Libri Plureos GmbH
in Hamburg, Germany